8 BILLION REASONS POPULATION MATTERS

The Defining Issue of the 21st Century

Valorie M. Allen

To Susan
Thanks for all your work on environment
Val Allen

One Printers Way
Altona, MB R0G 0B0
Canada

www.friesenpress.com

First Edition — 2022

ISBN
978-1-03-910805-9 (Hardcover)
978-1-03-910804-2 (Paperback)
978-1-03-910806-6 (eBook)

1. SOCIAL SCIENCE, WOMEN'S STUDIES, ENVIRONMENTAL STUDIES, POLITICAL SCIENCE

Distributed to the trade by The Ingram Book Company

Advance Praise for Eight Billion Reasons Population Matters

Valorie Allen's research and writing are essential to understanding how human society can settle down into a sustainable path. We are not just in a climate crisis; we are in a full-scale ecological crisis. Ms. Allen writes about the global effects of human population growth, one of the driving factors of our ecological dilemma. All pathways out of our ecological catastrophe require the contraction of human enterprise. In this book, Ms. Allen clearly articulates the importance of addressing both twin towers of our crisis—consumption and population. Fortunately, there are non-coercive ways to reverse human population growth that offer other social gains, primarily universal women's rights and universally available contraception. Ms. Allen's book is part of a third important focus: public education.

REX WEYLER — A FOUNDER OF GREENPEACE INTERNATIONAL; AUTHOR OF *GREENPEACE: THE INSIDE STORY*, *BLOOD OF THE LAND*, AND OTHER BOOKS. HE WRITES THE ECOLOGY BLOG, DEEP GREEN.

****

Population growth is not a victimless crime. It hurts those already here and further damages a planet reeling from over-exploitation by too many people wanting and exhorted by unscrupulous leaders to acquire and consume too much stuff. Every hectare of farmland paved over with more houses and every bit of added energy demand makes it harder for future generations to ensure food security and renewable energy resilience. Val Allen's books are treasure troves of information and perspective serving up a veritable researchers' bonanza while still carrying any interested reader through the whys and wherefores of the biophysical threats we face and up to a new and well-grounded awareness.

JOHN ERIK MEYER — AUTHOR; PRESIDENT OF CANADIANS FOR SUSTAINABLE SOCIETY

****

Val Allen's book is a crash course on the population issue. It examines the history of human population growth, its connections to all other environmental problems, the factors that influence it, the people and the organizations trying to do something about it, the impediments to addressing it, the growth promoters and deniers, and much more. It is a trove of information both for novices and for those who have had a long-standing interest in the issue. It can serve as both an inspiration and tool kit for anyone who wants to do something about the problem.

DR. MADELINE WELD — PRESIDENT, POPULATION INSTITUTE CANADA

****

Val Allen's book is a great resource for students to learn about population and its connection with numerous vital issues. It makes those links admirably well.

BILL RYERSON — FOUNDER AND PRESIDENT OF POPULATION MEDIA CENTER; CHAIR OF THE BOARD OF THE POPULATION INSTITUTE

CONTENTS

Acknowledgments

The power of accurate vision is commonly called cynicism by those who do not have it.
—G. B. Shaw

Since writing a book is a collaborative endeavour, I would like to thank all those who shared my vision and helped me express it in an accurate and enthusiastic way—hopefully with a tolerable dose of cynicism. Thanks also to all those groups, institutions, and individuals whose quotes I have borrowed. I hope that I have accredited them properly, as they have greatly contributed to the credibility of this book.

A special thanks to Brian Horjsi, Madeline Weld, Steve Wilmot, Gregg Miklashek, and Vivian Pharis, who agreed to proofread, edit, or comment on parts of my book. Without your help I never would have been able to accomplish this goal. Thanks to Sean Arthur Joyce, Anne Champagne, and Julia Caceres for their editing, research, and diligence in adding footnotes and references, and Irene Carlson for her assistance with the title. You are all an integral part of this book.

I would also like to extend a special and heartfelt thank you to those respected experts who have read and endorsed my book: John Meyer, author of *The Renewable Energy Transistion*; Rex Wyler, founder of Greenpeace International and author of *Greenpeace: The Inside Story*; Madeline Weld, president of Population Institute Canada; Bill Ryerson, founder and president of Population Media Center and chair of the board of the Population Institute.

My publishing crew at Friesen Press deserves recognition for the role they played, and a special thanks to Emily LeGrand Indexing Services for her excellent work. If I have neglected to mention anyone, please forgive me.

This book is a Population in Sync project. The goal of Population in Sync is to bring the human population back into harmony with the natural world. www.populationinsync.net

PART I
HUMAN HISTORY

Introduction: For the Greater Good

Once it was necessary that the people should multiply and be fruitful if the race was to survive. But now to preserve the race it is necessary that people hold back the power of propagation.
—Helen Keller

In the 21st century, we should all be enjoying the best life our amazing planet has to offer. With the trillions of dollars that humankind has spent to improve the human condition, develop modern technology, and protect the environment, we should have achieved a utopian lifestyle by now. We should, in fact, be the healthiest, happiest civilization that ever existed.

With humankind's most noble efforts from trillions of hours of volunteer time put towards reducing human suffering and protecting wildlife, we should surely have eliminated poverty and secured safe wildlife habitat and a healthy life support system.

Instead, the opposite is true. We have created a world reeling from environmental, social, political, and economic turmoil—a world plagued with decision-makers in denial of the true root cause of our failures, citizens exhausted by apathy and hopelessness, and animals disappearing at hundreds of times the normal rate, primarily because of shrinking habitats. Their biggest threat: humans.

So where did we go wrong? Never before on this planet have so many inhabitants been so misinformed about something that is so important and urgent as the population issue. Far too many of us have been duped into believing that simple population strategies do not exist, or even that population isn't really an issue, at least not in developed nations. But the biggest myth of all is that population growth is good for our planet, and economic growth is needed for humanity to prosper. The purpose of this book is to expose these myths and explain the many practical solutions that point the way to a healthy and sustainable future. This will require that every human being make this a priority.

I was one of those volunteers who spent thousands of hours trying to make this planet a better place to live, and I won the Canadian Volunteer Award for my efforts. It was winning this award that made me look back over all the work I had done, and what improvements my efforts had made. I was horrified and extremely disappointed to realize that all of the issues I had been working on had become worse instead of better. Massive and escalating overpopulation had undermined all of my hard work, and that of millions of others trying to reduce suffering and environmental degradation.

For example, I had been working for decades with environment groups to promote the reduction of fuel consumption of vehicles in order to reduce climate change. And in fact, over a thirty-year period, this fuel consumption had been reduced by about half, which many of my generation celebrated as a huge success. However, in that same time, the population had doubled, causing the number of cars to double. So actually, we had made no progress at all in the big picture.

If environmental groups and governments had been promoting a sustainable population level during that same time, this could have truly been a victory. This scenario of escalating population undermining our efforts is true of all of the critical world issues threatening our planet. One only needs to look at the increasing climate crisis, poverty, and world insecurity over resources to see these effects.

This realization changed the direction and focus of my life, and I decided to devote my efforts to population from that time forward. I published my first book, *Growing Pains – A Planet in Distress*, in 2010 and began working with Scientists in the Classroom and doing population presentations in schools and for community groups. In 2013, I was invited to be keynote speaker at a Critical World Issues conference in England, where I had the opportunity to share my concerns with other presenters from around the world. Many of these distinguished thinkers were eager to make the population connection to issues like climate change, the slave trade, food shortages, national security, and water wars.

You may ask, "Why should population matter to me personally or any of my immediate family?" In the 21st Century, people may not realize that in our global community, as long as there is population growth anywhere on the planet, **it will impact every living being every day of their lives**—but it most **often comes in disguise**. Every year, as population increases by 80 million people, your cost of food will go up, your need for ever more expensive and extensive health care will increase, your taxes will increase, your air and water will become scarcer and more polluted, and pandemics will increase and become more problematic **for you**. All of these problems will continuously be in your face and personal for the rest of your life, but you may never recognize them as being **caused by overpopulation.**

Also, as you have your morning coffee, use bananas for your smoothie, or buy a beef burger sourced from a foreign land, you are robbing developing countries of their resources and wildlife habitat. As we send our recycled plastic to other countries for disposal and dump our foul wastes in the oceans, all of it will come back to haunt us—and sooner than we think. As population increases, more people from other countries will end up on our doorsteps wanting us to share our food and other necessities with them as well. We are all working harder for less real income, and we are sacrificing our wonderful wildlife and wilderness to do it. Population density stress, caused by severe overcrowding that is driving our increasingly overactive stress responses, contributes towards your inner turmoil and disease, your conflict over resources and space, your everyday stresses of earning a living and caring for your family, and your vulnerability to pandemics that will increasingly torment humanity if we continue to increase our numbers. This will all be in your face and personal, and every animal—including humans—will be affected.

Watching our children suffer and die from their myriad and increasing childhood and adult diseases hits us right in the gut and demands our immediate attention. We are a very, very short-sighted and self-absorbed species and will only act, including using contraception, if we absolutely must to prevent the horror of our sick and dying loved ones in our arms (or behind a Plexiglas panel) right now, today, this very moment. For me, the health consequences of human overpopulation are just as real as running before a tsunami on a beachfront in Thailand.

The purpose of this book is to highlight ways we might prevent much of this suffering and turmoil that our species is creating on this planet. This book was also written to shed light on the extraordinary efforts being made by a select few organizations and individuals to finally acknowledge the damage done by unchecked population growth and work to reverse it. It is my hope that as you read through this book, the realization will finally set in that population growth is a problem that deeply impacts every one of us, and that we all must be part of the solution if we are to survive and have anything to pass on to future generations.

CHAPTER 1
POPULATION HISTORY AND LANDMARKS

History is a race between education and catastrophe.
—H.G. Wells

It was in 1971 that the population bomb rocked my world. Just three years earlier, Paul Ehrlich had published his book, *The Population Bomb*, which was creating quite a stir. I recall our social studies teacher discussing overpopulation in class, and that was a lightbulb moment for me! Suddenly, everything started to make sense. All of the poverty, extinction of species, water pollution, wars over resources, and a myriad of other problems began to take a logical place in the global scenario that had previously been such a puzzle.

I decided that very day to remain **childfree** and have never regretted that decision. I wanted to be part of the solution instead of part of the problem. I also made a commitment to focus on population and environmental issues in order to make this planet a better place to live for all forms of life. Little did I realize what a challenge that would prove to be, or that one day childfree groups would be starting up all over the world.

The 1970s ushered in a new era of freedom of expression that manifested in various ways. Modern art, psychedelic music, the "flower power" lifestyle, and a new awakening to issues of the day were all part of the modern scene. For many, these issues included the environment and population, as the hippie community searched for more meaningful and long-term solutions to world problems.

Protesters were demonstrating against the Vietnam War, and John Lennon's songs "Imagine" and "Give Peace a Chance" became part of the new peace movement that swept the planet. It was a time when anything seemed possible. During this time, the first heart transplant was performed, the first test-tube baby was conceived, the first Boeing 747 jumbo jet was introduced, and the first man walked on the moon. There was a new realization, after the Woodstock phenomenon, that by working together, the '70s generation had the power to create its own destiny. Today, there are many people looking back at that time and wondering what happened to derail the momentum of the hippie era. Alan Kuper, in his article "From Sentience to Silence," had this to say about it:

> In the heady days of the new environmental awareness, at the first big Earth Day celebration in April 1970, the ecological threat posed by U.S. population growth was part of every discussion. David Brower, executive director of the Sierra Club, had encouraged Paul Ehrlich to write *The Population Bomb*, which became a runaway

bestseller. The educational work of the new organization Zero Population Growth, or ZPG, became familiar to American schoolchildren. In 1972 the Sierra Club, the nation's premier conservation organization, adopted a ZPG platform, declaring as one of its objectives to "...bring about the stabilization of the population first of the United States and then of the world." Other groups made similar commitments.[1]

Unfortunately, by the mid-'80s most of these efforts had started to lose their momentum, and there was no new population bestseller to inspire a disillusioned planet. Similarly, David Nicholson-Lord, in his *Internet Forum* article "Whatever Happened to the Teeming Millions?" feels that when it comes to dealing with population issues, we are going backwards:

> In the 1960s and 70s, concern about population growth was a mainstream environmental issue. Paul Ehrlich wrote *The Population Bomb*; pressure groups flourished – Population Countdown in the UK, Zero Population Growth in the US. In 1973 a government-appointed panel declared that Britain's population could not "go on increasing indefinitely" and the government should "define its attitude" to the issue. The newly founded Ecology Party debated sustainable population levels, Oxfam publicly supported zero population growth and Greenpeace's slogans included "Stop at Two." Three decades later, the mainstream has largely abandoned the topic. Greenpeace says it's "not an issue for us," Oxfam doesn't list it on its website A-Z, the Greens didn't even mention it in their (2005) election manifesto. Population Countdown, worried about alienating funders, became Population Concern and, more recently, Interact Worldwide. Zero Population Growth, for similar reasons, morphed into Population Connection. As an issue, population is, in short, off the radar.[2]

So what happened? Well, there were a number of factors involved in this disappearing act:

- Many activists were distracted by the Vietnam War.
- Paul Ehrlich was too "time specific" in his premature predictions of when food shortages would cause mass starvation and was therefore dismissed by many. Advances in agriculture instigated the unsustainable and devastating "Green Revolution" that is now poisoning our soil and water.
- The Baby Boomers became less idealistic about achieving a sustainable population as they became preoccupied with raising their families.
- Many were distracted by the HIV epidemic, and much of our attention and funding was channeled into addressing this dire emergency instead of family planning.
- The rise of the religious right and political left in North America put a damper on both environmental concerns and birth control.

---

1 Alan Kuper, "From Sentience to Silence: How the Environmental Establishment Changed Its Tune on U.S. Overpopulation," archived article, Georgetown University Library online: https://repository.library.georgetown.edu/handle/10822/988984

2 David Nicholson-Lord, "Whatever Happened to the Teeming Millions?'" NPG Internet Forum, archived article: https://npg.org/forum_series/iforums/teemingmill.html

- Population myths arose to cloud the issue, which I will discuss throughout the book.
- The growth lobby, "green" agriculture advocates, space explorers, and modern economists became very active in supporting our unsustainable growth ethic, and shifting public focus to unrealistic and idealistic solutions.

Yes, it is off the radar, but can't you hear a faint bleep off in the distance? Do you think it could possibly be making a comeback? Well, from all indications, there is a movement afoot to erase that taboo stigma and make population the buzzword it was in the '70s, so watch for further updates ahead!

Understanding Our Capacity for Radical Change

I think we can all agree that 2019 was an extraordinary year for life-changing events, with Greta Thunberg emerging on the world stage in the spring, Australia's wildfires grabbing our attention in the summer, and COVID-19 making the world rethink what is really important to our future in the winter. This pandemic will be discussed in detail in Chapter 9 on food. COVID-19 showed us that we are capable of making radical changes if necessary, an awareness that we must utilize when dealing with the overpopulation and climate change emergencies in the 2020 decade. We must shed the growth ethic and allow logic to prevail.

Steven Chu, a former U.S. Secretary of Energy, 1997 Nobel Prize winner in physics, and president of the American Association for the Advancement of Science, argues that the world economy is a pyramid scheme. This model of constant economic growth is based in part on ever-increasing population, Chu explained in a lecture at the University of Chicago in 2018. It's a scheme that economists don't talk about and that governments won't face, a scheme that makes sustainability impossible and that is likely to fail.

> *It is now evident that our present growth ethic is in fact a Ponzi scheme doomed to failure.*
> **— Steven Chu, former US Secretary of Energy**

"The world needs a new model of how to generate a rising standard of living that's not dependent on a pyramid scheme," Chu said.[3] We need a model based on the greater good, and possible solutions to this problem will be outlined throughout the book. The common misconception that all other species must die so that the economy can live will also be challenged. This book will outline why Earth will be in serious jeopardy if we do not stop population growth at EIGHT BILLION, and I will offer scenarios that could make this possible.

I invite anyone to meet the great challenge of Professor Al Bartlett (1923–2013): "Can you think of any problem, in any area of human endeavour, on any scale, from microscopic to global, whose LONG-TERM solution is in any DEMONSTRABLE [emphasis added] way aided, assisted or advanced by further increases in population, locally, nationally, or globally?"[4]

---

3 Jeff McMahon, "The World Economy Is A Pyramid Scheme, Steven Chu Says," *Forbes*, April 5, 2019: https://www.forbes.com/sites/jeffmcmahon/2019/04/05/the-world-economy-is-a-pyramid-scheme-steven-chu-says/#7a2d8e5b4f17

4 Al Bartlett, "Reflections on Sustainability, Population Growth and the Environment," January 1998: https://www.albartlett.org/articles/art_reflections_part_2.html

No one has been able to meet Bartlett's challenge since he set it out in 1996 at the University of Colorado, where he was an emeritus professor of physics. Instead, we have added almost two billion more people to the planet since that challenge was put forth. Professor Bartlett explained how "sustainable growth" is a contradiction. His view was based on the fact that even a modest percentage of growth will equate to huge escalations over relatively short periods of time.

Bartlett regarded the failure to understand the laws of the exponential equation as "the greatest challenge" facing humanity. He was an early advocate on the topic of overpopulation and has been highly influential in shaping our view on the present population dilemma.

Considering How Overpopulation Threatens Us

The majority of people have failed to consider the fact that overpopulation is unravelling our web of life (see Figure 1.1). Most politicians, corporations, media, and even most environmental groups have been in denial that our alarming numbers are the root cause, decimating our planet and all life forms that exist here. One would think that with the climate change issue gaining so much attention in the media recently, people would feel compelled to connect the very few dots to the underlying human overpopulation crisis. Yet, until recently, this elephant in the room has disturbingly remained ignored.

According to the Australian Academy of Science, "Despite population increase being such a serious issue, the United Nations (UN) has held only three world conferences on population and development (in 1945, 1974 and 1994)."[5] The fourth only took place in 2019, after a gap of 25 years, and I will be discussing it in Chapter 10. This failure undermined the work of millions of volunteers trying to combat the population crisis.

The lack of concern for the greater good in a world dominated by self-centred thinking and individualized image-based social media is causing mass destruction of our planet at unprecedented levels. The alarming level of **ecocide** taking place today is in turn causing an unprecedented level of compassion fatigue, hopelessness, and depression. Ecocide is defined as "loss or damage to, or destruction of ecosystems of a given territory, such that peaceful enjoyment by the inhabitants has been or will be severely diminished."[6]

Those of us who dare to open our eyes to this travesty are mourning the loss of species and Indigenous peoples taking place at an alarming rate, as well as the destruction of the life support system we all depend on. We are struggling with a feeling of guilt as we pass on a planet greatly diminished to the next generation and see that they often can't cope with this overwhelming burden. But we can also see the many solutions that must be implemented in the next ten years, according to UN warnings.

The continued dismissal of the fundamental right to universal family planning resources is causing an increase in unwanted children, which in turn often leads to divorce, family abuse,

---

5 Prof. Stephen Dovers, Australia National University, *Population and Environment: A Global Challenge*: https://www.science.org.au/curious/earth-environment/population-environment

6 Ecocide Law, https://ecocidelaw.com/the-law/restorative-justice/; Climate Ecocide website: https://www.earthlaw.org/climatecrime_pr/

increased addictions, homelessness, and suicides (discussed in later chapters). Now, in the 21st century, Pregnant Teen Help notes that about half the babies born in the U.S. are unplanned, consistent with statistics globally.[7] According to Population Speak Out, seven in ten pregnancies among single women in their twenties in the U.S. are unplanned.[8] Perhaps Family Planning 101 is a class that should be included in all high schools to reduce this unacceptably high number, as has been done in Iran and other countries.

Yet, with advances in family planning options and availability, this appalling situation is most often unnecessary. Every parent has the right to be a willing parent, and every baby has the right be a wanted baby. What would it take to persuade every potential parent to actually plan their family size and spacing, consider the impact to our planet, and be provided the opportunity to prevent unintended pregnancies? Check out Chapter 13 for answers.

Tackling Urgent Challenges Together

This book will help reveal the disturbing chain of events that brought us to this critical state of affairs that is causing so much distress today. Since every global citizen is responsible to some extent, I will suggest ways that we can all help to achieve a sustainable population and reduce our impact on the planet. For example, the United Nations advises, "Family planning could bring more benefits to more people at less cost than any other single technology available to the human race."[9] Also, the private interests of wealthy insiders, the flawed gross domestic product (GDP), and the deceptive Vatican have all played a role. Solutions to these problems and more will be highlighted.

We will look at reports from many universities, think tanks, population groups, and other distinguished thinkers that indicate that a sustainable population level would be between one and three billion, rather than our present level of almost EIGHT BILLION. This reduction in population would be necessary in order to leave some space for tall timber and the other critters we share this planet with.

> *There are 7.7 billion of us today, and by 2050, the UN predicts there will be 9.7 billion. It is no wonder people have despaired. But I believe we have a window of time to have an impact.*
> **—Dr. Jane Goodall (2019)**

As Population Institute Canada stated in 2003, "We must provide a compelling vision of the many benefits resulting from a much lower global human population of, for example, one billion, which was approximately the population of the Earth in 1850. The quality of life of all people would soar. We would have all the advantages of modern technology but little, if any, environmental deterioration."[10]

This movement will require a bold and provocative global effort, since every country is contributing towards climate change and other critical world issues in its own way. While developed nations are responsible for overconsumption to a great extent, out of desperation, developing

7 Pregnant Teen Help: www.PregnantTeenHelp.org
8 Population Speak Out: www.populationspeakout.org
9 NFPA (United Nations Population Fund): https://www.unfpa.org
10 Population Institute Canada: https://populationinstitutecanada.ca

countries are killing endangered species for food and to sell the animal parts. Also, as their standard of living increases, they are striving to match our highly consumptive lifestyles, often leading to further environmental destruction.

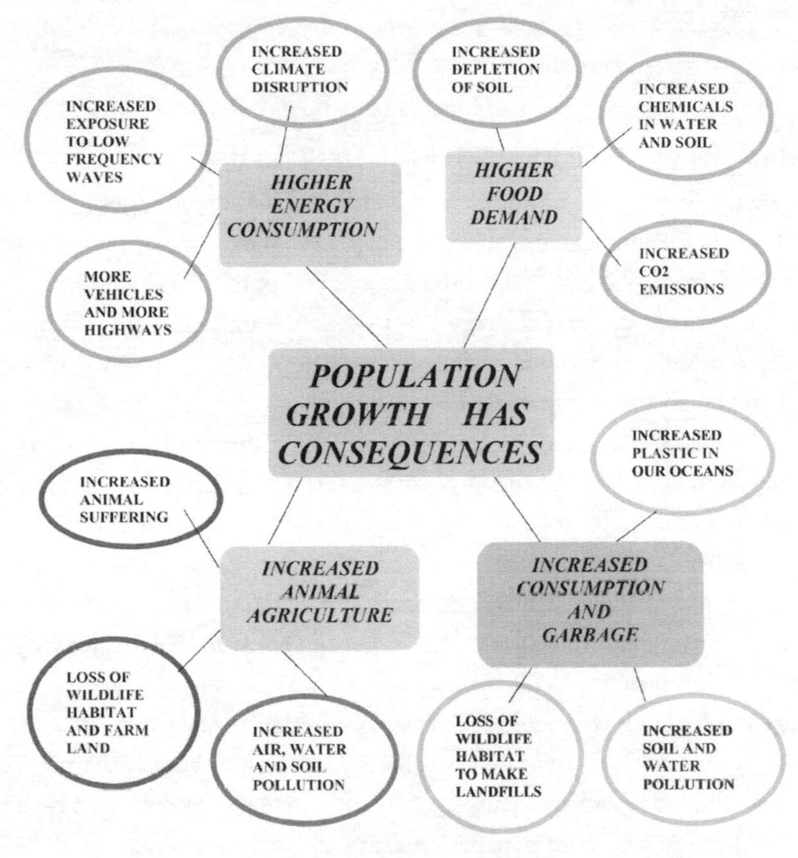

Figure 1. Population in Sync

People who believe that humans at the lowest end of consumerism do not contribute toward climate change should consider this fact: when you consider the deforestation necessary to provide their firewood alone, the destruction of wildlife habitat and carbon sinks is staggering.

Twenty years ago, Cornell University Professor of Ecology and Agricultural Science Dr. David Pimentel gave this warning: "If the human population increases dramatically over the next several decades, as it is projected to do, the strains on these limited resources will grow as well. Some people are starting to ask just how many people the Earth can support?... There is no solid answer yet, but the best estimate is that Earth can support about 1 to 2 billion people with an American standard of living, good health, nutrition, prosperity, personal dignity and freedom."[11]

Dr. Pimentel recognizes that "a drastic demographic adjustment to two billion humans will cause serious social, economic and political problems," but insists that "continued rapid population

11 Dr. David Pimentel, US professor on population growth, "Earth's carrying capacity, biofuel, big money & politician ignorance," archived web article: https://sites.google.com/site/biofuelgenocide/pimentel-david

growth will result in even more severe social, economic and political conflicts – plus catastrophic public health and environmental problems."[12] Humane scenarios for accomplishing this will be discussed throughout the book.

Hundreds of distinguished thinkers have dared to speak out about the population crisis that has caused them so much hand-wringing over the last century, and many of their quotes can be found on my website at www.populationinsync.net.

Getting Started: Where the Rest of This Book Will Take Us

My previous book, *Growing Pains*, took an in-depth and eye-opening look at Earth's greatest threat: that of too many people living on one small planet. It exposed the myths and taboos that are holding us back from addressing this critical issue and pointed us to a path of sustainability. In this book, I will take a more inclusive and up-to-date look at the global impact of EIGHT BILLION people. I will further demystify the population puzzle, explore emerging and more subtle impacts (eco-anxiety, compassion fatigue, etc.), and highlight the brave and rare efforts being made to bring positive solutions into the world. As a **unique feature**, each critical world issue will be followed by **solutions and success stories**, recognizing that so many books of this type fail to include them.

I will invite you to look at our family values and traditions. Are they really true to who we want to be as a society? Each country must ask if their traditional way of thinking is adequate to solve the enormous global problems facing us today. We will consider the **greater good** and then take a look at our present lifestyle choices. Do they match up? Will it take a small, furious Swedish girl to wake us up?

When Greta Thunberg addressed the United Nations in 2019, she let a room of world leaders know exactly what they were doing by refusing to act swiftly and decisively on climate change. "You have stolen my dreams and my childhood with your empty words," she told them. "We are in the beginning of a mass extinction and all you can talk about is money and fairy tales of eternal economic growth – how dare you!"[13] This small girl's presence and message are powerful because she reminds everyone that the threat of climate change is not only to the planet, but the victims are humans; more specifically, they are today's children. And they are not the only living beings who are victims. Africa's elephants have been in dangerous decline, from an estimated 10 million in the early 1900s to about 400,000, both due to population pressures and the ivory trade. Despite the presence of elephant reserves across the continent, they are not safe from poachers. According to Professor Carl Safina, "The rising tide of humanity has cut these reserves into isolated islands in the stream of time. In just the last 40 years in Kenya, humans quadrupled. Meanwhile, elephants dropped by four-fifths."[14]

When Paul Hawken and his team investigated and ranked carbon-reduction solutions for their Drawdown project, they found that, of all potential solutions, the combination of family planning and the education of girls (call it the female-empowerment package) carried the most potential to

12 *Ibid.*

13 Valerie Volcovici and Matthew Green, "'How dare you': Greta Thunberg gives powerful, emotional speech to the UN," Reuters, September 23, 2019: https://globalnews.ca/news/5940258/greta-thunberg-speech-un/

14 Carl Safina, *Beyond Words: What Animals Think and Feel* (New York: Picador, 2015), pp. 118, 122.

reduce greenhouse gases later this century. Together they could prevent 120 gigatonnes of GHGs by 2050—more than on and offshore wind combined.

We will find hope in emerging new groups like Fair Start Movement and BirthStrike, and in the Family Planning 2020, Thriving Together, and Self Care global initiatives discussed in Chapter 13. The Interfaith initiative discussed in Chapter 11 will surely give hope for Africa and set an example for the global community. Yes, there is a light at the end of the tunnel!

But most importantly, LET'S TALK! I want to get everyone talking about population again, and make population the buzzword it was in the '70s. We should make it part of every conversation from climate change to the economy.

The scientific facts revealed in this book will leave no doubt in anyone's mind about the implications and consequences of human population growth. We are predicted to add another billion people to this small crowded rock we live on in the next twelve years. Where will they live? What will they eat? But even more importantly, my book will leave no doubt as to the many practical solutions available. My earnest hope is that the information presented here will motivate us all to take this message to heart while we still have a choice.

In 1993, Population Connection summed it up aptly: "Imagine taking a journey into uncharted territory without a compass or a guide and with no idea whether you have enough provisions for everyone who is going with you. Crazy, right? But that's exactly what our country is doing – lumbering into the future, poorly prepared, with the sketchiest of plans and perhaps not even headed in the right direction."[15]

It took over 200,000 years to add our first billion people to the planet around 1804. It was another 123 **years** before it **reached two billion** in 1927. But it took only thirty-three **years** to **reach** three **billion** in 1960. It took only twelve years to add our last billion in 2011, and it will take less than twelve years to add the next billion (see Figure 1.2). This kind of growth is not sustainable on a finite planet if we are to leave room for old growth forests and wild critters.

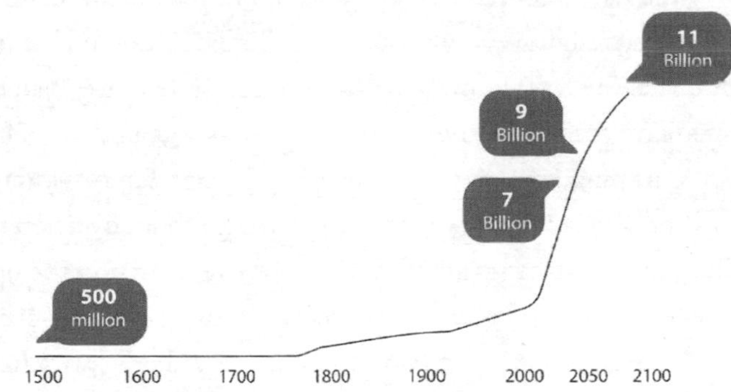

Figure 2. Population Growth, Population Institute Canada

The dilemma is that humanity's self-proclaimed dominance over Earth is really a 10,000-year experiment we have unleashed that is now threatening to destroy our planet. We are muddling our way through history in trial-and-error mode with little more understanding of the big picture than we had in the Stone Age. What is desperately needed at this time is a plan of action that acknowledges the folly of our ways and the urgency of changing our course.

According to Population Institute Canada, "For most of history, human population had little impact on our planet. In fact, for the last 200,000 years of human history, our population was less

15 Zero Population Growth (now Population Connection), "Beyond the Green Revolution: Singin' the Population Blues," *ZPG Reporter* 24, no. 4, September 1992.

than a quarter million. When Columbus came to America 500 years ago, global population was still only 425 million. If it had remained at that level, we would not be facing the high level of species extinction, global warming, or water shortages that are so worrisome today."

Humans quadrupled in the twentieth century. Although this is an average growth rate of just 1.39 percent per year, this innocent-sounding figure resulted in an exponential growth rate, soaring from 1.65 billion in 1900 to six billion in 1999.[16]

That rate of growth is referred to as "plague phase" by biologists. If any other species had begun to multiply at this rate, governments would have instructed biologists to take drastic measures to bring them under control as an "invasive species." Carrying capacity would have been calculated, and limits determined to bring the species into balance with its environment. I wonder, is it possible that these same limits could also apply to the human species, known scientifically as the "wise hominids" (*Homo sapiens*)?

Instead, anthropocentrism has guided our stewardship of the planet. The *Oxford English Dictionary* defines anthropocentrism as "regarding humans as the central element of the universe, and interpreting reality exclusively in terms of human values and experience." We have chosen to deny our natural place in the scheme of things, or our role in mismanaging the world's resources.

At one time humans actually thought that Earth was the centre of the universe. It wasn't until 1543 that Copernicus disproved this theory, causing quite a stir.[17] This revolutionary new thinking greatly influenced a re-emergence of atheism that lasted for almost two centuries. So the idea that humans are not intended to be the central element of our planet, but rather an integral part of an ecocentric (ecology-centred) life system, is equally incredible today.

The Earth-centered solar system is like the human-centred ecology system that persists because of strong religious dogma and an economic system out of sync with today's environment. We need to make the transition to the Copernican system in which we throw off the dogma and look at the facts. We are part of the environment, not independent of it.

Carl Jung, one of the fathers of psychology, famously remarked, "People cannot stand too much reality." Perhaps that explains why warnings about overpopulation from many of our most respected authorities are being ignored. To find the real reasons, the operational reasons, you must look to the power of some religions and corporate economists and their "growth at any cost" ideology.

Throughout the last two centuries of our history, we have chosen to deny that overpopulation is the missing link in achieving sustainable lifestyles and bringing humanity back into harmony with our environment. Yet, television brings us graphic scenes of human suffering and landscapes being savaged by the effects of ever-growing numbers of people every day. As populations in these developing nations become unmanageable, their citizens are encouraged to emigrate to find greater opportunities elsewhere. However, similar to the lifeboat scenario, this often causes additional problems in already overpopulated countries that allow entry. In addition, the immigrants often adopt these high-consumption lifestyles, causing even greater strains on the world's resources. It

---

16 World Population Growth, Our World in Data, "How has world population growth changed over time? World population from 10,000 BC to today [Graph]": https://ourworldindata.org/world-population-growth

17 Stanford Encyclopedia of Philosophy: https://plato.stanford.edu/entries/copernicus/

is hard for them to resist, with the combined appeal of higher wages, greater variety of products available, and tremendous advertising and societal pressures.

Often, this is a situation of people from tropical climates migrating to countries with colder climates, where they will require far more resources just to survive. *But before we in colder climates pound ourselves on the head for our high energy consumption, we must remember that it takes far more resources to survive in a cold climate than in a hot one.* The basic necessities of food, clothing, and shelter needed to survive varying and often extremely cold seasonal temperatures are far greater. Growing seasons are far shorter. Also, if we want people in developing countries to have the basic necessities they require for a humane existence, their standard of living will have to improve dramatically. This, of course, will increase their consumption rate and eventually bring it up to par with developed countries, factoring in their energy requirements. We must realize that the greatest rate of **growth in consumption** in the 2020 decade will be in **developing countries**. The best scenario would be to help people live a humane and sustainable existence in their own country, rather than taking them away from their families and homeland in order to help them.

However, in the last half-century, the migration of people has often been dictated by the wealthy elite or by force through the slave trade, military, and sex trade. It is often left up to the most powerful decision-makers on the planet—those that prey on the vulnerable. It is mainly our present global economic system and elite dictators that have allowed this travesty to occur, and they are actually in control here. I will talk about this in more detail in Chapter 10.

Landmarks in Population History

Baby Boomers will no doubt remember the findings of the Club of Rome, published in 1972 from their Project on the Predicament of Mankind. The technical part was written by a Massachusetts Institute of Technology project team, which built an elaborate computer model to track a number of variables from 1900 to 1970 and project their paths to the year 2100. The results were disturbing, to say the least. The variables in the model were resources, population, pollution, industrial output per capita, and food per capita. The model gave rather similar pictures under all of the scenarios where population was not controlled. Population, industrial output per capita, pollution, and food per capita all grew exponentially. Natural resources declined with the growth of population and industrial output.

Since the time of the Club of Rome project in 1970, there have been numerous landmarks in population history, some of which are listed below:

1970: Two months after the first Earth Day, the First National Congress on Optimum Population and Environment convened in Chicago. Religious groups—especially the United Methodist Church and the Presbyterian Church—urged, for ethical and moral reasons, that the federal government adopt policies that would lead to a stabilized U.S. population. President Nixon addressed the nation about problems it would face if U.S. population growth continued unabated.

1970: Paul Ehrlich's book, *The Population Bomb*, which was written in 1968, was being widely read and discussed.

1970: Title X of the *Public Health Service Act* was established as the single largest source of federal U.S. funds for domestic family-planning programs. Title X supported approximately 4,000 family-planning clinics serving low-income individuals and teenagers.

1972: Stockholm Conference on Population

1972: The Rockefeller Commission issued its 186-page report concluding that further population growth would do more harm than good.

1973: ***Roe v. Wade*** Supreme Court decision prohibited states from restricting access to abortion and protected a woman's right to choose to terminate a pregnancy prior to fetal viability.

1974: The United Nations designated 1974 as World Population Year. The aim of this was to focus the attention of governments, organizations, and individuals on the possible effect on various aspects of life that the size and growth of the world's population would have.

1974: President Nixon, recognizing the gravity of the overpopulation problem, directed in National Security Study Memorandum 200 (NSSM) that a new study be undertaken to determine the "Implications of World Population Growth for U.S. Security and Overseas Interests."

1974: International Population Conference in Bucharest, Romania

1975: Our planet was host to four billion people.

1975: President Ford completed the 1974 NSSM study, which recommended that America should try to reach population stability by 2000. This would have required a policy of encouraging one-child families.

1980: The abortion pill is invented in France and in 1987 is made available to the French public. It would take until 2000 for it to be made available in the U.S., and until 2015 for Health Canada to approve the pill.

1981: The Global 2000 Report, from the U.S. Department of State and Council on Environmental Quality, presented to President Carter the following recommendation: "The United States should develop a national population policy which addresses the issues of population stabilization as well as just, consistent, and workable immigration laws."

1984: At the United Nations International Conference on Population held in Mexico City, President Reagan initiated the "Mexico City policy" (also known as the "global gag rule"). This policy withheld U.S. funding from any private organization that provided information, counselling, or health care related to abortion, such as the International Planned Parenthood Federation.

1986: The Reagan administration withheld its contribution to the United Nations Population Fund (UNFPA) over allegations that UNFPA was participating in the management of China's controversial One Child Policy, despite verification of the fact that UNFPA was not involved in any coercive activities.

1987: The Brundtland Report of the World Commission on Environment and Development was released, promoting family planning.

1987: Global population reached five billion.

1989: World Population Day was established on July 11 by the Governing Council of the United Nations Development Program. It was inspired by the public interest in Five Billion Day on July 11, 1987, approximately the date on which the world's population reached five billion people.

1992: One of Bill Clinton's first acts as president was to sign a Presidential Memorandum **overturning the "Mexico City policy"** of 1984.

1992: UN Conference on Environment and Development, Earth Summit at Rio de Janiero.

1993: President Clinton created the President's Council on Sustainable Development and charged it with helping to "grow the economy and preserve the environment," objectives that some see as conflicting today.

1994: International Conference on Population and Development was held in Cairo. The Vatican was successful in its efforts to veto proposals that would have addressed the overpopulation issue. Therefore, little progress was made.

1994: South-South Initiative: Countries of the south formed the Partners in Population and Development (PPD). This is an intergovernmental initiative created specifically for the purpose of expanding and improving south-to-south collaboration in the fields of reproductive health, population, and development. PPD was launched at the 1994 International Conference on Population and Development (ICPD), when ten developing countries from Asia, Africa, and Latin America formed an intergovernmental alliance to help implement the Cairo Program of Action.

1995: Fourth World Conference on Women in Beijing marked a significant turning point for the global agenda for gender equality. It sets strategic objectives and actions for the advancement of women and the achievement of gender equality in numerous critical areas of concern.

> *The Beijing Declaration and the Platform for Action, adopted unanimously by 189 countries, is an agenda for women's empowerment and considered the key global policy document on gender equality.*
> **—UN Women**

1999: World population reached six billion.

1996: The President's Council on Sustainable Development called for a "move toward stabilization of U.S. population." Yet, funds for population programs were slashed by nearly 90 percent.

1999: President Clinton and Congress ended a three-year standoff and agreed to pay nearly $1 billion in back dues owed to the UN, which included funding for family planning programs. A concession the Clinton administration was forced to make to the Republican-dominated House of Representatives was to restrict funding to programs that advocated abortions in foreign countries.

1999: France becomes the first country in the world to distribute a brand of emergency contraception in pharmacies without prescription or parental consent needed.

1999: *Catholics for a Free Choice* launched the See Change campaign to change the status of the Holy See (Vatican) at the United Nations. This has not been successful, and the Vatican remains the only religious entity allowed to influence the United Nations at conferences through veto power.

2000: On July 14, Governor Gray Davis signed a proclamation declaring the week of October 22–28, 2000, as World Population Awareness Week (WPAW) in California. WPAW is an intense educational campaign designed to create public awareness about the trends in world population growth, the detrimental effects they have on our planet and its inhabitants, and the urgent need for action in order to change this situation.

2001: President Bush cut funding to family planning programs, promoting abstinence as the most effective form of birth control. He reinstated **enforcement of the Mexico City "global gag rule."**

2009: On his third day as president, **President Obama reversed the "global gag rule'** for family planning, which had been imposed by President Bush eight years earlier. He also proposed that family planning provisions be included in his stimulus bill that would have expanded eligibility for Medicaid-funded family planning services. However, the Democrats caved in to Republican outrage, and this proposal was not adopted.

2011: **World population reached seven billion.**

2012: The Family Planning 2020 initiative was launched as a global partnership to empower women and girls by investing in rights-based family planning.

2015: The abortion pill, mifepristone, was approved by Health Canada in 2015 and became available in Canada in January 2017. Mifepristone, also known an RU-486, is considered the "gold standard" for inducing safe, effective medical abortions and is on the World Health Organization's list of essential medicines.

2017: President Trump **reinstated the "global gag rule"** for family planning, a giant step backwards for women's rights and family planning.

2018: "World Scientists' Warning to Humanity: A Second Notice": This document was renewed and widely distributed. Twenty-five years previously, in 1992, the Union of Concerned Scientists, consisting of more than 1,600 scientists, including the majority of living Nobel Laureates in the sciences, penned the "Scientists' Warning to Humanity." Now, the authors of the second notice are renewing that call on humankind to curtail environmental destruction and caution that "a great change in our stewardship of the Earth and the life on it is required, if vast human misery is to be avoided." They warn that fundamental changes are urgently needed to avoid the consequences of our present course of continued human population growth.

2018: Our population crisis has prompted a new movement initiated by the youth of the planet who are demanding action: Birthstrike'is a group refusing to have kids because of the ecological crisis overpopulation has created. Fair Start Movement is a group that believes that better family planning is the most effective and comprehensive way to protect kids, animals, and the environment, and to make our communities stronger.

2019: On World Population Day, Thriving Together launched its campaign. Among its signatories are Dr. Jane Goodall, Greenpeace, Marie Stopes International, the United Nations Population Fund, and the Bill & Melinda Gates Institute for Population & Reproductive Health.

> *Now 151 organizations have joined forces to create the Thriving Together campaign, which is spearheaded by the London-based Margaret Pyke Trust. Together they spend £8 billion each year on family planning and environmental work in 170 countries globally.*
> **—Thriving Together**

The campaign will focus on getting the environmental organizations to talk about family planning and vice versa in a coordinated approach.

2019: The Self Care Global Coalition, an initiative of Population Services International, was formed. At least half of the world's population cannot obtain essential health services. It is time to recognize the importance of self-care as an integral component of the global health system.

2019: This year marks the 25th anniversary of the groundbreaking International Conference on Population and Development (ICPD), which took place in Cairo in 1994. At that conference, 179 governments adopted a Programme of Action, recognizing that reproductive health, women's empowerment, and gender equality are an important pathway to sustainable development. The 2019 International Conference on Population and Development in Nairobi highlighted women in developing countries without access to contraception but unfortunately was silent on the need to specifically address population growth.

2021: President Joe Biden overturns the "global gag rule" to restore a full range of family planning services.

2022: World population predicted to reach EIGHT BILLION.

And what else does the future hold for us? When will India's population exceed China's—predicted by 2027? When will Africa reach two billion—predicted by 2050? It was under 200 million in 1950, a ten-fold increase in 100 years. Will global population in 2050 be nine billion or eleven billion? Or dare we hope for six billion? This will depend on which of the vastly differing scenarios we set as a goal in 2020.

If we are willing to actually take population reduction seriously and reverse our engines, perhaps we could even stop population growth by 2027, and it could fall back to six billion by 2050. This scenario would greatly benefit all life on the planet.

Each of the above events, conferences, or reports greatly affected the population movement. Many supported the notion that population and consumption could not keep on growing indefinitely, and that change was both urgent and mandatory. However, over the years, support for the "global gag rule" went up and down like a yo-yo with each new president. In 2021 perhaps citizens will realize that we must turn Martin Luther King's dream for universal family planning programs into a reality.

I have outlined some of the circumstances and people who have brought us to the present population dilemma. If we are to learn anything from our history, I hope that it will be to rectify our past mistakes. We must insist that our leaders deal with the topic of population in an honest and meaningful way, just as Greta Thunberg has with the climate crisis.

In *A Short History of Progress,* Ronald Wright outlined the high stakes we face: "The future of everything we have accomplished since our intelligence evolved will depend on the wisdom of our actions over the next few years."[18]

18 Ronald Wright, "Can We Dodge the Progress Trap?," *The Tyee,* September 20, 2019: https://thetyee.ca/Analysis/2019/09/20/Ronald-Wright-Can-We-Dodge-Progress-Trap/?utm_source=daily&utm_medium=email&utm_campaign=200919

In the 15 years since Wright's prediction, our population has risen by over a billion (we've added another China or 40 more Canadas—to the world), plastic is set to outweigh all fish in the ocean by 2050, human trafficking and sexual exploitation are at epidemic levels, and political tip-toeing is still the rage. A child-centric approach to family development, which aims for a smaller family size, is so critical at this point in our evolution, yet remains just out of reach! Consequently, those in extreme poverty and hunger still number at least a billion. Fortunately, there are many solutions yet to come.

The commanding height of this group, the billionaires' club, has more than 2,200 members with a combined known worth nearing $10 trillion....
—Ronald Wright

Wright cautions: "The wealthiest billion—to which most North Americans and Europeans and many Asians now belong—devour an ever-growing share of natural capital. This super-elite not only consumes at a rate never seen before but also deploys its wealth to influence government policy, media content, and key elections. Such, in a few words, is the shape of the human pyramid today."[19]

While the climate change issue and the human footprint concept have triggered an attempt to reduce our consumption rates, population rarely gets a mention, as if the staggering number of people consuming our resources had no relevance to how great our footprint is on this planet.

At this critical crossroads, we need to decide what kind of world we want for our children and grandchildren. Do we want a balanced healthy society, living in harmony with the natural world? Do we want sustainable economies and communities? In order to achieve sustainable lifestyles, it is essential to develop a strategy to achieve a sustainable population level first and foremost.

Humankind's most noble objectives, such as human rights for all—as endorsed in the Universal Declaration of Human Rights in 1948—will not be attainable in a world in which overpopulation is unravelling the web of life. If all the world's refugees and displaced people were to come together as a single nation, they would collectively create one of the largest countries on Earth.[20] So eradicating poverty and reducing the number of wartime and environmental refugees would go a long way toward reducing human suffering and establishing these basic human rights.

In this book, scientific facts and expert opinions will demystify the causes and consequences of our disturbing population dilemma. I will explore the solutions, some surprisingly simple and practical, that everyone can begin implementing right now. If we fail to recognize the important role overpopulation is playing in our ecological, social, and financial predicament, we may soon find that the destruction is irreversible.

19 ibid

20 According to the most recent UNHCR figures available, there are a total of 70.8 million "displaced people" worldwide, which includes 25.9 refugees and 3.5 million asylum seekers: https://www.unhcr.org/figures-at-a-glance.html

CHAPTER 2
REASONS FOR HIGH POPULATION LEVELS

Family planning, to relate population to world resources, is possible, practical and necessary.
—Martin Luther King Jr.

In 1963, Reverend Martin Luther King Jr. gave his "I Have a Dream" speech, which inspired a generation and changed the course of history. Now, decades later, we are still celebrating his legacy as we pause to honour his message of hope. Every year on Martin Luther King's birthday (January 15), celebrities and working folk alike pay tribute to this legend of a man. Oprah dedicated a show to him, noting the sacrifices he had made and the significance of her hero's life. He was the youngest person ever awarded the Nobel Peace Prize and was remembered for his supreme intelligence and his dedication to the civil rights movement.

However, what most people have neglected to remember is one of his most captivating and stirring speeches given in 1966, the one about family planning. He believed that overpopulation was a special and urgent concern even at that time, when our population was far less than today's 7.7 billion. He pointed out that we spend paltry sums for population planning, even though our spontaneous growth is an urgent threat to life on our planet. It seems to me unfortunate that his most important message has been forgotten by our society at a time when it so desperately needs to be heard.

As King passionately stated that year, "There is no human circumstance more tragic than the persisting existence of a harmful condition for which a remedy is readily available. Family planning, to relate population to world resources, is possible, practical and necessary. Unlike plagues of the dark ages or contemporary diseases we do not yet understand, the modern plague of overpopulation is soluble by means we have discovered and with resources we possess."[21]

> *What is lacking is not sufficient knowledge of the solution but universal consciousness of the gravity of the problem and education of billions who are its victims.*
> **—Martin Luther King**

It is this senseless refusal to acknowledge overpopulation that is at the very core of all of our critical world issues and is causing so much unnecessary suffering. POPULATION: Surely this one innocent

21 Population Media Center, *Martin Luther King, Jr. Discusses Family Planning and Population* January 15, 2018: https://www.populationmedia.org/2018/01/15/martin-luther-king/

word should not have the power to cause the terror, confusion or irrational behaviour it is alleged to have. At this urgent time in our evolution, we desperately need a bold and powerful movement that will start a meaningful conversation about population. This book will provide many positive steps that we can take to make this happen.

Factors for High Birth Rates

The unfortunate truth is that it's not only the political, environmental, media, and health communities that are failing to initiate dialogue about population, but sadly parents are also failing to talk to their children about family planning or sex education. Further, this seems to indicate a need for better communication within families in both developed and developing nations.

Let's Talk About It at Home

GeoPoll is the world's largest mobile survey platform, with a network of 200 million users in Africa and Asia. By asking people questions on their mobile phones without the need for data plans or Internet access, GeoPoll is able to help customers understand trends, preferences, and conditions in countries where data has traditionally been difficult and expensive to obtain.

A 2019 Geopoll report shows that **parents have failed to educate their children**, especially girls, on sexual reproductive health, leaving the responsibility to the Internet. This report also shows that youths are more worried about contracting sexually transmitted diseases than unplanned pregnancies or being sexually assaulted. While a majority of the youth would be interested in sexual reproductive health education, parents play a minimal role in imparting this knowledge to them.

For example, 62 percent of the 1,125 African youths polled said they learnt about menstruation from school, with just 12 percent saying they were informed of the same by their parents. When these youths between the ages of 18 and 24 were asked what the most important source of information on sexual reproductive health was, parents also came a distant second at 17 percent, the same level with social media, the Internet, books, and magazines. The most important source was TV and radio at 42 percent for non-educated youths. For educated youth, the picture was even worse as parents did not appear among the listed top nine preferred sources of sexual reproductive health. Polls in other countries have shown similarly disturbing results. (More about this in Chapter 13.)

It would be a serious mistake to continue to ignore the population crisis that is swallowing up and fouling our natural ecosystems. Unfortunately, human population size and growth remains a taboo topic in mainstream media and even in the environmental community. I believe most environment groups are betraying the interests of their members by refusing to address the population issue, as if it didn't impact local issues in any way.

Let's Talk About It in the Media

In his recent book *The Treason of the BBC*, Jack Parsons describes how in 40 years of correspondence with the BBC, he failed to get the network to include a mention or discussion of the probable role of overpopulation in exacerbating many of the political, economic, social, and environmental troubles it was vividly portraying in its documentary programs.[22] Similar attempts have been made by myself, Population Institute Canada, and many others to get the CBC to include a meaningful discussion on population, with little success. It has only been in the last couple of years that the CBC has allowed any conversations on the topic.

Of course, there have been many brave souls over the years, like Paul Ehrlich and Thomas Malthus, who have defied the powers that be and tackled this seemingly overwhelming issue. I will be discussing the contributions of these pioneers in Chapters 10 and 11. Then there are those who recognize the population problem, but prefer to tackle it without actually ever mentioning the taboo word "population." This **population phobia** likely stems from the disturbing amount of misinformation out there, conflicting positions on abortion, as well as the coercive measures taken in China, Peru, India, and other countries in the past that have now been discontinued.

In 2018, environmental journalist David Roberts explained why he never writes about overpopulation, although he believes reducing population would have more impact than virtually any other climate policy. It would simply be a matter of choosing the correct terminology, such as using "female empowerment," "family planning," and "education of girls" instead of "population planning." He points out that tackling population growth can be done without the enormous, unnecessary risks involved in actually talking about population growth. We would be contributing to the most effective solutions to the problem without stepping on any moral landmines, since it is now a universal goal to achieve gender equality. Also, these are goals shared by powerful pre-existing coalitions.[23]

Roberts adds that the mere mention of "population" raises all sorts of ugly historical associations, such as racism, xenophobia, or eugenics. Unfortunately, this is all the more reason that we need to make this a priority conversation and get this issue sorted out once and for all. The future of humanity depends on this dialogue.

Let's Talk About It with Open Minds

It has been my objective to maintain a balance of statistical and lay information for a wider perspective. The purpose here is not to be morbid, but to be realistic, and some may find this information shocking or distasteful, as reality so often is. There is no malice, racism, or discrimination intended. The only purpose of this book is to look at the numbers globally, and how we can bring the human numbers in sync with our natural world and reduce human suffering.

The points outlined below are fairly brief, as most of them will be discussed in more detail later in the book in the appropriate chapters. They are listed here only to give a glimpse of the countless implications in this complex topic of high birth rates.

---

22 The Center for Media and Democracy, SourceWatch, "Jack Parson's Archive," Optimum Population Trust: https://www.sourcewatch.org/index.php/Jack_Parson%27s_Archive. Parsons helped found the Conservation Society in Britain in 1967 and gave the keynote address to the 1993 World Congress on the Optimum Population.

23 David Roberts, "I'm an environmental journalist, but I never write about overpopulation. Here's why," *Vox*, November 29, 2018: https://www.vox.com/energy-and-environment/2017/9/26/16356524/the-population-question

Natural Instinct, Tradition, or Societal Norm?

In the past, tradition and ignorance dictated what the family norm should be. These conditioned value systems used when having a family often went unquestioned. Girls had been conditioned to believe that their only purpose was to grow up, have children, and be good mothers and wives. However, with changing times, the Internet, busy lifestyles, a rapidly growing population, and globalization, it is imperative that we rethink this premise. In fact, many women now are choosing to stop at two or to be childfree. In these troubled and more knowledgeable times, not everyone sees having children as a blessing.

Now, far more women do have a choice, and herein lies the solution to population growth. Birth control is more reliable, people are marrying later, marriages seem less stable, our world is a scarier place to raise children, and women are more reluctant to give up their careers. According to the National Center for Health Statistics, in 1975 about one in eleven women was childless by the age of 44. By 1993, that number had risen to about one in six. In 2018, 15 percent of women between the ages of 40 and 44 were childless, according to the 2014 U.S. Census Bureau Current Population Survey.[24] This is an encouraging step, but we must do much better if we are to divert a global catastrophe and meet the UN Population Goals.

> *Family planning could bring more benefits to more people at less cost than any other single technology available to the human race.*
> **—United Nations**

Just Say No to Coerced Parenthood

Many more couples are weighing all the pros and cons and deciding to follow the old Greenpeace slogan from the '70s and "Stop at 2" or be childfree. However, this is only possible where family planning programs are available and where couples are encouraged to recognize the benefits and options of small families or a childfree lifestyle. After a baby is born, marital satisfaction declines, says Ralph LaRossa, a professor of sociology at Georgia State University. When researchers chart this decline, it's a U-shaped curve that rises only as children get older and leave home. (More about this in Chapter 13.)

Many childfree couples claim that they give their decision to be childfree more thought than most parents give to having children. It is apparent, given that over 40 percent of pregnancies are unintended globally. Also, the numerous cases of abandoned babies, child neglect, and abuse would indicate that far too few people give adequate thought to having and raising children.

Some people feel that childfree couples are being selfish, but in reality the opposite is true. In 2009, Dr. Bonnie Bukwa, retired chemistry instructor at College of the Rockies, pointed out:

> Those without children make more opportunities and resources for the children of those who do breed. In effect the non-breeders are denying their own genes for the benefit and opportunities of the genes of others. It actually is a **VERY selfless act**,

24 Emma Gray, *A Record Percentage of Women Don't Have Kids. Here's Why That Makes Sense*, *Huffington Post*, April 9, 2015: https://www.huffingtonpost.ca/entry/childless-more-women-are-not-having-kids-says-census_n_7032258?ri18n=true

genetically equivalent to taking a bullet to protect another or throwing oneself on a grenade in combat! The really selfish ones are those having many children on purpose!

We often hear parents say that they both work because they want to give their children all the things they could never have. What they are really talking about is material possessions. But in order to do that, they are sacrificing the truly important things like spending time with their children, not the least of which is sharing experiences of nature with them.

The vast majority of today's children suffer from "nature deficit disorder," which can render them disconnected from the land and unsympathetic to the plight of our disappearing wildlife. Unfortunately, many mothers have bought into the misconception that they can do it all without any consequences.

Lack of Access to Family Planning Services

> *We believe that the Catholic tradition supports a woman's moral and legal right to follow her conscience in matters of sexuality and reproductive health.*
> **—Catholics for Free Choice**

According to the UN, over 220 million women around the world would like to limit their family size but are denied access to safe, affordable family planning programs.[25] The United Nations and participating countries have failed time and again to provide adequate funding for these family planning programs. It seems that governments can always find funding to continue senseless wars or pricey space missions, but not for this basic necessity that should be every woman's human right. Now this burden is placed to a great extent on volunteer organizations like Planned Parenthood Federation and new initiatives like Thriving Together. The costs of unintended pregnancies are often placed on taxpayers through medical treatment, child care, social services, and adoption programs.

Of course there is one aspect of family planning that is seldom brought up these days. Dare we mention it? Abortion must be part of this family planning services package, especially to deal with unwanted pregnancies from rape, teenage sex, child brides, and human trafficking, to be discussed in later chapters.

Unintended Pregnancies

Unintended pregnancies are pregnancies that are mistimed, unplanned, or unwanted at the time of conception. Sexual activity without the use of effective contraception through choice or coercion is the predominant cause of unintended pregnancy. Worldwide, the unintended pregnancy rate is approximately 40 percent of all pregnancies, but rates of unintended pregnancy vary in different geographic areas and among different socio-demographic groups. Efforts to decrease rates of unintended pregnancy have focused on improving access to effective contraception through improved counselling and removing barriers to contraception access.[26]

---

25 UN, *Family Planning*: https://www.unfpa.org/family-planning

26 Wikipedia, *Unintended Pregnancies*: https://en.wikipedia.org/wiki/Unintended_pregnancy

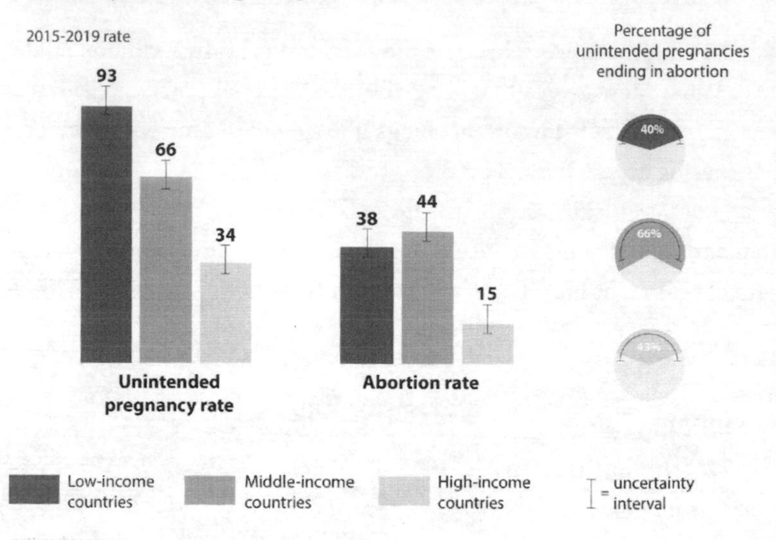

Figure 3. Guttmacher.org Besides the lack of funding for family planning services, there are numerous other reasons for unplanned pregnancies. Human trafficking, not giving up until a boy child is conceived, appeasing grandparents, mothers plagued with addictions or AIDS who are not capable of taking precautions, child brides, rape in its many forms, and religious terrorism are a few. These issues will all be discussed in later chapters.

Having children that are unplanned or unwanted is increasingly becoming the focus of child-centric conversations happening around the world—conversations about child-bearing choices versus protecting the environment. In an attempt to be more Earth-friendly, our reproductive choices are now increasingly inciting eco-anxiety. Today it is so very important, in this time of climate chaos, to **consider ways that we can prevent the millions of unplanned pregnancies from occurring. For every extra child that is born, a piece of nature will have to be sacrificed to make room for it.** Species are disappearing at an alarming rate as our population continues to grow and encroach on wildlife habitat. These are all genuine concerns.

The Global Rape Epidemic

Around the world, sexual abuse and rape are everyday violent occurrences, affecting close to a billion women and girls over their lifetime. Organizations like Equality Now have been working with the survivors of rape and sexual assault to get justice, strengthen laws, and increase enforcement. It has recently released a global report titled *The World's Shame: The Global Rape Epidemic.*[27]

The numerous forms of rape, from war rape to date rape, are the cause of millions of unwanted pregnancies every year, as will be discussed in Chapter 6. The US National Library of Medicine points out, "Rape-related pregnancy occurs with significant frequency. It is a cause of many unwanted pregnancies and is closely linked with family and domestic violence. As we address the

27 *The World's Shame: The Global Rape Epidemic*, Equality Now: https://www.equalitynow.org/the_global_rape_epidemic_campaign

epidemic of unintended pregnancies in the United States, greater attention and effort should be aimed at preventing and identifying unwanted pregnancies that result from sexual victimization."[28]

Globally, governments have committed and recommitted to ending all forms of violence against women and girls, including sexual violence. In September 2015, the UN General Assembly adopted "Transforming Our World: The 2030 Agenda for Sustainable Development" ("Agenda 2030").[29]

> *If this sexual violence were a medical disease, it would be getting serious attention and the funding to address it from governments and independent donors alike.*
> **—Equality Now**

The goals of this document were to achieve gender equality and empower all women and girls, as well as eliminate all forms of violence against women and girls in the public and private spheres. However, most governments have put little effort into implementing these goals since the UN has no power of enforcement. Also, the global elite who benefit from overpopulation have a great deal of influence and power over decisions regarding family planning.

Technology, in particular the Internet, has enabled sex trafficking and sexual exploitation to become the fastest growing criminal enterprise in the world. We must work together to develop new ways to address this illegal industry. (See also Chapters 10 and 11.)

Religious or Moral Reasons

To surrender to ignorance and call it God has always been premature, and it remains premature today.
—Isaac Asimov

Religious compliance is often a form of religious terrorism, as will be discussed in great detail in Chapter 11. Poor and vulnerable people are often persuaded to follow the unfounded and outdated beliefs of the representatives of that religion, not necessarily the religion itself. Often, foreign aid is linked to agreement to these religious orders, and options are very limited. In many of these cases, family planning programs are denied, even for the prevention of AIDS or in the case of rape. However, in 2018 the Thriving Together initiative was launched, and it is tackling this issue head-on. So this is a huge step forward, and this initiative promises to make some fundamental changes in the role religion has played.

The most powerful religious entity that opposes family planning is the Vatican. Although the Catholic religion itself, including many priests, nuns, and practising Catholics, often supports family planning, the Vatican does not. As a representative of the Roman Catholic Church, the Pope refuses to give up this outdated and unpopular stance.[30]

---

28 Rape-related pregnancy: estimates and descriptive characteristics from a national sample of women, PubMed.gov, Aug. 1996, https://pubmed.ncbi.nlm.nih.gov/8765248/

29 "Transforming our world: the 2030 Agenda for Sustainable Development," UN Sustainable Development Goals Knowledge Platform: https://sustainabledevelopment.un.org/post2015/transformingourworld

30 Catholics for Free Choice: http://www.catholicsforchoice.org/about-us/

Reasons for Low Death Rates

Although the greatest increase in population is due to high birth rates, low death rates also play a role in maintaining a high population. In some instances, we will go to any length or expense to keep a patient alive, often against his or her will. It is unfortunate that the heroic measures we often take to extend human life are not similarly taken to prevent unwanted births.

"Like knowing the time the first nuclear device exploded (5:30 a.m., July 16, 1945), it is known when the fuse was lit on today's population bomb: February 1935," said plant pathologist William Paddock in 1990.[31] A common theme to Paddock's research and writings was the belief that famine and population growth were linked and that the Green Revolution was instigated to address overpopulation. He cautioned that optimism about humanity's ability to feed itself, as today's rate of population growth continues, is precisely what we do not need and cannot afford in the race with the population bomb.

> *Because of our expertise and historical background, plant pathologists have an obligation to send an accurate message to world leadership.*
> **—William Paddock, plant pathologist**

In February 1935, Gerhard Domagk, a pharmaceutical chemist, discovered prontosil, the first of the sulfa drugs. It was a turning point in medicine when it was learned that bacteria could be attacked within the living body. This was followed by the discovery of penicillin in 1939.

Simultaneously in 1939, noted Paddock, another technological breakthrough was the discovery by Paul Hermann Müller of the insecticidal properties of DDT. Again, World War II rushed the product into production. DDT became a major factor in the increase in food production during the next twenty years. As Paddock pointed out, "The results of the wonder penicillin and the miracle DDT were 20th century equivalents of Ireland's 19th century potato. Müller, Chain, Florey, and Fleming all received the Nobel Prize for discoveries that ironically helped to turn loose the unprecedented human growth rate now threatening the survival of much of humankind."[32]

In addition to the exponential increase in population fostered by the Green Revolution, it's interesting to note that it was in 1964 that the Vatican gained veto status in the United Nations. In that capacity, the Holy See has been able to influence the decisions and recommendations of the United Nations with its relentless campaigns against family planning programs. Also during that period, there was a shift to petroleum-based energy for agriculture, allowing for increased food production.

Within 15 years after the inauguration of the United States foreign aid crusade, it became abundantly clear it wasn't working. Billions were spent to alleviate hunger and reduce poverty, but every year there was more hunger and more poverty. Paddock advises, "Governments must be told, in a way they can hear, that they are condemning their citizens to greater poverty when they leave to the

---

31 William Carson Paddock, *Washington Post* obituary, March 13, 2008: https://www.washingtonpost.com/wp-dyn/content/article/2008/03/12/AR2008031204230.html. Paddock, who died at 86 in 2008, developed a disease-resistant strain of corn high in vitamin A and wrote extensively on world food issues, including the book *Hungry Nations* (1964).

32 *Our Last Chance to Win the War on Hunger*, William Paddock, Carrying Capacity Network Handbook, 1994

future action on the sticky problem of exponential population growth and that political tar baby: motivating their citizenry to have fewer children."[33]

The Right to Die with Dignity

> *Dying is not a crime. How could anyone think it is a crime to help a suffering human end his or her agony?*
> —Dr. Jack Kevorkian

Dr. Jack Kevorkian (born May 26, 1928) is a controversial American pathologist. Nicknamed "Doctor Death," he is most noted for publicly championing a terminal patient's "right to die," aiding those who wanted to die to reach their goal without suffering. Unfortunately, Kevorkian was convicted for his beliefs about the right to die and served eight years in prison.

This is how Ronald Sokol, former lecturer in law at University of Virginia, puts it:

> A generation ago, in 1980, a number of people in France formed an Association for the Right to Die with Dignity (ADMD), which now has over 40,000 members. As medical care improves and people live longer, one can expect to see more such associations around the world, and eventually a change in perspective. At present, the law focuses on the act of the physician or nurse, and not on the rights of the patient. As that focus shifts so that the right of the patient to die with dignity becomes paramount, one can expect to see the law proclaim a fundamental right.[34]

This new perspective is shared by other advocacy groups around the world, including the British Columbia Civil Liberties Association (BCCLA). The BCCLA has been at the forefront of championing critically ill patients who wish to see Canadian law changed to allow them a medically assisted death. Since February 2015, Canada has had medical assistance in dying legislation for people with a "grievous and irremediable condition."[35] However, the legislation requires that a patient's natural death must be "reasonably foreseeable," that they must be in a state of advanced decline. Some patients with conditions that are debilitating—literally a life sentence—feel this doesn't go far enough.

In 2018 the BCCLA represented Julia Lamb in the Supreme Court of British Columbia.[36] Lamb, who was only 25, suffered from spinal muscular atrophy. That means she could have lived for years, even decades, with her condition. However, she was fighting for the right to end her suffering on her own terms. "My biggest fear is that if my condition suddenly gets much worse, which can happen

---

33 *Our Last Chance to Win the War on Hunger*, William Paddock, Carrying Capacity Network Handbook, 1994 ibid

34 Ronald Sokol, "The Right to Die," *New York Times*, March 21, 2007: https://www.nytimes.com/2007/03/21/opinion/21iht-edsokol.4978393.html

35 Government of Canada website, *Medical assistance in dying*: https://www.canada.ca/en/health-canada/services/medical-assistance-dying.html

36 BCCLA online, *Lamb v. Canada: the Death with Dignity Case Continues*: https://bccla.org/our-work/blog/lamb/

any day, I will become trapped," she told a news conference in Vancouver. "If my suffering becomes intolerable I would like to be able to make a final choice for how much suffering to endure."[37]

In court, the Government of Canada's expert witness admitted that Lamb would now qualify for an assisted death if she requests the procedure.[38] In a fall 2019 BCCLA update on the case, they reported, "It confirms that people who are seriously ill and face enduring and intolerable suffering have the right to die with dignity, even if they are not at or near the end of life and their deaths are not considered 'reasonably foreseeable.'"[39]

The World Federation of Right to Die Societies presently consists of 37 member organizations from 23 countries. Assisted suicide has been legal in Switzerland since 1942, providing a doctor has been consulted and the patient is aware of the consequences of their decision.

It is after a long-fought battle that more options are becoming available for the terminally ill in some countries. But what about those in other countries and those with diminished life quality, such as the elderly, the sick, or persons with a disability? Many people in these conditions have given a great deal of thought to how they might wish to die. They say that control, peace, and dignity are what count.

---

37 Laura Stone and Sean Fine, *B.C. woman, rights group file legal challenge against assisted-dying law*, May 16, 2018, *Globe and Mail*: https://www.theglobeandmail.com/news/politics/rights-group-launches-legal-challenge-of-assisted-dying-law/article30623211/

38 Canadian Press, *B.C. woman wins right to doctor-assisted death*, City News Radio, September 18, 2019: https://www.citynews1130.com/2019/09/18/cp-newsalert-b-c-woman-wins-right-to-doctor-assisted-death-lawsuit-adjourned/

39 BCCLA newsletter, *Victory for Julia*, fall 2019 (print).

CHAPTER 3

HOW MANY IS ENOUGH?

The first rule of intelligent tinkering is to save all the parts.
—Aldo Leopold, Sand County Almanac

Scientists tell us that we have only a small window of time to take action on issues like climate change and population before destruction is irreversible. So a revolutionary shift in our thinking is required. Perhaps the world's most creative sales pitch is in order for the world's most urgent challenge. If we could find within ourselves the same passion we conjure up for buying cellphones, watching Star Wars movies, and posting photos on Facebook, we would be off to a good start indeed! It would launch us into a millennium of hope, which is rather rare today, with population pressures causing a growing state of discontent and COVID-19 a constant threat.

The public needs to appreciate that when "pro-lifers" talk about protecting life, they mean human life only. They don't really care about the gorillas, grizzly bears, or elephants sentenced to death by our frenzy of human procreation. When they talk about "pro-life," they are talking about quantity of life, not quality. They don't care in the least that often these children will be delivered into a hostile world, unwanted and unloved. Nor do they lose sleep over the billions of children born into poverty, AIDS, rat-infested slums, abusive homes, the slave trade, drug-addicted families, the sex trade, or many of the other hopeless situations that exist today. Do we really want this self-centered and inhumane ethic of conduct to dictate our progress?

Setting Goals

I suggest that we set a target date to set our compass by, and launch our humane population objectives on that course for 2030. I propose stopping population growth at EIGHT BILLION, and starting to reduce our human numbers by 2025. If we don't seize the momentum that has been built by finally connecting those few dots from population to climate change, we will condemn future generations to a world greatly diminished.

Redefining How We Talk About Overpopulation

Population growth in Canada is a result of both natural population growth and immigration. If we were to reduce our immigration to equal our emigration (i.e., balanced immigration), we would be in the enviable position of population stabilization and—eventually—a slow decline in Canada. Scaling back Canada's immigration programs could provide much needed funds for family planning services both here and in other countries. This strategy would begin to address many of our most critical national and global problems, such as the increasing food and energy costs and urban expansion onto agricultural land and wildlife habitat. Studies also indicate that urban density, housing costs, and traffic congestion would start to decrease if we reduced our numbers.

In order to restore a sustainable population level, we must remind citizens of how far we have strayed from our basic human rights due to overpopulation. In 1948, the United Nations developed the Universal Declaration of Human Rights. This declaration has served to guide organizations like Zero Population Growth (now Population Connection in the U.S.), Canadians for a Sustainable Society, and the Sierra Club to endorse the intent of this historic document with the following resolution: "Every human being, present and future, has a right to a world with a healthy environment, clean air and water, uncluttered land, adequate food, sufficient open space, natural beauty, wilderness and wildlife in variety and abundance, and an opportunity to gain an appreciation of the natural world and people's place in it through firsthand experience."[40]

Rather than "population control," a term too easily linked to the unsavory past of eugenics, this is how Population Connection defines its mission: "We must ensure that people are fully able to choose their own paths—including deciding for themselves if and when they want to have children. We must address the unmet need for family planning of the 214 million women in the developing world who want to prevent pregnancy but have an unmet need for modern contraception."[41]

> *It's never about population control. It's always about empowering people—especially women and girls.*
> **—Population Connection**

How attainable are these human rights in a world so overpopulated that our natural web of life is unravelling and rapidly destroying our life-support systems? And what of the wildlife and natural beauty declared as a universal right? Who will be our champion to defend this declaration if not the United Nations who created the document? How can we get there from here?

In fact, at this point I don't think we want to stabilize population growth. It's now slightly over 1 percent per year. This is the last thing we want to stabilize; we must be aiming at a growth rate of below zero. We want to stop population growth and decrease the size of the population. One misconception I hear all the time is that the problem has been resolved because fertility rates are decreasing. Yes, it is true that the number of children per family in most countries has decreased, but the growth in actual numbers continues to increase. This is similar to taking your foot off the accelerator when driving down the highway. The car continues to go forward for some time, just due to momentum.

Population growth hit a peak of about 2 percent in the late 1960s, and had fallen to 1.08 percent in 2019.[42] According to Chris Rapley, former director of the British Antarctic Survey, this decline

40 Sierra Club, Official Sierra Club Population Policy/History: https://www.susps.org/history/scpolicy.html

41 Population Connection, Mission and Programs: https://www.populationconnection.org/us/mission/

42 Worldometers, *World Population, Past, Present and Future*: https://www.worldometers.info/world-population/

is mainly a result of increased use and availability of birth control.[43] Population growth is calculated by subtracting the number of deaths from the number of births per year. However, world population continues to increase by an estimated 82 million people each year. So, while the rate of increase is slowing, in absolute numbers, world population growth continues to be substantial. According to the latest figures compiled by the UN, "global population could grow to around 8.5 billion in 2030, 9.7 billion in 2050, and 10.9 billion in 2100."[44]

While developed countries are expected to see a continuing decline, some of the hardest hit nations will be those least able to absorb the excess population. For example, sub-Saharan Africa is expected to add a billion people to that continent alone, and India is projected to surpass China as the world's most populous country around 2027 at 1.7 billion. Several countries in sub-Saharan Africa continue to experience high levels of adolescent fertility, contributing to an estimated 62 million babies born to mothers aged 15 to 19 worldwide. These are precisely the women who will have to struggle hardest to raise healthy children, since developing countries are likely to be hardest hit by desertification and climate change.[45]

To reach a zero level population growth rate, a family would have to limit the number of children to replacement level, which would be a fertility rate of two children. The growth rate means the number of children over the replacement level.

The growth of human population has been, is now, and in the future will be almost entirely determined in the world's less developed regions—Africa, India, and Asia. Africa is projected to overtake Asia in births by 2060. Half of babies born worldwide are expected to be born in Africa by 2100, up from three-in-ten today. Nigeria is expected to have 864 million births between 2020 and 2100, the most of any African country. The number of births in Nigeria is projected to exceed those in China by 2070.[46]

Of course, there are exceptions to this trend. In Thailand, "Mr. Condom" (Mechai Viravaidya) has made a tremendous impact by addressing overpopulation at every level imaginable. With this innovative and non-coercive approach to family planning, Thailand reduced its population growth rate from 3.3 percent (7 children per family) in 1974 to 0.5 percent (1.5 children per family) in 2019. He stressed that it was a case of everyone joining in, that it is everybody's job to change their attitude and behaviour. Hopefully his example should lead those in developed countries to conclude that we are also part of the problem, as many of us have yet to achieve this level of success. Consequently, developed countries should be making every effort to achieve a declining population rate as well. Japan, Tunisia, and parts of Europe have similar success stories to tell, yet to come.

43 Chris Rapley, *The Independent, This planet ain't big enough for the 6,500,000,000,* June 27, 2007: https://www.independent.co.uk/environment/climate-change/this-planet-aint-big-enough-for-the-6500000000-5333634.html

44 United Nations Department of Economic and Social Affairs World Population Prospects 2019 report: https://population.un.org/wpp/

45 Anthony Cilluffo and Neil G. Ruiz, Pew Research Center, *World's population is projected to nearly stop growing by the end of the century,* figures based on UN report, June 17, 2019: https://www.pewresearch.org/fact-tank/2019/06/17/worlds-population-is-projected-to-nearly-stop-growing-by-the-end-of-the-century/

46 Anthony Cilluffo and Neil G. Ruiz, Pew Research Center, *World's population is projected to nearly stop growing by the end of the century,* figures based on UN report, June 17, 2019: https://www.pewresearch.org/fact-tank/2019/06/17/worlds-population-is-projected-to-nearly-stop-growing-by-the-end-of-the-century/ , ibid

Understanding How Population Relates to Climate Change

Published in time for the Fourth United Nations Environmental Assembly, UN Environment's sixth Global Environment Outlook (2019) "calls on decision makers to take immediate action to address pressing environmental issues to achieve the Sustainable Development Goals as well as other Internationally Agreed Environment Goals, such as the Paris Agreement."[47]

The first Global Environment Outlook (GEO) was launched by UN Environment in 1997 and was prepared by about 390 experts. It points out that population growth combined with unsustainable consumption has resulted in an increasingly stressed planet. The report makes an urgent call for action, explaining, "Changing drivers—such as population growth, economic activities and consumption patterns—have placed increasing pressure on the environment."[48]

This report points out that humanity's survival will be largely determined by the decisions individuals and society make now to reduce population in all countries of the world. Time for delaying action has run out if we are to defuse the population time bomb and avoid even more catastrophic consequences than we are already witnessing.

Michael Shanks makes a compelling argument about population as it impacts the environment: "What's less frequently discussed, however, when it comes to personal contributions people can make, is also one of the most effective actions on the climate front: a smaller family size...."

Shank adds, "And at a time when we need to keep more and more carbon in the ground in order to slow global warming, this is a serious contribution to climate mitigation. To put this in perspective, each American produces an average of 16.5 metric tons, or 36,376 pounds, of CO_2 emissions annually. (For comparison: In China, it's an average of 7.5 metric tons per person per year; in the United Kingdom, it's 6.5 tons; in France, it's 4.6 tons.) Now, multiply this annual carbon footprint by the life expectancy of a baby born in this decade, which is 78.6 years. You get this: The average American produces roughly 1296.89 metric tons of CO_2, or 2,859,153.6 pounds, over their lifetime. That's equivalent to burning 1,417,794 pounds of coal or driving 3,170,881 miles in a passenger vehicle."[49]

If we are to aspire to a truly sustainable population level, we need strong leadership and international cooperation. Derrick Jensen's 2009 article, "Forget Shorter Showers," in *Orion* magazine makes this point quite well:

> Why now, with all the world at stake, do so many people retreat into these entirely personal "solutions"? Part of the problem is that we've been victims of a campaign of systematic misdirection. Consumer culture and the capitalist mindset have taught us to substitute acts of personal consumption (or enlightenment) for organized political resistance. *An Inconvenient Truth* helped raise consciousness about global warming.

---

47 UN Environment Programme, Global Environment Outlook (GEO) 6: https://www.unenvironment.org/resources/global-environment-outlook-6

48 UN Environment Programme, Global Environment Outlook (GEO) 4 report, Chapter 1, 'Main Messages' (see links): https://www.unenvironment.org/resources/global-environment-outlook-4

49 Michael Shanks, "This Controversial Way to Combat Climate Change Might Be the Most Effective," *Newsweek*, October 29, 2019: https://www.newsweek.com/this-controversial-way-combat-climate-change-might-most-effective-opinion-1468410

> But did you notice that all of the solutions presented had to do with personal consumption – changing light bulbs, inflating tires, driving half as much—and had nothing to do with shifting power away from corporations, or stopping the growth economy that is destroying the planet?[50]

Jensen went on to say in another *Orion* article that "more than 50 percent of the children who are born into this world are unwanted. We demand that all children be wanted. The single most effective strategy for making certain that all children are wanted is the liberation of women."[51]

Population Numbers Outpace Science

Each day we share Earth and its resources with a quarter million more people than the day before. With 300,000 people being born each day, and 50,000 dying, that leaves us with an additional 250,000 bodies to feed, clothe, and shelter every day. Our population is increasing by three people per second, with a growth rate of over 82 million each year. Today's grand total is about 7.7 billion. Just 350 years ago, there were only half a billion people on Earth.[52] During the 20th century alone, global population grew from 1.65 billion to 6 billion! Although our numbers will grow at a slower rate in the 2020 decade, what is desperately needed is to reach a state of **degrowth** in the next ten years.

The UN warns that we are on course to reach 10 billion by 2050—that is, unless the world unites to address this colossal problem now and agrees that we STOP AT EIGHT BILLION, and then reduce to a sustainable number.

Our phenomenal growth has prompted the National Academy of Science and Royal Society of London to release a second joint "Warning to Humanity" in 2018, updating the first 1993 warning.[53] It restated that if population continues to grow as predicted, science and technology cannot be expected to rescue the environment or society at large.

The Science-Policy Platform on Biodiversity and Ecosystem Services (IPBES) 2019 report is a definitive new global synthesis of the state of nature, ecosystems, and nature's contributions to people. It is the first such report since the landmark Millennium Ecosystem Assessment published in 2005, and the first ever that is inter-governmental. It was presented to 132 government representatives for consideration of approval in 2019. Prepared by 150 leading international experts from 50 countries, balancing representation from the natural and social sciences, with additional contributions from a further 250 experts, the IPBES intends to promote stronger policies and recommend more urgent action in the coming decade..[54]

---

50 Derrick Jensen, "Forget Shorter Showers," *Orion*, July 7, 2009: https://orionmagazine.org/article/forget-shorter-showers/

51 Derrick Jensen, "Self-Evident Truths," *Orion*, July 2, 2012: https://orionmagazine.org/article/self-evident-truths/

52 World Data Info, "Population growth worldwide": https://www.worlddata.info/populationgrowth.php

53 "World Scientists' Warning to Humanity: A Second Notice," Oxford Academic Bioscience: https://academic.oup.com/bioscience/article/67/12/1026/4605229

54 IPBES 2019 Global Assessment Report on Biodiversity and Ecosystem Services: https://elementaldigest.com/ipbes-2019-global-assessment-report-on-biodiversity-and-ecosystem-services/

This report indicates that we have every right to be alarmed about overpopulation, and millions of people are in fact starving, just as Ehrlich predicted. It confirms earlier reports that humans have far exceeded Earth's carrying capacity and population limits. It also confirms that growing populations and growing economies are associated with higher consumption and increased pressure on ecosystems. The following statistics are clear indicators of these findings:

- In 2020 almost a billion humans are malnourished worldwide, and this number is rising.
- Water demands already far exceed supplies in nearly 80 nations of the world; one-third to one-half of the world's population lacks access to clean water, a situation expected to worsen with climate change.[55]
- Food production per capita started declining in 1980, and continues to decline.
- Forests have been almost completely eradicated in 25 countries; in another 29, the area covered by forest has fallen by more than 90 percent.
- Crop fertilization has doubled the availability of nitrogen worldwide since the mid-19th century and tripled the availability of phosphorus since 1960. This leads to eutrophication of lakes and rivers and creates dead zones on the ocean floor due to oxygen depletion.
- Around 3.2 billion people worldwide are suffering from degraded soils. That's almost half of the world population, and soils deliver ecosystem services that enable life on Earth. We are losing from the soil the organic carbon, and this undermines agricultural productivity and contributes to climate change. We absolutely have to restore the degraded soil we've got.
- Over the past few hundred years, humans have increased the species extinction rate by as much as 1,000 times over background rates typical over the planet's history.[56]

Redefining How We Tackle Poverty and Malnutrition

Although great strides have been made in recent decades to reduce global poverty rates, malnutrition across the world remains unacceptably high. The World Health Organization's 2018 "Global Nutrition Report" warns that malnutrition is a serious disease because it increases susceptibility to other diseases like malaria and AIDS. In addition, it reduces the quality of life and often takes these people out of the work force.[57]

> *Finding out that 1 million species face extinction without radical corrective changes in human behavior is akin to finding out you have a fatal disease. One day you have a thousand problems; the next, you have just one. Nothing in today's headlines compares to the catastrophic potential posed by climate change and the decimating effects of careless consumerism around the globe.*
>
> **—Kathleen Parker,** *Washington Post*

Unfortunately, some of the gains in poverty reduction have come at a high environmental

55 "Vulnerable People and Places," Chapter 6, Millennium Ecosystem Assessment reports, p. 146: https://www.millenniumassessment.org/documents/document.275.aspx.pdf

56 "Ecosystems and Human Well-Being: Synthesis: Summary for Decision Makers," Millennium Ecosystem Assessment reports, p. 4: https://www.millenniumassessment.org/documents/document.356.aspx.pdf

57 "Global Nutrition Report," Executive Summary, p. 11, WHO: https://www.who.int/nutrition/globalnutritionreport/en/

cost, explains the Millennium Ecosystem Assessment: "The changes that have been made to ecosystems have contributed to substantial net gains in human well-being and economic development, but these gains have been achieved at growing costs in the form of the degradation of many ecosystem services, increased risks of nonlinear changes, and the exacerbation of poverty for some groups of people. These problems, unless addressed, will substantially diminish the benefits that future generations obtain from ecosystems."[58]

Overpopulation and Its Impact on Nature

People are slow to recognize how much population stresses affect each one of us individually, socially, and environmentally. This growth is taking its toll as natural habitats are being paved over, polluted, logged, developed, and mined—all to "benefit" encroaching human civilization. Of the many environmental issues, population is the one most neglected, yet the most crucial for the well-being of future generations. At the Rio Earth Summit, Secretary General Maurice Strong put it this way: "Population must be stabilized, and rapidly. If we do not do it, Nature will, and much more brutally."[59]

So, it seems to me that however you look at it, if our human population increases by another two to three billion by mid-century, our plant and animal populations will have to decrease by a similarly large amount. Once again the Millennium Ecosystem Assessment makes this clear: "**Over the past 50 years, humans have changed ecosystems more rapidly and extensively than in any comparable period of time in human history**, largely to meet rapidly growing demands for food, fresh water, timber, fibre, and fuel. This has resulted in a substantial and largely irreversible loss in the diversity of life on Earth."[60] Our drain on the biological resources will drastically reduce populations and diversity in native wildlife. Is that really what we want? More importantly, do we have any right to do this? After all, our deep-rooted feelings toward nature compel us to surround ourselves with nature's bounty.

We have windows in our homes to bring the sun into our private world, and we often surround ourselves with houseplants. Many of us bring cats and dogs into our environment and place bird feeders in our backyards. Those who can afford it have a pool on their property, and living near a park is considered an asset. If we can't be a part of nature, we try to capture a piece of it to be part of us. Deep down, we know that there is something special about this relationship. If at all possible, we try to create our own little rainforests in our concrete jungles.

Yet overpopulation threatens this very nature that we crave and value as part of our lifestyles. It's a paradox that so many people think of population growth as non-threatening and inevitable—even

58 "Ecosystems and Human Well-Being: Synthesis: Summary for Decision Makers," Millennium Ecosystem Assessment reports, p. 5. See also Kathleen Parker, "Nothing in today's headlines compares to the coming catastrophe," *Washington Post*, May 7, 2019: https://www.washingtonpost.com/opinions/the-end-of-the-everything-may-be-what-weve-been-needing/2019/05/07/902027ac-7101-11e9-9f06-5fc2ee80027a_story.html

59 Strong organized the Stockholm Conference in 1972, subtitled "Only One Earth," and the Rio Earth Summit in 1992, subtitled "Our Last Chance to Save the Earth." The Stockholm Conference is recognized as a major landmark.

60 "Ecosystems and Human Well-Being: Synthesis: Summary for Decision Makers," Millennium Ecosystem Assessment reports, p. 2.

desirable—in a world in which our very success as a species threatens our present-day and future survival.

So, are we really making this planet a better place to live by increasing our population? Indeed, there are a lot of us who are working very hard to improve our quality of life and save what remains of our wildlife and specialplaces. We certainly wouldn't want to think that it's all been in vain. Unfortunately, this may very well be the case, for our growing numbers are quickly undermining any progress we do make. That is why, despite our best intentions, all indicators show that **global degradation continues to increase.** The problems of widespread deforestation, disappearing species, sprawling cities, escalating crime rates, third world poverty, pollution, environmental health problems, and wars over natural resources continue to escalate. Russell Mittermeier, chief conservation officer for Global Wildlife Conservation, makes Shank's point more succinctly: "If steps are not taken soon to stabilize population growth, all our efforts in conservation will have little or no impact."[61]

> *Environmentally speaking, by choosing to have two kids instead of three kids, for example, an American family avoids nearly 3 million pounds of carbon dioxide emissions.*
> **—Chris Rapley, British Antarctic Survey**

Granted, the climate crisis is a complex problem with many factors besides population, as explained by experts such as Kathleen Mogelgaard, a consultant on population dynamics and climate change and an adjunct professor at the University of Maryland: "It is a very complicated, multifaceted relationship. Population issues certainly are an important dimension of how society will unfold, how society will be able to cope with this crisis over the course of this century. And fully addressing population growth is not, on its own, going to be able to solve the climate crisis. But **it is an important piece of the puzzle.**"[62] In 2020, I would venture to say that population is by far the most essential piece of the puzzle, and it is paramount that we set goals today for the future we want in 2030.

Introducing Ecological Overshoot

The hungry world cannot be fed until and unless the growth of its resources and the growth of its population come into balance. Each man and woman—and each nation—must make decisions of conscience and policy in the face of this great problem.
—Lyndon Baines Johnson, former U.S. president

A recent concept known as "ecological overshoot" has provided yet another warning of humanity's impact on the planet. Our annual "take" from the natural world has, since the 1970s, exceeded what Earth can renew. Ecological overshoot has continued to grow over the years, reaching a 50 percent deficit in 2008.[63] A similar concept, "ecological footprint" was introduced in 1990 by

61 Global Wildlife Conservation team bios: https://www.globalwildlife.org/team/russell-mittermeier-ph-d/
62 Nicole Martillero, BBC News, "Is population control the answer to fixing climate change?" CBC News online, October 25, 2019: https://www.cbc.ca/news/technology/population-climate-change-1.5331133 (emphasis mine)
63 "What does ecological overshoot mean?" World Wildlife Federation, https://wwf.panda.org/knowledge_hub/all_publications/living_planet_report_timeline/lpr_2012/demands_on_our_planet/overshoot/

Mathis Wackernagel and William Rees at the University of British Columbia. Far more than just a catchphrase, it's a "resource accounting tool that measures the amount of the Earth's regenerative capacity demanded by a given activity." It is used by scientists, businesses, governments, individuals, and institutions working to monitor ecological resource use and advance sustainable development.[64]

A National Academy of Sciences report by an international who's who of ecologists and economists calculated that humans have been borrowing for four decades against the ecological production of future years. In 1961, humans were using 70 percent of the planet's yearly potential for biological productivity; by 1999, it was 120 percent.[65]

One reason is that native biological capacity of large parts of the planet is being degraded by narrowly focused agriculture. And demands on that biological capacity are increasing as the world population grows. This study does not factor in any land being set aside for animal habitat. Many scientists have calculated that 12 percent of the planet's land and water ecosystems should be protected as parks and reserves for species that are not human. More recently, distinguished thinkers believe we need to give back 50 percent of the land to wildlife.[66]

> *We are preparing for ecological bankruptcy.*
> **—Mathis Wackernagel**

Personal correspondence with wildlife and forest ecologist Brian Horejsi reveals that this notion of protecting only 12 percent of ecologically functional land is a serious miscalculation and that it continues to have damaging political and industrialization repercussions. He adds, "For grizzly bears, for example, at least 50 percent of the land base must be roadless," and that land base must be large enough to contain 2,000 bears in a population to insure genetic viability (could be four areas of 500 bears each, as long as there is unrestricted exchange between). Over a 100-year horizon, protecting 50 percent of the landscape still only provided a 50 percent probability of survival. This would be more in line with the Half Earth project, striving to protect half of Earth for wildlife—to be discussed in Chapter 5.[67]

A species may greatly overshoot the long-term carrying capacity of its environment, at least in the short term! Humans are not likely to get away with much more than 50 years overshoot before severe consequences destroy the world as we know it. Already we have climate disruption impacting us, and now COVID-19, plus all the threatened and endangered wildlife populations around the world.

---

64 Global Footprint Network, "Ecological Footprint": https://www.footprintnetwork.org/our-work/ecological-footprint/ Science Direct, 'Learn More About Ecological Footprint': https://www.sciencedirect.com/topics/agricultural-and-biological-sciences/ecological-footprint

65 Mathis Wackernagel, Niels B. Schulz, et al, "Tracking the ecological overshoot of the human economy," edited by Edward O. Wilson, Harvard University, 2002: https://www.pnas.org/content/pnas/99/14/9266.full.pdf

66 Stephen Leahy, *National Geographic, Half of all land must be kept in a natural state to protect Earth,* April 19, 2019: https://www.nationalgeographic.com/environment/2019/04/science-study-outlines-30-percent-conservation-2030/

67 E.O. Wilson, *Half-Earth Project,* www.half-earthproject.org

What Is Ecological Overshoot?

Overshoot becomes possible when a species encounters a rich and previously unexploited stock of resources that promote its reproduction. A huge stock of resources may be available for millions of years before it encounters a species that can exploit it easily. After such an encounter, only predation and disease limit reproduction of the species. Without these two limiting factors, the population of a species can grow to a size hundreds of times that which can be supported at long-term viability levels by the resources.[68]

> *In 2019 Earth Overshoot Day was July 29th. In seven months humanity had burned through the resources it takes the earth a full year to replace*
> **—Global Footprint Network**

As overpopulation accelerates, individuals increasingly, then desperately, compete for the remaining resources. They resort to alternative resources of lower and lower quality. This incrementally degrades the ability of their environment to restore resources or produce alternative resources. Eventually, most of the population dies. Ecologists call this a crash or die-off; it results in carrying capacity for the overshot species being reduced to below moderate levels.

The question is, "How much longer can we continue in this overshoot consumption?" Annual Earth Overshoot Day reports the date when we've used more resources than Earth can regenerate in a year.[69] Beyond that date we operate in ecological deficit. This critical situation is caused by only one of the 25 million species that inhabit this planet—humans! The UN tells us that we have only ten years to reverse this unmanageable rate of growth.

We are facing a titanic-like catastrophe. In the movie *Titanic*, a few of the crewmembers realized they were heading directly for an iceberg, and sure disaster, and gave the order to reverse the engines. At this point they had waited too long, causing great chaos and then enormous loss of life. Those who the passengers had entrusted to take care and look out for them had failed miserably. Much as we see in our leadership today, the captain and those who influenced him had personal gain, and not survival, in mind in their race to reach their destination.

How Can We Reverse the Overshoot Clock?

Should humans not be reversing the overpopulation engines? Every ecological, social, and economic indicator points toward population growth as the engine propelling our planet to destruction. As young climate activist Greta Thunberg has said, "Our political leaders' legacy will be the greatest failure of human history."

Just as on the *Titanic*, the wealthy first-class passengers may survive as they sacrifice the poor people and animals in steerage—the less advantaged and the middle class of today— to ensure the safety of the rich.

---

68 David M. Delaney, "Overshoot in a Nutshell," October 2003; no longer extant on the web but archived as a PDF by the Population Institute of Canada at: https://populationinstitutecanada.ca/wp-content/uploads/2018/03/Overshoot-in-a-nutshell-David-M-Delaney.pdf

69 Earth Overshoot Day: https://www.overshootday.org

Humans are very good at adapting to almost any situation. The problem is that our adaptability is now working against our long-term survival as a species. Mark Hume outlined the problem quite well in the *Vancouver Sun*, "In laboratory experiments with Norway rats, behavioural scientists have observed a disturbing phenomenon. Allowed to breed at will, the rats multiply until they reach a point where they just can't stand it any more. When the overcrowding becomes too great, they stop breeding and start killing each other."[70]

Aprodicio Laquian, the scientist who studied the rats, pondered their behaviour and the alarming rate at which the human population is growing. "What happens with the rats is not too far-fetched a scenario for humans. We are just biological creatures, after all," he says.[71]

Isn't this already occurring, especially in our overcrowded cities? A person can be killed for a pair of $50 sneakers, gangs are fighting for their turf, and violent road rage has become a common occurrence. Governments perpetuate wars all over the world to gain power over land, water, or natural resources. Millions of women are raped or killed every year in a struggle for power and dominance. We are a species at war with each other and the planet, yet the critical role that population undoubtedly plays is rarely mentioned.

"The October 2018 **bombshell report** from the UN's Intergovernmental Panel on Climate Change, with its dire warning that we have some 11 years to make the huge systemic changes necessary to save ourselves from ecological destruction, only **compounded my reproductive anxiety**," wrote Katie O'Reilly for the Sierra Club. "The UN's assessment in May predicting an 'unprecedented' and 'accelerating' decline in global biodiversity—alerting us all that more than a million animal and plant species are on track to go extinct within the next few decades—didn't help the existential spiral."[72]

How could I look my hypothetical child in the eye and acknowledge that I willingly brought them into a chaotic, increasingly uninhabitable world, that I knew all their favorite picture-book animals were going extinct?
—Katie O'Reilly, Sierra Club

This all must sound very grim, but the good news is that it doesn't have to be this way. It is clear that the time has come for a new consciousness, a global revolution! Humans are at the point of no return, and if we don't take action now to stop human population from reaching EIGHT BILLION and then reduce our numbers to two to three billion, we deliberately impose a bleak future on forthcoming generations.

---

70 Mark Hume, *Vancouver Sun*, "Like Too Many Rats," February 19, 1994

71 Aprodicio Laquian, Professor Emeritus, University of British Columbia, Wilson Center: https://www.wilsoncenter.org › person › aprodicio-laquian

72 Katie O'Reilly, "To Have or Not to Have Children in the Age of Climate Change," *Sierra Club* magazine, November 1, 2019: https://www.sierraclub.org/sierra/2019-6-november-december/

CHAPTER 4

SETTING THE LIMITS

Now, two hundred years after Malthus, humans have multiplied their numbers far beyond any sustainable limit...
—David M. Delaney

Transition to an Optimum Population

Many people argue that the planet is not overcrowded, that there's plenty of land, as if the concern about population is one of physical overcrowding alone. I've heard overpopulation deniers claim that all of the people on the planet standing side-by-side could fit into a state the size of Texas or some other ridiculous confined area.

But as Madeline Weld with Population Institute Canada explains, "The evidence is strong, if not overwhelming, that we are already exceeding the Earth's regenerative capacity. In many areas of the world, lakes and rivers are shrinking and water tables are falling precipitously. Around the world, arable land is in short supply and topsoil is being eroded at a worrisome rate. Deserts are expanding and forests are contracting. Extinction of species and indigenous peoples is increasing, as our hope and security are decreasing."

But it's never been a question of how many people the world can contain; it's how many people the world can sustain at what standard of living and for how long.
—Madeline Weld, Population Institute Canada

What We've Learned from the Fate of Easter Island

The puzzling thing is that it isn't as if we haven't seen this coming. Our history, our science, and our common sense have told us for decades, if not centuries, that this growth ethic we are so attached to is a path to destruction. Let's look at the cautionary tale of Easter Island, for example. This South Pacific island with the famous stone statues tells a story with a very sobering message for our present circumstance. This isolated piece of volcanic rock witnessed the collapse of Polynesia's most advanced megalithic culture. Archaeologists have uncovered a chilling tale of how overpopulation impacted this once-pristine tropical island.

Polynesians first set foot on this isolated island, roughly 2,250 miles northwest of Chile, around 1200 CE. This small island boasted a lush palm forest, abundant population of birds and seafood, and a fertile coastal plain sufficient for their taro and yam staples. It was the subtropical forest that played a central role in the rise and fall of the Rapa Nui society. Cut down for timber, the trees were used for building their homes, for fruit, clothing, canoes, and firewood. The intact forest also provided homes for many of the nesting birds on the island. Writing on the fate of Easter Island, Peter Tyson notes, "The parallels between what happened there and what is occurring today in the world at large—albeit more slowly and on a much vaster scale—are, the evolutionary biologist Jared Diamond says, 'chillingly obvious.'"[73]

By the middle of the second millennium CE, the islanders had developed a complex society of from 10,000 to 20,000 people. However, they were blind to the impacts of their large population, and began exhausting all of their resources. This shortsighted thinking managed to cause one of the most extreme examples of forest destruction in history, and by 1600 CE, all the island's tree species had become extinct. The supply of seafood, birds, and crops also began to dwindle, and civil war broke out as they squabbled over the remaining resources. At that time, the Rapa Nui civilization had begun to fall apart, and by the mid-1800s, it had all but disappeared. By 1872, the number of Rapa Nui had plummeted to just 111 individuals. This once-thriving society had been reduced to a handful of starving islanders. By the time they realized their mistake, it was too late.

The parallel of this haunting scenario, and what is happening today in our world, should act as an urgent call to action. The only difference is the scale—from a small island in the Pacific to the entire planet.

"Against great odds the islanders had painstakingly constructed, over several centuries, one of the most advanced societies of its type in the world," explains the Sustainable Footprint Education Project. "It was in many ways a triumph of human ingenuity and an apparent victory over a difficult environment. But in the end the increasing numbers and cultural ambitions of the islanders proved too great for the limited resources available to them."

Recent archaeological evidence suggests that colonization of Easter Island started much later than previously estimated—1200 CE instead of 400 CE—making the lesson all the more poignant. Collapse of that civilization from ecological overshoot occurred in just four centuries. "Like Easter Island the Earth has only limited resources to support human society and all its demands."[74]

But the globalization of the Western lifestyle, with its high rate of consumption, is speeding up our planetary overshoot just as it did on Easter Island. As Jared Diamond explains, "Thanks to globalization, international trade, jet planes, and the Internet, **all countries on Earth today share resources and affect each other**.... Polynesian Easter Island was as isolated in the Pacific Ocean as the Earth is today in space."[75]

---

73 Peter Tyson, "The Fate of Easter Island," World in the Balance, *NOVA*, April 19, 2004: https://www.pbs.org/wgbh/nova/article/fate-of-easter-island/?ncid=txtlnkusaolp00000619. The editor notes, "Since this article appeared in 2004, new studies posit a later arrival for the first Easter Islanders, a smaller maximum population, and a more complex explanation for what transpired there."

74 "Easter Island—A Lesson for Us All," *Sustainable Footprint*, http://sustainablefootprint.org/teachers/theme-lessons/easter-island-a-lesson-for-us-all/

75 Jared Diamond, *Collapse: How Societies Choose to Fail or Succeed*, Penguin Books, 2005, p. 119.

Defining the Thresholds

So what is the best course of action for Canada to be a good global citizen and a model of a sustainable society in an overpopulated world? Population Institute Canada believes that Canada should strive to stabilize its own population and should encourage other countries to do the same. The institute says, "Given Canada's low birth rate, this is an achievable goal.... Instead of pursuing a policy that drives population growth by about 1% annually, Canada could pursue a balanced migration policy, taking in about as many immigrants as the number who leave. Economists have calculated that the high annual intake of newcomers costs the government **between $18 and $23 billion more in services** than it generates in taxes every year. Not only does Canada's rapid growth through immigration fail to provide economic benefits for most Canadians, the money saved could be directed to support family planning in those countries also trying to stabilize their populations, thereby helping a far greater number of people and making a real difference in helping those countries achieve sustainable population levels."

According to Andrew Revkin, an environmental reporter for the *New York Times*, Americans, the world's greatest per-capita emitters of greenhouse gas emissions, produce about 20 tonnes of the stuff per person, per year. If they were to cut that in half, as emissions rose with the quality of life in developing countries, and everyone on the planet met around 10 tonnes per person, per year, simple multiplication says we'd collectively emit 90 billion tonnes of carbon dioxide annually come 2050.[76] That's three times the already problematic current number.

> *The single most concrete, substantive thing a young American could do is not turning off the lights or driving a Prius. It's having fewer kids.*
> **—Andrew Revkin,** *New York Times*

So if we were to think about coping with nine billion people, a lot of our options would suddenly disappear. At that point, most of our life support systems would have crumbled, like the disappearance of our bumblebees and other pollinators. We would be in a constant state of war over our remaining resources, like water and food, and climate change would be irreversible.

So how should we tackle this problem? First we need to address the politically touchy roadblock of family planning, and point out that population planning is a distinctly different animal than population control. The coercive measures used in China in the past are no longer on anyone's radar screen, as we have **evolved beyond that mindset** and only non-coercive measures can be tolerated.

In fact, we have evolved to the point of already beginning the decline in fertility rates so desperately needed, so now the UN prediction of ten billion by 2050 has been reduced to nine billion—and that is GOOD NEWS! Now if we could just continue this trend and make a global effort to stop growing at EIGHT BILLION, we may still have a chance to start reducing our numbers by 2025. In that case we could be down to six billion by 2050. Then we would have something to cheer about! By the end of the century, we could be living a healthy and happy lifestyle with two to three billion people sharing the planet with a host of other critters.

76 Emily Badger, "Let's Try Cap-and-Trade on Babies," May 3, 2017: https://psmag.com/news/let-s-try-cap-and-trade-on-babies-3405

So what do distinguished thinkers believe is the magic number for an ideal population? Unless we make an effort to determine optimum population numbers, we will have no benchmark against which we can measure success. Here are a few recommendations from world experts, and their views on how to transition to an optimum population:

One, Two, or Three Billion?

1. The World Wildlife Fund, in its publication *Living Planet,* has this to say: "The world optimum population is **between 2 and 3 billion**—less than a third of its forecast mid-century peak."[77] The human population has become a force at the planetary scale. Collectively, our exploitation of the world's resources has already reached a level that, according to the World Wildlife Fund, could only be sustained on a planet 25 percent larger than our own.
2. "Very few writers seem to recognize that growth cannot continue forever in a limited space, and that mathematical truism applies to the real world, today," writes former National Security Council member Lindsay Grant. "Dr. Smail is one of those few who do. Moreover, he suggests that human numbers have already passed the long-term capacity of the Earth to sustain us, and that an optimum world population lies perhaps in the range of **2–3 billion**. So short is memory that the proposal sounds revolutionary—almost blasphemous—to most ears. Humans' ability to accommodate to change is both our strength and our peril. People have learned to consider six billion 'normal,' and presumably they will try to adjust to 10 or 12 billion, desperate as their situation may be by that time."[78]
3. The College of Agriculture and Life Sciences at Cornell University has been giving population limits a great deal of thought and in 1999 produced a paper entitled "Will Limits of the Earth's Resources Control Human Numbers?" This is the number that they came up with:

   The human population has enormous momentum for rapid growth because of the young age distribution both in the U.S. population and in the world population. If the whole world agreed on and adopted a policy so that only 2.1 children were born per couple, more than 60 years would pass before the world population finally stabilized at approximately 12 billion. On the other hand, a population policy ensuring that each couple produces an average of only 1.5 children would be necessary to achieve the goal of reducing the

The population problem concerns us, but it will concern our children and grandchildren even more. How we respond to the population threat may do more to shape the world in which they will live than anything else we do.
—Lester Brown, Worldwatch Institute, State of the World 2001

77 World Wildlife Fund, Living Planet Report, 2018: https://www.worldwildlife.org/pages/
78 Lindsay Grant, "In Support of a Revolution," 1997: https://doi.org/10.1017/S0730938400024576

world population from the current six billion to an optimal population of **approximately two billion.**[79]

"If this policy were implemented," adds another researcher, "more than 100 years would be required to make the adjustment to 2 billion people. Again, the **prime difficulty** in making the adjustment is **the young age distribution and growth momentum** in the world population."[80]

4. Another supporter of setting optimum population at about **two billion** comes from the Sustainable Scale Project.[81] They provide educational material and resources to assist government decision makers, civil society organizations, and students. They warn that current global consumption is unsustainable, since it exceeds our planet's ability to regenerate itself. The amount of bioproductive land and sea available to supply human needs is limited to 11-plus billion acres of productive earth; divided by 6.3 billion people, this results in an average of about 1.8 hectares per person as the "equal earth share" available. Collectively, we are currently using approximately 2.2 hectares per person, or over 20 percent more than replacement levels.

   Since current population policies encourage increasing consumption in countries dealing with poverty, this will increase the "equal earth share." On the other hand, many developed countries have set goals to decrease their "equal earth share." So for purposes of discussion, a balanced consumption rate of six hectares per person is used. What would the population have to be to allow the "equal earth share" to be six hectares per capita? Divide the 11-plus billion bioproductive hectares by six hectares per person to get a total population of **1,830,000,000**, which is close to 1930 population levels. The Sustainable Scale Project estimates that this would be achievable by around the year 2100 if we implemented a 1.5-child-per family-policy now. They warn that the alternative would be to leave billions of people currently alive in misery and chaos and to continue to degrade ecosystems for all future generations.

Can We Shift the Paradigm?

So, regardless of whether the optimum population is one billion or three billion, I think it is safe to say that at nearly eight billion, we are at least double a sustainable level and accelerating. Would it not be logical, then, to assume that we need to address this crisis as soon as possible? Or will we continue to debate the numbers until, as we did with global warming, the problem is irreversible? It is this generation that has unfairly been given the task of telling our political leaders that we no longer want growth, and that we must first stabilize the population in all countries, so that we can begin to reduce it to a sustainable level.

---

79 D. Pimentel, O. Bailey, & P. Kim, College of Agriculture and Life Sciences at Cornell University, *Will Limits of the Earth's Resources Control Human Numbers?*, 1999, https://link.springer.com/article/10.1023/A:1010008112119

80 Bartlett and Lytwak, PRB, 1996, Bartlett, 1997–1998.

81 Sustainable Scale Project, http://www.sustainablescale.org/

> *Present need to reduce numbers is greatest in wealthy countries where per capita use of energy and earth materials is highest. A reasonable objective is the reduction to population levels as they were before the widespread use of fossil fuels, that is, to* ***one billion or less****. This will be accomplished either by intelligent policies or inevitably by plague, famine, and warfare.*
> **—Ted Mosquin and Stan Rowe,**
> **A Manifesto for Earth**

Otherwise, what will we tell future generations? That we knowingly chose to continue to rob future children of their right to necessary resources, healthy environment, and diversity of nature? That we were so preoccupied with our self-interests that we ignored the warnings of disaster for the planet? Several Baby Boomers, such as Jane Goodall, Margaret Pyke, and Oprah, have already spearheaded this movement, so if we could recruit some youth activists like Greta Thunberg, we would be off to a great start towards a sustainable population.

Most of the world's policymakers have backgrounds in law, business, or economics. Very few have backgrounds in biology or ecology. We cannot expect them to make the right decisions for us. We must let them know that there is a quality of life that cannot be measured in economic terms, and that we want this lifestyle to be available to our children and grandchildren.

Achieving the Limits

In order to stay within our planetary limits, several steps need to be taken:

- The economy must stop growing.
- Our consumption must stop growing.
- The human population must stop growing.

These are all issues that bring up a lot of moral and ethical debate. One's gut reaction when looking at these monumental problems might be to say that they are just far too big to tackle. Not so. They can all be achieved by stopping population growth at EIGHT BILLION and then reducing to two to three billion by 2100.

1. The Economy Must Stop Growing

Economists, policymakers, reporters, and the public rely on the gross domestic product (GDP) as a shorthand indicator of progress, but the GDP is merely a sum of national spending with no distinctions between transactions that add to well-being and those that diminish it. America's leading policy institute dedicated to smart economics, Redefining Progress, develops solutions that create a just society, protect the environment, and shift public policy to achieve a sustainable economy. This is done by using the genuine progress indicator (GPI), a system of accounting that is far more ethical and logical than the GDP presently being used (see Figure 4-1). Redefining Progress, a San Francisco–based group, created the GPI in 1995 as an alternative to GDP. The

GPI enables policymakers at the national, state, regional, or local level to measure how well their citizens are doing both economically and socially.[82]

Honest national accounting would inject a large dose of accountability to the political process. It would stop politicians and interest groups from hiding the implications of bad policy behind what amounts to a rigged set of books. So how would using the GPI help control population? Well, the GPI counts resource depletion as a cost, and as the population increases, resource depletion increases. This becomes more evident as the standard of living rises in countries like China and India and as more citizens from undeveloped nations immigrate to developed nations. Both of these situations increase resource depletion and highlight the impacts and costs of population growth.

Redefining Progress explains that if today's economic activity depletes the physical resource base available for tomorrow, then it is not creating well-being; rather, it is borrowing it from future generations. The GDP counts such borrowing as current income. The GPI, by contrast, counts the depletion or degradation of wetlands, forests, farmland, and nonrenewable minerals (including oil) as a current cost.

Long-term environmental damage arising from the use of fossil fuels, chlorofluorocarbons, and atomic energy are unaccounted for in ordinary economic indicators. On the other hand, the GPI treats environmental damage as a cost to society. Interestingly, while the GDP has been rising for decades, the GPI has actually been falling since the mid-1970s.

Redefining Progress warns that with both individuals and governments around the world borrowing and spending well beyond their means, we are accumulating a huge load of debt on future generations. With each additional person added to this planet, both economic and environmental debt increases, causing loss of biodiversity and scarcity of resources.

If every country adopted the GPI accounting system, statistics would warn us that matters are growing worse much faster than expected. We would have a better picture of the costs of climate disruptions, extreme weather events, relentless population growth, and disappearance of wildlife habitat. We would have more incentive to take action to remedy these problems and improve our well-being. Presently, a huge gap exists between the fraud of economic growth as defined today and what a healthy economy really is.

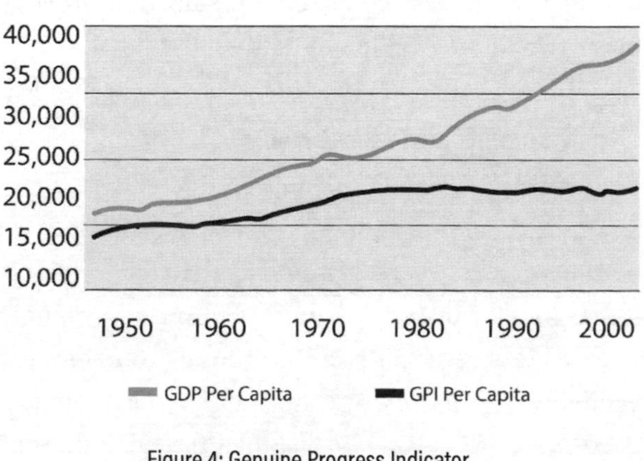

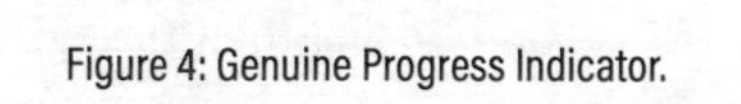
Figure 4: Genuine Progress Indicator.
Graph by permission of Redefining Progress

Something is fundamentally wrong with the way we assess economic performance and social progress, says Nobel economist Joseph Stiglitz. In 2019, Stiglitz called on humanity to

82 Community Indicators Consortium (CIC) cites Redefining Progress as the lead organization on the GPI: https://communityindicators.net/indicator-projects/redefining-progress-the-genuine-progress-indicator

end our obsession with growth and the widely used GDP accounting system. He cautions that if we don't end this obsession now, there will be little chance of adequately fighting back against the triple threat of climate destruction, the scourge of financial inequality, and the crises of democracy now being felt around the globe. "If our economy seems to be growing but that growth is not sustainable because we are destroying the environment and using up scarce natural resources, our statistics should warn us," explains Stiglitz. "But because GDP didn't include resource depletion and environmental degradation, we typically get an excessively rosy picture. If we measure the wrong thing, we will do the wrong thing."[83]

Despite continuous increases in GDP and the 2008 economic crisis being well behind us, social and environmental conditions continue to worsen. "We see this in the political discontent rippling through so many advanced countries," writes Stiglitz. "We see it in the widespread support of demagogues, whose successes depend on exploiting economic discontent; and we see it in the environment around us, where fires rage and floods and droughts occur at ever-increasing intervals."

It seems increasingly clear that our present economic system has steered us off course and blinded us to the true impacts of our global and environmental overspending. We must urge economists around the world to adopt the GPI accounting framework, now used by the Pembina Institute in Canada, to provide an accurate assessment of the costs accrued every time money changes hands for destructive behaviour. This one essential act could **radically change the course of humanity** in the 21st century. This tool is something the United Nations should be using and promoting so that all countries could make this transition together.

2. Consumption Must Stop Growing

Contrary to many debates, **it is not consumption OR population**, but **both together** that are destroying our planet and our well-being. We must deal with both of these issues simultaneously. Yet, while reducing consumption seems to be the buzzword these days, population still remains a taboo issue in many circles. It is also essential that we **differentiate between growth and progress**. Growth has to stop, but in a resilient society, progress can continue forever. In fact, we cannot actually make any progress until both our economic and population growth begin to decrease.

There is no doubt that developed countries are consuming far more than their fair share of Earth's resources, partly due to the greater resources needed to survive in colder climates. However, there is no accounting for the excesses of the **wealthiest elite existing in all countries of the world**. It is essential that we all must do our part to reduce our footprint on Earth and live within our carrying capacity. It is our children's legacy that we are squandering along with an alarming loss of biodiversity.

83 Jon Queally, "Nobel Economist Calls on Humanity to End Obsession with GDP," *Salon*, December 1, 2019: https://www.salon.com/2019/12/01/nobel-economist-calls-on-humanity-to-end-obsession-with-gdp_partner/ See also Joseph Stiglitz, Jean-Paul Fitoussi and Martine Durand, *Measuring What Counts: The Global Movement for Well-Being*, The New Press, 2019.

It is not that we don't know better, but rather that we lack the willpower to live more sustainably. Even those who are willing are constrained by the narrow range of options available to the general public. Turning off lights, replacing bulbs with LEDs, and recycling are all great, but compared to readjusting national economic priorities or ending a system premised on planned obsolescence, these barely make a dent. This issue of consumption is a book in itself, and of course there are plenty on bookshelves already. **Ethically, we have no right to continue stealing from future generations and the other critters on the planet.**

As we begin to shift away from an unrealistic economic system that gives an excessively rosy picture of our situation, and begin to reduce consumption in a truly meaningful way, we will be on the course to recovery. Both of these actions will highlight the underlying population crisis knocking down the planet's door, and we will decide to stop at EIGHT BILLION simply because it is the only ethical thing to do. Then our population can finally begin the decline to a sustainable level and the healing can begin.

3. Population Must Stop Growing

Many decades before our present crisis, science-fiction author Isaac Asimov saw the problem clearly:

> Which is the greater danger—nuclear warfare or the population explosion? The latter, absolutely! To bring about nuclear war, someone has to DO something; someone has to press a button. To bring about destruction by overcrowding, mass starvation, anarchy, the destruction of our most cherished values—there is no need to do anything. We need only do nothing except what comes naturally—and breed. And how easy it is to do nothing.[84]

Because we have neglected this problem for far too long, there are no easy answers left when it comes to reaching a sustainable population. It is like dealing with logging practices that should have been controlled decades ago; we now have few choices left. Should we log the last of our old-growth forests, log our community watersheds, or resort to helicopter-logging to get the last few remaining trees on the mountainsides? The dilemma of overpopulation forces us to face similar moral and ethical choices, and we have to ask ourselves which of the often expensive and challenging choices are more acceptable and attainable.

Feeding People vs. Saving Nature?

When we must choose between feeding the hungry and conserving nature, people usually come first. This is how Holmes Rolston III explains it in the journal *Environmental Values*: "A bumper sticker reads: 'Hungry loggers eat spotted owls.' That pinpoints an ethical issue, pure and simple, and often one where the humanist protagonist, taking high moral ground, intends to

---

84 Isaac Asimov, Earth Citizen: https://earthcitizen.co/blogs/quotes/which-is-the-greater-danger-nuclear-warfare-or-the-population-explosion-dr-isaac-asimov

put the environmentalist on the defensive. You wouldn't let the Ethiopians starve to save some butterfly, would you? Humans win? Nature loses? After analysis, sometimes it turns out that humans are not really winning if they are sacrificing the nature that is their life support system. Humans win by conserving nature—and these winners include the poor and the hungry."[85]

> *Basically, then, there are only two kinds of solutions to the population problem. One is a "birth rate solution," in which we find ways to lower the birth rate. The other is a "death rate solution," in which ways to raise the death rate—war, famine, pestilence—find us.*
> **—Paul Ehrlich,** *The Population Bomb*

Rolston makes a good point about our lack of respect for nature. It is not simply a trade-off over hungry people versus nature. It is a matter of finding a win-win situation that will provide for people's needs in the short term without destroying the resource base that is essential to providing for those needs. If the alleviation of human suffering always results in an increase in human population, we are simply creating a larger problem—and more human suffering—in the future. This is exactly what we have done every time we have provided food aid in Africa with no effort to stem the rapid population growth. The number of hungry people in each crisis simply gets larger, and the environment becomes more degraded. The protection of nature and human well-being both depend on our acknowledging the population crisis.

In Part 1, we discussed population history, reasons for our high birth rates, an example of a society that did not have the foresight to remedy this, and why a sustainable population level would be approximately two billion. Our economic system, rate of consumption, and growing population were identified as deterrents to achieving this goal. Solutions to these problems were outlined and will be discussed in greater detail in future chapters. **There is so much more to be hopeful about**, so please check out all the solutions and success stories offered in Chapters 5 through 12.

85 Holmes Rolston III, "Saving Nature, Feeding People, and the Foundations of Ethics," *Environmental Values*, Vol.7, No.3, August 1998: https://www.jstor.org/stable/30301647

PART 2

CRITICAL WORLD ISSUES

The population problem concerns us, but it will concern our children and grandchildren even more. How we respond to the population threat may do more to shape the world in which they will live than anything else we do.
—Lester Brown, Worldwatch Institute

Since I became aware of the population crisis in 1971, I have taken notice of the disappearing wildlife around the world, starting in my own backyard. As a child growing up in the Crowsnest Pass in Canada, there was a special little pond near my home where I would marvel at the frogs and salamanders I became so attached to. After I moved away, I would return to that pond occasionally to check on those frogs and salamanders, and was horrified to see fewer and fewer of them each time, until one day not only had my fascinating friends disappeared, but the pond had disappeared as well. It had been filled in to make room for development as more people moved to the area. I was devastated and filled with grief over this loss.

At the same time, Jane Goodall witnessed a chimpanzee fishing for termites by manipulating blades of grass, and the line between humans and other animals was suddenly obscured. Tool use was no longer uniquely human, leading Dr. Louis Leakey, Jane's mentor, to famously say, "Now we must redefine tool, redefine man, or accept chimpanzees as human." Later, Dr. Goodall would witness the horrifying loss of gorillas and their habitat to development and the sale of bush meat, as Africa's growing human population put more and more pressure on wildlife. The '70s also witnessed the Penan people of Borneo fight logging companies to protect the forests that they relied on for food and survival. The planet's growing numbers were demanding more and more of our global forests, with no regard for the wildlife or the Indigenous peoples that called these forests home.

Women everywhere must be able to choose whether to have children, how many children, and the spacing between them. This is critical for their own wellbeing. But they also need to be equipped with the knowledge as to how their choice affects the health of the planet and thus the future of their own children. For we are part of the natural world and rely on its "services" for our very survival.
—Dr. Jane Goodall

Anthropocentrism is dominating the landscape, as humans put themselves on a pedestal and all other species are reduced to nil.

Anthropocentrism is defined as " a worldview that considers humans to be the most important thing in the Universe, or at least on the planet Earth. In contrast, the **biocentric worldview** considers humans to be a particular species of animal, without greater intrinsic value than any of the other species of organisms that occur on Earth."[86] Should it not be this latter view that we strive for if we want to find our true purpose and place on the planet?

Part 2, Critical World Issues, explores the many ways humans have impacted this spectacular world we live in. It takes an in-depth look at how our obsession with growth has impacted other species, how empowering women will truly bring us into the 20th century, how intricately climate change is linked to population, and how vital it is that we are able to supply enough healthy food and clean water to sustain a planet of almost EIGHT BILLION people. It becomes very clear in Part 2 that if we do not work to stop population growth now, we will become the victims of our own folly.

The chaos of climate change is coming at us from every direction. From melting glaciers to increasing storm intensity, climate change and population will be at the forefront of our thoughts for the rest of the 21st century. It will also have a colossal impact on water and food availability, forcing humanity into situations of conflict as demands from a growing population far outpace supply, just as Malthus predicted. Every chapter will end with solutions and success stories, so don't despair, as we still have a small window of time to turn things around.

86 "Anthropocentrism," Encyclopedia.com: https://www.encyclopedia.com/history/modern-europe/czech-and-slovak-history/anthropocentrism

CHAPTER 5
EXTINCTION OF SPECIES

You know, I have often thought that at the end of the day, we would have saved more wildlife if we had spent all WWF's money on buying condoms.
—Sir Peter Scott, founder of World Wildlife Fund (WWF).

The Hundredth Monkey Effect

You may have heard the story of the "Hundredth Monkey Effect." In 1952, on the island of Koshima, scientists were providing Japanese macaques with sweet potatoes dropped in the sand. The monkeys liked the taste of the raw sweet potatoes, but they found the dirt unpleasant. An 18-month-old female named Imo found she could solve the problem by washing them in a nearby stream. She taught this trick to her mother. Her playmates also learned this new way and they taught their mothers, too. This cultural innovation was gradually picked up by other monkeys and then **inexplicably transmitted to colonies of macaques on other islands.**[87]

The hundredth monkey story remains unproven, but as a metaphor it is still useful in recognizing the tipping point for real change. Today the capacity to **spread ideas around the world with lightning speed** in the age of email and the Internet is no myth. We even have a phrase for it—"going viral." If there has ever been an issue that needs to trigger that "hundredth monkey effect," our population crisis is it. We desperately need a critical mass of people to spur politicians and the media into action. Sitting at our computer, each one of us can be that hundredth monkey and help motivate citizens around the globe to make sustainable population of all species a goal.

> *The greatness of a nation and its moral progress can be judged by the way its animals are treated. I hold that, the more helpless a creature, the more entitled it is to protection by man from the cruelty of man.*
> **—Mahatma Gandhi**

LinkedIn contributor David Floyd states, "In a world where war is no longer the primary threat to our survival, but in which rampant consumerism and our inability to recognise and respond to the damage we're inflicting on our own planet are moving us closer to the brink of destruction, it's not too late to make amends."

Extinction of species is occurring at an alarming rate, so in this chapter, many causes and solutions to this tragedy will be discussed. From the loss of our last male white rhino to the Girl Guides'

87 David Floyd, Linkedin, *The Hundredth Monkey*, January 27, 2020: https://www.linkedin.com/pulse/hundredth-monkey-david-floyd-first-on-the-beach?trk=related_artice_The%20Hundredth%20Monkey_article-card_title

project to save the bees, this is an issue that is causing great debate all over the world. Ronald Wright, author of *A Short History of Progress,* points out, "Of all land mammals and birds alive today, **humans and their livestock make up 96 percent of the biomass; wildlife has dwindled to four percent**. This has no precedent. Not so far back in history the proportions were the other way round. As recently as 1970, humans were only half and wildlife more than twice their present numbers. These closely linked figures are milestones along our rush towards a trashed and looted planet, stripped of diversity, wildness, and resilience; strewn with waste. Such is the measure of our success."[88]

A Vanishing Biodiversity

The loss of genetic and species diversity is the folly our descendants are least likely to forgive us.
—Edward O. Wilson, professor at Harvard University

If trends continue, our grandchildren may live on a planet inhabited by less than half the species of plants and animals populating ours. Some people argue that extinction is a natural process, that species have always come and gone. But in the 21st century, this **sixth mass extinction** is causing species to vanish from Earth at a rate of dozens a day, surpassing even the mass extinctions 65 million years ago when the dinosaurs perished. And there is a major difference: today's extinctions are being caused by humans and our relentless resource extraction, animal agriculture, deforestation, and a host of other activities. What price will we pay for this travesty?

"What do the snow leopard, giant panda, whooping crane, green pitcher plant, noonday snail and the Queen Alexandria birdwing butterfly have in common? They are all endangered species."[89] These, and thousands of other plants and animals, share a common enemy—the human species. The phenomenal growth in human population is taking its toll as natural habitats are paved over, built on, polluted, lumbered and mined—all to "benefit" encroaching civilization. The clash between humans and wildlife is evident on every part of the planet.

Jerry Coyne and Hopi Hoekstra tell us that 250 million years ago, a monumental catastrophe devastated life on Earth. We don't know the cause—perhaps glaciers, volcanoes, or even the impact of a giant meteorite—but whatever happened drove more than 90 percent of the planet's species to extinction. After the Great Dying, as the end-Permian extinction is called, Earth's biodiversity—its panoply of species—didn't bounce back for more than ten million years.[90]

Aside from the Great Dying, there have been four other mass extinctions, all of which severely pruned life's diversity. Scientists agree that we're now in the midst of a sixth such episode. This new one, however, is different—and, in many ways, much worse. For unlike earlier extinctions,

88 Ronald Wright, "Can We Still Dodge the Progress Trap?" *The Tyee,* September 20, 2019: https://thetyee.ca/Analysis/2019/09/20/Ronald-Wright-Can-We-Dodge-Progress-Trap/

89 Zero Population Growth (now known as Population Connection): https://www.populationconnection.org

90 Jerry A. Coyne and Hopi E. Hoekstra, "The Greatest Dying," *The New Republic,* September 23, 2007: https://newrepublic.com/article/63245/the-greatest-dying

this one results from the work of a single species, *Homo sapiens*. We are relentlessly taking over the planet, laying it to waste and eliminating most of our fellow species. Moreover, we're doing it much faster than the mass extinctions that came before. Every year, up to 30,000 species disappear due to human activity alone. At this rate, we could lose half of Earth's species in this century. And unlike with previous extinctions, there's no hope that biodiversity will ever recover, since the cause of the decimation—us—is here to stay.

To scientists, this is an unparalleled calamity, far more severe than global warming, which is, after all, only one of many threats to biodiversity. Yet global warming gets far more press. Why? One reason is that, while the increase in temperature is easy to document, the decrease of species is not. Biologists don't know, for example, exactly how many species exist on Earth. Another reason is that biologists often fail to document the decline in biodiversity or argue for the public interest—avoiding criticism from politicians, industry, and lobbyists. But we do know some things.

Tropical rainforests are disappearing at a rate of 2 percent per year. Populations of most large fish are down to only 10 percent of what they were in 1950. Many primates, and all the great apes—our closest relatives—are nearly gone from the wild.[91] Human-induced habitat loss and degradation are the primary threats to gorilla survival. In the last 50 years, populations of bonobos, chimpanzees, gorillas, and orangutans have declined by 50 percent. For some subspecies, declines have been even greater. While human populations are in the billions, the great apes only number in the thousands.

According to the CITES Red List, only about 20 percent of gorillas in Africa live inside protected areas where, in theory, they are safe from habitat modification by encroaching humans. The remaining 80 percent are severely threatened by human-induced habitat modification. It is humanity's insatiable demand for Earth's resources that drives the precipitous decline in primates of all species. It is the remorseless hunting, logging, livestock grazing, fuelwood gathering, agriculture, food collection, mining, and civil wars fuelled by a growing human population that are having devastating effects on primates and their habitat. Africa's mountain gorillas now number only a few hundred.

> *There are more human babies born each day—about 250,000—than there are individuals left in all the great ape species combined, including gorillas, chimpanzees, bonobo and orangutans.*
> **—Richard Cincotta, ecologist, Population Action International**

You Think You're Stressed?

The last decade has seen an ominous spate of storms and disasters that we could never have imagined 50 years ago, leaving people reeling with disbelief and a feeling of foreboding for the future. A form of post-traumatic stress disorder (PTSD) sets in as we wait on the edge of our seats for the next disaster to strike. Where will it happen? When will it happen? Could I be the next victim?

91 Jerry Coyne and Hopi Hoekstra, "A Fate Worse Than Global Warming," *The New Republic*, September 24, 2007: https://www.unz.com/print/NewRepublic-2007sep24

For decades, medical experts have referred to this kind of chronic stress as a "silent cause of disease and killer of millions." Exposure to behaviour or situations that cause stress elevate the release of the stress hormone cortisol, the "flight or fight" hormone, with its myriad of physical, physiological, and neural impacts on the body. Stress felt by all species is the product of both environmental and social factors, some made by choice and others totally out of our control.

Wildlife biologist Dr. Brian Horejsi asks, "You think you're stressed?" He points out that in times of turmoil like this, it is not only people that are stressed out, but all the animals that people value so highly as well. Ancients and Indigenous peoples often tell stories of how animals are even more sensitive to changes taking place on the planet than people, as they are more in tune with Earth and often foretell of dangers to come. Dr. Horejsi notes, "People have always known that when a situation is not 'right' they feel uneasy, anxious, uncomfortable, they worry, they make mistakes, other aspects of their life don't progress 'normally,' they lose their appetite, they get ill; the list is extensive!"[92] Wildlife responds similarly.

Why does the "fight or flight" stress syndrome even exist? Dr. Horejsi explains that it evolved as a feedback system that allowed wild animals and early humans to sense (determine) that something wasn't "right" about what they were doing, who they were with, or where they were. It provides a valuable early warning signal: get out, change your circumstances! Those who reacted by modifying their behaviour or leaving, survived and reproduced; others weren't so lucky. That's how natural selection works.

Think of the consequences of human-induced stress on animals confined to very limited and constantly changing landscapes, in a heightened state of stress over loss of habitat to humans and their cattle, shrinking food supply, and extreme weather hazards. The disruptive and tumultuous insecurity imposed on animals by humanity is deplorable and cumulative, undoubtedly a factor in the die-offs occurring at record numbers worldwide.

Dr. Horejsi reminds us that today's wildlife populations survive in contracted landscapes surrounded by the ever-increasing threat of human encroachment, a mere remnant of what they were in pre-modern human society. They are constantly exposed to humanity's industrial and recreational extravagances (like snowmobiles). Even remote and supposedly protected places are routinely invaded by humans in a world where violent and selfish behaviour is not only tolerated but often encouraged. It's a social phenomenon psychologists Jean M. Twenge and W. Keith Campbell have described as a "narcissism epidemic," with negative repercussions that ripple out far beyond just human society.[93]

When an animal is not allowed to react to stress naturally and flee to escape a hazardous situation, the result could be significant or even deadly. When there is no safe place to flee to, there is nowhere else to find adequate food, or there are fires or storms raging all around, an animal is in a state of crisis and likely to perish. The state of an animal's health and behaviour plays an immense

92 Dr. Brian Horejsi, "Opinion: Common Ground," *Penticton Herald*, December 18, 2019: http://www.pentictonherald.ca/opinion/

93 Jean M. Twenge and W. Keith Campbell, *The Narcissism Epidemic: Living in the Age of Entitlement*, Atria, New York/London/Sydney/New Delhi, 2009.

part in successful reproduction as well; even the inherent capabilities of an animal's body, the "biological imperative" to produce young, cannot escape the burdens of stress. We can only speculate how many wild animals are NOT conceived, not born, or don't survive because of stress-induced complications, but it is safe to say the consequences for wildlife conservation are very real.

Half a Billion Animals Perish in Bush Fires

This fight for survival can be seen centre-stage in the heartbreaking inferno that ravaged Australia's landscape, as half a billion animals perished in bushfires in the four months between September 2019 and January 2020. According to News Corp Australia, it is likely that entire species have been wiped out.[94] More than 6.3 million hectares of land have been burned, nearly six times as that destroyed during last year's devastating Amazon fires. The fires were so prominent that they could be seen from space, and smoke reached nearby countries like New Zealand.

Ecologists from the University of Sydney now estimate 480 million mammals, birds, and reptiles have been lost in Australia since September. Record-breaking temperatures and months of severe drought have fuelled these infernos that have destroyed 30 percent of the koala's habitat and a third of all koalas, which are only found in Australia. As of 2020, millions of hectares of national park have burnt in this wildlife apocalypse.

Koalas were already on the brink in Australia. In May 2019, Deborah Tabart, chairman of the Australian Koala Foundation (AKF), argued that koalas may be "functionally extinct" in Australia and called on Scott Morrison, Australia's prime minister to protect koalas and their habitat by enacting the Koala Protection Act (KPA). She added, "The KPA has been written and ready to go since 2016. The plight of the koala now falls on his shoulders."[95] The Koala Protection Act is based on America's Bald Eagle Act, which works with both the federal Endangered Species Act and the Environment Protection Authority to protect bald eagles in the United States.

"The AKF thinks there are no more than 80,000 koalas in Australia. This is approximately 1% of the 8 million koalas that were shot for fur and sent to London between 1890 and 1927," Tabart added. Once the koala population falls below a critical number, it can no longer produce the next generation, resulting in definite extinction. Deforestation, disease, and the effects of climate change are the biggest threats to koalas, and the main reasons their numbers are rapidly decreasing.

Prime Minister Morrison argued that there was no direct link between the severity of the fires burning across Australia at alarming rates and the country's carbon emissions, which are among the highest per capita in the world. He told ABC AU during a radio interview on November 21 that there was no "credible scientific evidence" that cutting carbon emissions could reduce the severity of the fires.

---

94 Marnie O'Neill, "Half a Billion Animals Perish in Bushfires," *Townsville Bulletin*, News Corp Australia, January 1, 2020: https://www.townsvillebulletin.com.au/technology/half-a-billion-animals-perish-in-bushfires/news-story/b316adb4f3af7b1c8464cf186ab9f52c

95 Eric Todisco, "Australian Koalas Considered 'Functionally Extinct'," *People* magazine online, May 15, 2019: https://people.com/pets/australian-koalas-considered-functionally-extinct/

To climate experts it is blatantly obvious that climate change is magnifying the destructive effect of events such as wildfires and hurricanes, but politicians don't want to risk appearing at fault by admitting that climate change is a huge problem. If no one owns up, then it will not jeopardize their longstanding ties to oil, gas, and coal industries who have been the biggest deniers of climate change over the past few decades.

Australian author Julian Cribb warns us, "On any day between 10,000 and 30,000 wildfires blaze around the planet. Realms as diverse and distant as Siberia, Amazonia, Indonesia, Australia and California are aflame. The advent of the pyocene—'**The Age of Fire**'—is among the bleakest warnings yet that humanity has breached boundaries we were never meant to cross and that our tenure on Planet Earth is now in jeopardy. It is time not only to think the unthinkable, but to speak it: that the world economy, civilization, and maybe our very survival as a species, are on the line. And **it is past time to act**."[96]

Nature is declining at rates unprecedented in human history.
—Jane Smart, IUCN director

The IUCN "Red List"

Koalas are one of the many species listed as threatened with extinction on the Red List by the International Union for Conservation of Nature (IUCN). Created in 1948, it is now the world's largest and most diverse environmental network, harnessing the knowledge, resources, and reach of more than 1,300 member organizations and some 15,000 experts.[97] The Red List is the world's most comprehensive inventory of the global conservation status of plant and animal species. It uses a set of quantitative criteria to evaluate the extinction risk of thousands of species. These criteria are relevant to most species and all regions of the world. With its strong scientific base, the IUCN Red List is recognized as the most authoritative guide to the status of biological diversity.[98] The list now includes 112,432 species, with 30,178 threatened with extinction—nearly a third of all assessed species, from monkeys to sea turtles.[99]

The IUCN describes the Red List of Threatened Species as an invaluable resource to guide conservation action and policy decisions. It is a health check for our planet—a **Barometer of Life** (see Figure 5-1). It is based on an objective system for assessing the risk of extinction of a species should no conservation action be taken. Species are assigned to one of nine categories of threat based on whether they meet criteria linked to population trend, population size, and structure and geographic range.

96 Julian Cribb, "Time to Speak the Unspeakable," Surviving C21 blog, December 1, 2019, https://juliancribb.blog/2019/12/01/time-to-speak-the-unspeakable/

97 The International Union for Conservation of Nature: https://www.iucn.org/about

98 The IUCN Red List of Threatened Species: https://www.iucn.org/resources/conservation-tools/iucn-red-list-threatened-species

99 UCN Red List, *Species recoveries bring hope amidst the biodiversity crisis,* December 10, 2019: https://www.iucn.org/news/species/201912/species-recoveries-bring-hope-amidst-biodiversity-crisis-iucn-red-list

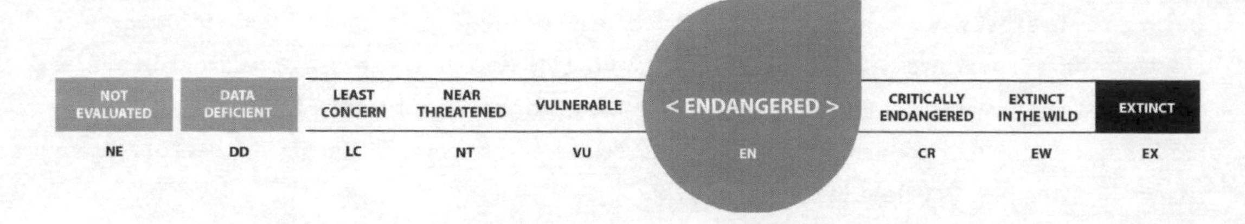

Figure 5: The IUCN Red List Categories

In May 2019, the Red List of Threatened Species included 9,000 new species, but this number doesn't account for the thousands of species that go extinct before scientists even have a chance to describe them. However, in December 2019, there was a hopeful landmark update reporting that conservation efforts had caused **ten species to improve in status**. This can be found in the SUCCESSES section of this chapter.

"With more than 100,000 species now assessed for the IUCN Red List, this update clearly shows how much humans around the world are overexploiting wildlife," said IUCN Acting Director General, Dr. Grethel Aguilar.

The Red List shows that most of the world's threatened species are suffering habitat loss where livestock are a factor. United Nations scientists point out that **the meat industry is one of the most significant contributors to the most serious environmental problems**, at every scale from local to global, including problems of land degradation, climate change and air pollution, water shortage and water pollution, and loss of biodiversity.

> *We must wake up to the fact that conserving nature's diversity is in our interest, and is absolutely fundamental to achieving the Sustainable Development Goals.*
> **—Dr. Grethel Aguilar, IUCN**

Wildlife in the Food Industry

Today it is estimated that 99 percent of threatened species are at risk from human activities. One animal that symbolizes this crisis is the Spix's macaw, a blue-grey Brazilian parrot of which only one remains in the wild. The macaw, which lives in the dry scrubland, has been driven to the brink of extinction by habitat destruction and the exotic pet trade. This trade also uses rare wildlife in meals and medicine, a centuries-old habit that is hard to break. Wild animals on the endangered species list are continuing to find their way to specialty restaurants for diners willing to pay.

The extravagant prices charged for wildlife dishes are no deterrent for those who have money and power. Medicine containing rhinoceros horn and tiger bone are still available in shops across China, as the government lifts its ban on commercial use of the products. An underlying problem is a general lack of compassion for animals and the difficulty of enforcing bans.[100] Ian Warkentin warns that humans are also eating frogs into extinction, as we harvest at least 400 million frog legs

100 World Wildlife Fund, *In a blow to wildlife, China lifts a ban on the use of tiger and rhino parts*, November 12 2018: https://www.worldwildlife.org/stories/in-a-blow-to-wildlife-china-lifts-a-ban-on-the-use-of-tiger-and-rhino-parts

each year. He says that frog populations, already under pressure from disease, climate change, and habitat loss, could go the way of Canada's East Coast cod stocks.[101]

In 2009, American Pulitzer Prize–winning journalist and author Chris Hedges warned that we were experiencing an accelerated obliteration of the planet's life forms because, simply put, there were too many people. Hedges cautioned that if the current rate of extinction continued, *Homo sapiens* would be one of the few life forms left on the planet, its members scrambling violently among themselves for water, food, fossil fuels, and perhaps air until they too disappeared. He was right.[102]

Species Declared Extinct in the 2010s

Scientists estimate that species are going extinct 1,000 times faster than they should be, and "literally dozens" go extinct each day.[103]

When it comes to extinct species, where does one begin? The last ten years should provide sufficient evidence of the treasures that we are losing. Some of the extinctions we have witnessed in the last decade, like **Lonesome George, the last of the Pinta Island tortoises**, died as recently as 2012. *Gizmodo* reporter Ryan F. Mandelbaum tells us that George's story is the perfect extinction story. It features a charismatic character with a recognizable face, an obvious villain, and the tireless efforts of naturalists.[104]

"But George's story is not a typical story," adds Mandelbaum. "Perhaps a better mascot of the extinction crisis is Plectostoma sciaphilum; a small snail, called a "microjewel" for its beautiful, intricate shell, that inhabited a single limestone hill in Malaysia. During the 2000s, a cement company wiped the hill off the map for its valuable resources, rendering the "microjewel" snail extinct."

Species today typically go extinct due to one or a combination of several factors: **humans** *clearing their habitat,* **humans** *purposely or inadvertently introducing invasive species to their habitats,* **humans** *polluting their habitats,* **humans** *over-harvesting the species for food or other uses, or* **humans** *indirectly harming habitats through the effects of climate change.*
—Ryan F. Mandelbaum, ***Gizmodo***

Declaring a species extinct itself can be an act of giving up, or worse—once a species is considered extinct, governments may no longer feel the need to fund protections for its habitat. But the trends that connect these 160 extinctions are true of the biodiversity crisis more generally, and the kinds of species it lists, plus how they went extinct, are even more relevant in the present day.

Stuart Pimm, Duke University conservation ecology professor, sounds a note of caution: "Nonprofits are attempting to fight the crisis. But this long list of species declared extinct should

101 Ian Warkentin, Society for Conservation Biology, *Eating Frogs to Extinction*, July 13, 2009: https://conbio.onlinelibrary.wiley.com/doi/abs/10.1111/j.1523-1739.2008.01165.x

102 Chris Hedges, TruthDig, *We Are Breeding Ourselves to Extinction*, March 09 2009: https://www.truthdig.com/articles/we-are-breeding-ourselves-to-extinction/

103 Ryan F. Mandelbaum, "Decade's End: All the Species Declared Extinct This Decade," *Gizmodo*, December 16, 2019: https://gizmodo.com/all-the-species-declared-extinct-this-decade-1840325660

104 *Ibid.*

show you just how serious the crisis is and how **the mere act of humans showing up someplace can precipitate ecological disaster**. Have hope, if you are able, but know that it's going to take serious, coordinated, international effort, some of which may be uncomfortably radical, in order to maintain the health of our planet and the species we share it with and thrive as a species ourselves."[105]

Losing the Great Migrations

Another issue where coordinated international effort will be critical is in the protection of migratory routes. Although governments and industry may see these migration corridors as hindrances to business as usual, they are essential in the long-lasting survival of a species. Not only do these corridors allow free movement to find alternative food and water sources, but they also provide links to ensure genetic diversity and ecosystem dynamics. This is especially critical in this time of climate disruption, when temperatures and food sources are in a state of flux and instability.

From bears to butterflies, migration routes play an important role in protecting and restoring endangered species. Like the locally extinct Chinook salmon that once accessed parts of the Mexico to Canada migratory corridor, the monarch cutterfly travels this historic international route each spring and fall. Since this traditional migratory route was at one time home to an abundant supply of milkweed and camas, the monarch butterfly could find sustenance and thrive. Celebrated worldwide for one of nature's great animal migrations, the monarch butterfly overwinters in Mexico and Coastal California, and returns to breeding sites in the United States and southern Canada for the summer. Their arrival in Canada is a welcome sign of the change in seasons from spring to summer.

However, without milkweed, this distinctively coloured butterfly cannot survive, and milkweed is disappearing at an alarming rate due to the heavy use of herbicides and habitat destruction. Since 2001, according to *Scientific American*, "Milkweeds in and around agricultural crop fields have gradually been eliminated, through a combination of spraying with Roundup (glyphosate) herbicide and increased planting of corn and soybeans genetically modified to be resistant to the herbicide.... This strongly suggests milkweed loss is the primary factor in the species' decline."[106] Other threats facing the monarch include climate change, severe weather events like wildfires, and encroachment by the growing human population. Ways to offset the decline will be discussed at the end of this chapter.

Queen Bee Farms, located on Canada's Vancouver Island, notes that as recently as the early 1990s, the eastern monarch population was estimated to be nearly 700 million making their epic journey across the northern plains of the U.S. and Canada to sites in the Oyamel fir forests just north of Mexico City. Meanwhile, western monarch populations were more than one million overwintering along the California coast. However, most recent estimates of monarch populations show an over 80 percent decline in the east, and 74 percent in the west.[107]

---

105 *Ibid.*

106 John Pleasants, "Monarch Butterflies Under Threat from Rising Herbicide Use," *Scientific American*, June 6, 2014: https://www.scientificamerican.com/article/monarch-butterflies-under-threat-from-rising-herbicide-use/

107 "The Monarch Butterfly," Queen Bee Farms, 2019: https://queenbeefarms.ca/pages/the-monarch-butterfly

In North America, the Yellowstone to Yukon migratory route is essential for wolves and bears. *National Geographic* reporter Jennifer Holland tells us that one herd of zebras is the new record-holder, migrating more than 300 miles across Namibia and Botswana—farther than any other known African mammal.[108] The Serengeti is the site of what most consider the most dramatic migration, with giant herds of millions of animals—some 750,000 zebras and 1.2 million wildebeests as well as gazelles and eland—travelling from the Ngorongoro area in southern Tanzania to the Masai Mara in lower Kenya and returning as the rains dictate.

Research with other migratory mammals has shown that generations of animals may follow the same corridors down to the nearest metre, says ungulate ecologist Mark Hebblewhite. "We have evidence that pronghorn antelope in the western U.S. have migrated over the same routes for more than 6,000 years. This is probably a product of both the landscape and cultural transmission of knowledge amongst social animals."

With the alarming global decline in migratory species, large-scale protected travel corridors are crucial to the survival of these animals and to the ecosystems that sustain them. Knowing their routes also gives conservationists a target when protecting land. The newly reported zebra migration takes place entirely within the boundaries of a complex of protected areas known as the Kavango-Zambezi Transfrontier Conservation Area (KAZA). At about 170,000 square miles (440,000 square kilometres), it's the **largest transfrontier conservation area in the world.**[109]

Scientists are now aware that just protecting isolated blocks of land doesn't work for many species; they need space to freely roam beyond these protected cores. We construct our environments, they do not, so they do not have our adaptive ability. Therefore, the unforeseen detriment to the population as a whole can be significant and the losses irreplaceable. In light of the established value of these migration corridors, it is ironic that the world's largest electrified fence has recently been constructed to encircle the forest reserve surrounding Mount Kenya. With the rapidly growing human population dominating the landscape in Kenya, there has been increasing conflict between wildlife and communities. This fence will surround the entire mountain, locking the animals in. The traditionally forest-dwelling Ogiek tribe has been thrown out, and say their legitimate rights have been ignored.

I fear, in today's world of shifting value, that these kinds of conflicts, and displacement, or replacement, are inevitable. As long as human population escalates, these unpleasant, crude decisions will continue to multiply! It's a duplicitous push–pull kind of politics, where politicians have never wanted to constrain the majority, but pay lip service to their expectations in places like Mount Kenya. It is inevitable that power will dominate some minorities, as the struggle continues for land and control.

One problem in Africa, as *Yale Environment 360* reporter Fred Pearce explains, is that, "wildlife and forest protection is still often seen as something imposed on communities by foreigners or remote government agencies, all this is a sea of change for conservation.... The rights and wrongs of

108 Jennifer S. Holland, "Longest Migration Among African Mammals Discovered," *National Geographic*, May 26, 2014: https://www.nationalgeographic.com/news/2014/5/140527-migration-zebra-mammal-africa-namibia-botswana-environment-conservation/

109 Kavango-Zambezi Transfrontier Conservation Area (KAZA) website: http://www.kavangozambezi.org/en/

electric fences that keep humans and wildlife apart have become a battleground in the debate about how best to preserve megafauna."[110] Pearce adds that there are now plans to create new migration corridors to other wildlife areas, which should alleviate problems such as inbreeding caused by isolating populations of large animals.[111]

One can't help but wonder if it might not have been more beneficial to both the wildlife and communities to make an equal investment in family planning programs in the area at the same time.

Loss of Forest Diversity

Another newly discovered scientific surprise comes in the form of a tree known in Maori mythology as the "whale of the forest." The Kauri is an iconic New Zealand native tree that has recently been classified as "threatened," and there are fears that it may go extinct over large areas of its current range. Historic logging had reduced up to 1.7 million hectares of kauri to 7,000 hectares, which had heightened the species' vulnerability to factors like disease to which it might have otherwise been buffered. When the "whales of the forest" start dying, you know there's something wrong with the forest.[112]

Of course the Kauri tree is not the only endangered tree we need to protect. Since trees are the lungs of Earth, they are essential for all life on the planet. In all their diversity, they are a source of food and shelter for countless animal species such as birds, primates, and squirrels, and play a key spiritual and economic role as well.

Of the approximately 6,750,000 square miles of lush forestry canopy that once blanketed Earth, only about 30 percent remains. In Madagascar, the Rosewood tree is also endangered, as this species is highly exploited for their precious wood, with over 90 percent of them now threatened. They are one of the world's most illegally trafficked wild products, so their continued survival is dependent on increased enforcement of local management plans, national laws, and international cooperation.[113]

Global Intergovernmental Report

The United Nation's Intergovernmental Science-Policy Platform on Biodiversity and Ecosystem Services (IPBES) report examined the rate of decline in biodiversity over the last 50 years and found that the adverse effects of human activities on the world's species is "unprecedented in human history."[114]

---

110 Fred Pearce, "In Kenya's Mountain Forests, A New Path to Conservation," Yale Environment 360, February 26, 2015: https://e360.yale.edu/features/in_kenyas_mountain_forests_a_new_path_to_conservation

111 Fred Pearce, "Kenya's electrified route to human-wildlife harmony," *New Scientist*, February 25, 2015: https://www.newscientist.com/article/mg22530104-200-kenyas-electrified-route-to-human-wildlife-harmony

112 Hannah Osborne, "Ancient Tree With Record of Earth's Magnetic Field Reversal in Its Rings Discovered," *Newsweek*, July 4, 2019: https://www.newsweek.com/ancient-tree-discovered-earths-magnetic-field-1447570

113 International Union for Conservation of Nature, "Over half of Europe's endemic trees face extinction," September 27, 2019: https://www.iucn.org/news/species/201909/over-half-europes-endemic-trees-face-extinction

114 Global Assessment Report on Biodiversity and Ecosystem Services, Wikipedia, 2019: https://en.wikipedia.org/wiki/Global_Assessment_Report_on_Biodiversity_and_Ecosystem_Services

Nature is declining globally at rates unprecedented in human history – and the rate of species extinctions is accelerating, with grave impacts on people around the world now likely.
—UN (IPBES) Global Assessment

"The overwhelming evidence of the IPBES Global Assessment, from a wide range of different fields of knowledge, presents an ominous picture," said IPBES Chair, Sir Robert Watson. "The Report also tells us that it is not too late to make a difference, but only if we start now at every level from local to global," he said. It is the first and most comprehensive intergovernmental report of its kind and builds on the landmark Millennium Ecosystem Assessment of 2005, introducing innovative ways of evaluating evidence. Compiled by 145 expert authors from 50 countries over the past three years, with inputs from another 310 contributing authors, the IPBES report assesses changes over the past five decades, providing a comprehensive picture of the relationship between economic development pathways and their impacts on nature.

The IPBES report further notes that since 1980, greenhouse gas emissions have doubled, raising average global temperatures by at least 0.7 degrees Celsius. With climate change already impacting nature from the level of ecosystems to that of genetics, impacts are expected to increase over the coming decades. "Biodiversity and nature's contributions to people are our common heritage and humanity's most important life-supporting 'safety net,'" said Professor Sandra Díaz, co-chair of the assessment. "But our safety net is stretched almost to breaking point."

With good progress on components of only four of the twenty Aichi Biodiversity Targets, most targets were not achieved by the 2020 deadline. Current negative trends in biodiversity and ecosystems have undermined progress towards 80 percent (35 out of 44) of the assessed targets of the Sustainable Development Goals, related to poverty, hunger, health, water, cities, climate, oceans, and land. It's scary to think that the total biomass of wild mammals has decreased by 82 percent, while **humans and their farm animals now make up 96 percent of all mammalian biomass on Earth.** Around 25 percent of the planet's ice-free land is being used to rear cattle for human consumption, so it is little wonder that the report recommended a massive move towards plant-based diets.

So what does this mean for the future of the planet? Without drastic action to conserve habitats, the rate of species extinction—already tens to hundreds of times higher than the average across the past ten million years—will only increase, says the analysis. Biodiversity should be at the top of the global agenda alongside climate, said Anne Larigauderie, IPBES executive secretary, at a press conference in Paris, France. "We can no longer say that we did not know."

***One million species**, including 40 percent of amphibians, almost a third of reef-building corals, more than a third of marine mammals, and 10 percent of all insects are threatened with extinction.*
—IPBES Global Assessment

Goodbye to Sudan

Journalist Ami Vitale began his career covering conflict in places like Kosovo and Afghanistan. "Those years in war zones led me to an epiphany: Stories about people and the human condition are also about nature," he said. "If you dig deep enough behind virtually every human conflict, you will find an erosion of the bond between humans and the natural world around them. These truths became personal guideposts when I met Sudan, a northern white rhinoceros and, eventually, the last male of his kind."[115]

Vitale explains in *National Geographic* that a century ago there were hundreds of thousands of rhinos in Africa. By the early 1980s, hunting had reduced their numbers to around 19,000. Rhino horns, like our fingernails, are simply keratin, with no special curative powers, yet they've long been valued by people around the world as antidotes for ailments from fever to impotence.

In 2009, Sudan was living at a zoo in the Czech Republic when Vitale met him for the first time: "I can recall the exact moment. Surrounded by snow in his brick and iron enclosure, Sudan was being crate trained—learning to walk into the giant box that would carry him almost 4,000 miles south to Kenya. He moved slowly, cautiously. He took time to sniff the snow. He was gentle, hulking, otherworldly. I knew I was in the presence of an ancient being, millions of years in the making (fossil records suggest that the lineage is over 50 million years old), whose kind had roamed around much of our world."

Those final moments were quiet—the rain falling, a single goaway bird scolding, and the muffled sorrow of Sudan's caretakers. These keepers spend more time protecting the northern white rhinos than they do with their own children. Watching a creature die—one who is the last of its kind—is something I hope never to experience again. It felt like watching our own demise.
—Ami Vitale, *National Geographic*

"How did we arrive at the point where such desperate measures were necessary?" Vitale asks. "It's astonishing that a demand for rhino horn based on little more than superstition has caused the wholesale slaughter of a species. But it's encouraging that a disparate group of people came together in an attempt to save something unique and precious, something that once lost would be gone forever."

Nine years after the airlift, Vitale received a call to hurry to Kenya. At 45, Sudan was elderly for his species. He had lived a long life, but now he was dying. In his last years he experienced again his native grasslands, although always in the company of armed guards to keep him safe from poachers. And he had found stardom—he'd been affectionately dubbed the "most eligible bachelor in the world."

The northern white rhinos may not survive human greed, yet there is a tiny sliver of hope. Today only two females are left in the world, but plans are in place to try in vitro fertilization to breed them.

Vitale commiserates, "This is not just a story to me. We are witnessing extinction right now, on our watch. Poaching is not slowing down. If the current trajectory of killing continues, it's entirely possible that all species of rhinos will be functionally extinct within our lifetimes. Removal of a keystone species has a huge effect on the ecosystem and on all of us. These giants are part of a complex world created over millions of years, and their survival is intertwined with our own. Without rhinos and elephants and other wildlife, we suffer a loss of imagination, a loss of wonder, a loss of beautiful

115 Ami Vitale, "What I learned documenting the last male northern white rhino's death," *National Geographic*, October 2019: https://www.nationalgeographic.com/animals/2019/09/life-changing-lessons-of-the-last-male-northern-white-rhino/

possibilities. When we see ourselves as part of nature, we understand that saving nature is really about saving ourselves.... Sudan taught me that."

Solutions and Successes

Family planning provision is the best way to respond to critical conservation challenges. But it is an issue which is almost entirely ignored.
—Jane Goodall

Extinction of wildlife is occurring at a rate never envisioned before—as the human population has doubled in the last 50 years, the number of animals has halved. Those who feel a kinship to other species are grieving this regrettable loss, a feeling of *environmental melancholia*. We often try to replace this loss with stuffed replicas of animals, which now are said to outnumber the real ones on the planet. This causes even more environmental harm, as they are usually made from fossil fuels and, like the genuine animals, are seen as disposable.

Humans can only hope that evolution will look favourably upon our last-ditch attempts to reverse the damage we have done on this planet. Over the last 50 years, there have been many extraordinary initiatives dedicated to protecting nonhumans and their habitats, and one recently launched was the *Born To Be Wild* multimedia presentation in 2019 by John Kay. Baby Boomers will recall that he is the legendary singer and guitar player from the band Steppenwolf. Now, 50 years after their first album was released, John Kay has transformed from rock star to wildlife advocate.[116] *Born To Be Wild* documents Kay's travels and conservation work with his wife Maue Kay, leading them to form the Maue Kay Foundation in 2004, and now bring an end to an era of music with this new conservation presentation.

There are conservation groups all over the world working to garner government cooperation in a last-ditch effort to protect endangered species from the brink of extinction. The **giant guitarfish is now the most imperiled marine fish in the world**, with all but one of its 16 species assessed as critically endangered, and efforts are now underway to reverse this predicament.

Involving the people who live in the Grevy's habitat area in the conservation process includes educating the next generation of land managers.
—Grevy's Zebra Trust

The Grevy's zebra is also officially listed as an endangered species by the International Union for Conservation of Nature. Conservation organizers, like the Grevy's Zebra Trust, are working with governments and communities in Kenya and Ethiopia to formulate an action plan to protect the endangered Grevy's zebra from extinction. Zebra hides can fetch big bucks, and both zebra fat and bone are claimed to have medicinal value. We have seen their numbers plummet over the years—in Kenya, the population has gone from around 15,000 in the 1970s to only 2,812 known left today.[117]

A primary threat to the species is land degradation from the overgrazing of livestock. Repeated droughts in East Africa in recent years have only added to the environmental pressures on the

116 Maue Kay Foundation: https://mauekay.org/
117 Andrew Peacock and Morgane Croissant, "Saving Africa's Grevy's zebra from extinction," Matador Network, 2018: https://matadornetwork.com/view/africa-grevys-zebra-extinction/

survival of this species, as they search for water and grazing land. Young grassland warriors are taught to recognize when herders are not sticking to agreed boundaries for livestock grazing and the Grevy's zebra ambassadors monitor the plains to protect the species from being hunted for its meat and act as conservation messengers and peacekeepers between ethnic groups.

Efforts like this make us realize that it isn't all doom and gloom. There's a lot we can do to stave off a species' extinction. With dedicated, coordinated international effort, we have the means to combat the extinction crisis one species at a time. According to a 2019 UN Sustainable Development Report, there has been a **29 percent average reduction in the extinction risk for mammals and birds in 109 countries**, thanks to conservation investments from 1996 to 2008. The extinction risk of birds, mammals, and amphibians would have been at least 20 percent greater without conservation action in recent decades.[118]

Fortunately, it is possible to save most endangered wildlife species—as long as new population groups like Roots & Shoots and Fair Start Movement are working along with conservation groups to promote a sustainable human population. It is this combined effort that can build the momentum that is required for this colossal task.

> *Human population growth and overconsumption are at the root of our most pressing environmental issues, including the species extinction crisis, habitat loss and climate change.*
> **—Center for Biological Diversity**

Fair Start Movement is one of the new youth-powered groups on the block, and they are making a big splash in the social media scene. "**Child-first family planning** provides a clear human rights-based solution and path forward," they say. "It's sustainable and equitable, and reflected in the *worldwide trend towards smaller families* that can invest more in each child and ensure a safer future."[119]

The Center for Biological Diversity has also found an innovative way of addressing the extinction of species crisis. They have acknowledged that the connection between human population growth and the decline of other species is well documented but little discussed. So they have identified a unique solution—condoms. "It's a conversation that needs to happen, but talking about how a person's sex life can affect the survival of wildlife can be difficult. Endangered Species Condoms are the perfect icebreaker."[120]

The Center maintains that taking responsibility for our reproductive decisions is part of ensuring plant and animal diversity around the globe. Since the Endangered Species Condoms Project was launched in 2009, they have handed out hundreds of thousands of free condoms, relying on thousands of volunteers for distribution. Other solutions for overpopulation that they recommend are supporting the empowerment of women, ensuring access to birth control as a fundamental human right, and educating women regarding the impacts of overpopulation on biodiversity and the health of our planet.

118 UN Sustainable Development Goals, 2019: https://www.un.org/sustainabledevelopment/blog/2019/05/nature-decline-unprecedented-report/

119 Fair Start Movement: https://havingkids.org/

120 Endangered Species Condoms, A Project of the Center for Biological Diversity: https://www.endangeredspeciescondoms.com/condoms.html

Birds—Our Sentinel Super Heroes

National Geographic reporter Alanna Mitchell maintains that birds are the planet's superheroes, built for survival. "The ice of Antarctica doesn't faze them," she says. "Nor does the heat of the tropics. They thrive in the desert, in swamps, on the open ocean, on sheer rock faces, on treeless tundra, atop airless mountaintops, and burrowed into barren soil. Some fly nonstop for days on end. With just the feathers on their backs, they crisscross the hemisphere, dodging hurricanes and predators along the way, arriving unerringly at a precise spot, year after year. They have penetrated nearly every ecosystem on Earth and then tailored their own size, habits, and colors to each one, pollinating, dispersing seeds, controlling bugs, cleaning up carrion, and fertilizing plants. But for all their superhero powers, birds are in trouble."[121]

Globally, one in eight—more than 1,300 species from the tropics to the poles—are threatened with extinction, and the status of most of those is deteriorating, according to BirdLife International.[122] **Despite all of this, birds are helping to save our lives.** Studies on how endocrine-disrupting chemicals like DDT, which still persist in our environment years after being banned, can have implications for both avian and human health. Birds truly are the canary in the coal mine when it comes to linking the fate of birds to that of humans. As Mitchell confirms, "They prophesy what might happen to us as the load of carbon-based, planet-warming gases in the atmosphere and oceans climbs ever higher."

Fortunately, nature is amazingly resilient: places we have destroyed, given time and help, can once again support life, and endangered species can be given a second chance. And there is a growing number of people, especially young people, who are aware of these problems and are fighting for the survival of our only home, Planet Earth. We must all join that fight before it is too late.

—Dr. Goodall speaks about threats to biodiversity following UN Report, 2019

Mitchell reminds us that humans have relied on birds' superpowers for millennia: "Imagine forest-dwellers of ancient times, anxious to avoid snakes and jaguars, listening for the alarm calls of sharp-sighted, high-flying, omnipresent birds." In many Indigenous cultures, birds are messengers sent by the creator, or protectors and healers. Today they play that role in a non-spiritual sense: They send warnings to tribes about the health risks of eating fish tainted with industrial pollutants, and herald the presence of pathogens, such as avian influenza and West Nile virus.

Birds also provide us with sheer joy—from their songs and striking colours, to the thrill of watching them swoop through the air. "Which of us has not wished we could do that?" asked John Fitzpatrick, director of the Cornell Lab of Ornithology in Ithaca, New York. "They move with the seasons. It's a major annual heartbeat we feel."

---

121 Alanna Mitchell, "The 1,300 Bird Species Facing Extinction Signal Threats to Human Health," *National Geographic*, August 26, 2014: https://www.nationalgeographic.com/news/2014/8/140825-bird-environment-chemical-contaminant-climate-change-science-winged-warning/

122 Martin Fowlie, "Saving the world's most threatened birds," BirdLife International, June 20, 2013: https://www.birdlife.org/worldwide/news/saving-worlds-most-threatened-birds

Amazing People Are Building and Planting Things!

In these ominous times of disturbing species decline and EIGHT BILLION people squabbling over the planet's remaining wealth, Dr. Jane Goodall is very busy proving that one person can make a huge difference. For decades, Goodall has been a force to reckon with, building Africa's largest chimpanzee sanctuary and transforming landscapes to benefit wildlife and people around the world.[123] Dr. Goodall first set foot in what is now known as Gombe Stream National Park in 1960 when she launched her pioneering research with wild chimpanzees. She was only 26 years old.

In 1977, Jane founded the Jane Goodall Institute, which continues to support the research at Gombe. The institute is committed to saving this iconic species—our closest relatives—who share 98.6 percent of our DNA. With 31 offices around the world, Jane and the institute are widely recognized for effective community-centred conservation and development programs in Africa and the protection of wild chimpanzees in Africa's largest chimpanzee sanctuary. Thanks to these efforts, Kabi, a very special chimpanzee, now calls the Jane Goodall Institute's Tchimpounga sanctuary his home. Jane pointed out that happiness, safety, and love were not always a part of Kabi's life. At the age of two, Kabi was rescued by EcoGuards who raided a poaching camp in far north Congo. When they found Kabi, he was alone, tied up alongside animal skins, victims of the illegal bushmeat and wildlife trafficking trade.

In 1991, after meeting with a group of Tanzanian teenagers to discuss community problems, Goodall created Roots & Shoots. This program is dedicated to inspiring young people to take action in their communities, and it has since grown to include approximately 150,000 individuals in more than 50 countries. Jane continues her work today by speaking out about the threats facing chimpanzees. Everywhere she goes, **Goodall urges audiences to recognize their personal power and responsibility to effect positive change** through consumer action, lifestyle change, and activism.[124] "Young people, when they understand the problems, are empowered to take action. When we listen to their voices (they) actually are changing the world and making it better for people, for animals, and for the environment because everything is interconnected."

In her interview with *Time* magazine, Jane shared the four seemingly unsolvable problems we must tackle to slow down climate change and the solutions:

1. Eliminate poverty.
2. Change the unsustainable lifestyles of so many of us.
3. Abolish corruption.
4. Think about our growing human population.

She added, "There are 7.7 billion of us today, and by 2050, the UN predicts there will be 9.7 billion. It is no wonder people have despaired. But I believe we have a window of time to have an

123 Mary Bagley, Jane Goodall biography, LiveScience, March 29, 2014: https://www.livescience.com/44469-jane-goodall.html

124 David Gelles, "Jane Goodall Keeps Going, With a Lot of Hope (and a Bit of Whiskey)," *New York Times*, September 12, 2019: https://www.nytimes.com/2019/09/12/business/jane-goodall-corner-office.html

impact.... but we must act now. What you have to do is to get into the heart. And how do you get into the heart? With stories."[125]

This compelling story Goodall tells about the resilience of nature does get to the heart of the matter: "Habitats and species on the brink of extinction can recover if given a chance. When I realized the plight of the people living around Gombe, the Jane Goodall Institute started a program called Tacare to help them find ways to make a livelihood that did not involve devastating the environment.... Today we have Tacare in six other African countries, and the hills in Gombe aren't bare anymore."

In 1968, the United Nations made it a basic human right for each woman to decide the number and spacing of her children. In the last 50 years since that decision, it has become clear that improving access to birth control is the key in the fight against environmental devastation.[126]

On 2019 **World Population Day**, 150 organizations joined forces to recognize the "ignored issue" that better family planning is crucial to securing the future of the planet. They are pushing for greater **links to be made between health and overpopulation** and **more funding** to be made available for women, especially in developing countries. Dr. Jane Goodall, one of the signatories to the Thriving Together campaign, said, "One challenge that we have to face, if we care about the future of Planet Earth, is the impact of our own species. The planet only has finite natural resources and in some places we are plundering them faster than Mother Nature can replenish them."

The Thriving Together campaign, which is spearheaded by the London based Margaret Pyke Trust, has spent £8 billion each year on family planning and environmental work in 170 countries globally. Among the organizations are Greenpeace, Marie Stopes International, the United Nations Population Fund, and the Bill & Melinda Gates Institute for Population & Reproductive Health. The campaign will focus on getting the environmental organizations to talk about family planning and vice versa in a coordinated approach.

Reviving an Iconic Migration

The Kootenay Native Plant Society has partnered up with the Xerces Society in southern British Columbia, Canada, to launch the Milkweed for Monarchs project. This call to action aims to rebuild a migration corridor and revive an iconic migration of monarch butterflies from Mexico to southern Canada that is on the verge of collapse. The western monarch butterfly, a symbol of international cooperation and a natural wonder, was listed as a species of special concern under Canada's Species at Risk Act in 2003. If nothing is done to preserve their habitat, western monarchs **could face extinction in the next decade**.

The Xerces Society points out that at one time, millions of monarchs overwintered along the Pacific coast in California and Baja, Mexico—now they number in the thousands. Therefore, it is a common goal of Canada, the US, and Mexico to help migrating monarch butterflies by planting

125 Jane Goodall, "These 4 Issues May Not Seem Related to Climate Change. But They Are and We Need to Solve Them Now," *Time* magazine, September 12, 2019: https://time.com/5669043/jane-goodall-climate-change/

126 Jane Wharton, "Environment crisis won't be stopped unless we improve access to contraception," July 11, 2019: https://metro.co.uk/2019/07/11/environment-crisis-wont-stopped-unless-improve-access-contraception-10187089/

milkweed, recovering habitat, and reducing pesticide use. Showy milkweed is a perennial plant native to western North America, and is the only native milkweed species known to occur in the West Kootenay region of Canada. Monarch butterflies depend on milkweed to survive. Adult butterflies lay their eggs only on milkweed, and milkweed is the only food caterpillars eat.

For a long time, milkweed has been considered a "noxious weed." The fact that it can make livestock sick is seen as especially problematic to the **influential livestock industry**, so we have lost much of this native wildflower across our landscape. As a result, **western monarch** butterflies are becoming a rare sight in southwestern BC.[127]

The David Suzuki Foundation has announced that their *Monarch Watch* will also promote the concept of the Milkweed Village, recognizing the importance of native milkweeds to local biodiversity, including monarchs. In the process, they will also engage communities to build a network of pollinator freeways to help butterflies and wild pollinators find food and shelter wherever they land. The endangered western bumblebee has also been seen on local milkweed, along with honeybees, milkweed beetles, cobalt beetles, and hummingbirds.[128]

Solution: So, what can you do? You can become an advocate for monarch habitat protection through your local group, or plant milkweed in your backyard to attract them. Also, contact your local elected official to ask that monarch migration corridors in your area be protected from destructive land use practices.

"Bee" the Change

Global Citizen's reporter Helen Lock tells us about another effort to restore our environment. "The bees are back in town," she says, because London is building a 7-mile "Bee Corridor" to help boost insect numbers. Of the 353 species of pollinating insects, bees are especially important for pollinating the vast majority of plants we rely on for food, from apples and vanilla to almonds and squash. Wildflower meadows are set to be planted in 22 parks in London to attract not just bees, but also other declining pollinators.[129]

> *If you plant it, they will come!*
> **—B.C. Butterflyway Rangers**

We know that protecting the diversity of bees and other pollinating insects is critical for our food production. But there has been a devastating decline in their numbers since 1980, mainly due to loss of habitat, pesticide use, and, some scientists say, **microwave frequency electromagnetism** generated by wireless transmissions. Although more study remains to be done to conclusively tie this radiation to the colony collapse disorder (CCD) responsible for bees' decline, what has been observed so far in studies is disturbing. "Recently, a study suggested that cellphones and cellphone towers near beehives interfere with honeybee navigation: in one experiment, it was found that

127 Milkweed for Monarchs, Kootenay Native Plant Society, 2019: http://kootenaynativeplants.ca/our-work/milkweed-for-monarchs/

128 "Where are all the monarchs?" February 4, 2019, David Suzuki Foundation: https://davidsuzuki.org/story/where-are-all-the-monarchs-b-c-butterflyway-rangers-to-use-citizen-science-in-2019-to-find-out/

129 Helen Lock, Global Citizen, "London's Getting a 7-Mile 'Bee Corridor' to Help Boost Insect Numbers," May 8, 2019: https://www.globalcitizen.org/en/content/london-bee-corridor-stop-insect-decline-uk/

when a mobile phone was kept near a beehive it resulted in collapse of the colony in 5 to 10 days," wrote one researcher.[130] In 2018, the EU launched a program to totally ban bee-harming pesticides, so there are hopes that these efforts combined will see our insects thrive once again.[131]

Grist reporter Rachel Ramirez tells us that in Colorado, the task of saving bees from the consequences of climate change has fallen to the girls who sell us the best cookies—the Girl Scouts—with their new motto, "**Bee All That You Can Be**!" These Girl Scouts are saving wild bees, by building one bee "hotel" at a time.[132] Like birdhouses for wild bees, these tiny bee hotels help fight the depopulation of bees across the country. Since wild bees don't make honey, they don't live in hives, but they're always in need of a suitable habitat. Out in the wild, these bees often nest in holes in fallen logs or dead trees. But natural habitats can be hard to come by in developed areas, which is where bee hotels come in. The Girl Scout troops repurposed cardboard boxes, old paper straws, toilet paper rolls, and other materials to create homes for bees in their local community.

Ramirez adds that the U.S. has more than 4,000 wild bee species, and 40 percent of them are now facing extinction. Dennis van Engelsdorp, an associate professor of entomology at the University of Maryland, says that bee hotels aren't only habitats but also a safe haven for bees to lay eggs. He adds that it's important for kids to learn that every effort counts. "What you're seeing is that you need bees to survive, and so who better to be concerned about that than the people who are going to inherit the next generation?" he said. "These efforts are really good because hopefully they set up a lifelong commitment to preserving biodiversity."

Planting Trees and Protecting Water

The World Wildlife Fund tells us there were once six trillion trees on the planet, yet now there are only three trillion—and ten billion trees per year are still being lost. This leads to a changing climate, shrinking habitat for wildlife, and harder lives for billions of people. So their project is to restore a trillion trees by 2050 to reverse these trends.[133]

Global Citizen reporter Joe McCarthy tells us that New Zealand has a similar plan, as they intend to plant one billion trees in their efforts to mitigate climate change. He adds that this initiative enjoys broad support, and the government is considering public–private partnerships that would provide grants to landowners to plant trees on their properties.[134] This sort of all-hands-on-deck approach to environmental rehabilitation is happening in other parts of the world, too. In Pakistan,

130 Daniel Favre, Apiary School of the City of Lausanne, Switzerland, "Mobile phone-induced honeybee worker piping," Vol. 42, Issue 3, May 2011: https://link.springer.com/article/10.1007/s13592-011-0016-x

131 Imogen Calderwood, Global Citizen, "The EU Will Totally Ban These Bee-Harming Pesticides," April 27, 2018: https://www.globalcitizen.org/en/content/bee-harming-pesticides-european-union-neonicotinoi/

132 Rachel Ramirez, Grist, "These Girl Scouts are saving wild bees, one 'hotel' at a time," October 21, 2019: https://grist.org/science/these-girl-scouts-are-saving-wild-bees-one-hotel-at-a-time

133 World Wildlife Fund, "Protecting and restoring a trillion trees," 2019: https://www.wwf.org.uk/end-of-year-successes

134 Joe McCarthy, Global Citizen, "New Zealand Wants to Plant 1 Billion Trees to Fight Climate Change," August 14, 2018: https://www.globalcitizen.org/en/content/new-zealand-wants-to-plant-1-billion-trees/

Prime Minister Imran Khan announced a plan to plant 10 billion trees,[135] and India has backed similarly ambitious projects.[136]

In contrast, in Florida's everglades, the plan is actually to prevent building something. Scientific studies have highlighted how important it is to conserve forests and not just continue harvesting mature timber and planting new growth. The Florida Department of Environmental Protection will purchase 20,000 acres of wetlands in the Everglades in an effort to save the area from oil drilling, the governor announced.[137] The wetland acquisition will protect the wildlife habitat of more than 60 endangered and threatened species.

> *We're phasing out single-use plastic bags so we can better look after our environment and safeguard New Zealand's clean, green reputation.... It's also the biggest single subject schoolchildren write to me about.*
> **—New Zealand Prime Minister Jacinda Ardern**

When it comes to protecting our precious water, there are also shreds of good news. In 2019, New Zealand joined the movement to ban plastic bags, as Prime Minister Jacinda Ardern followed the trend of New Zealand's two largest grocery stores. This bold new law could find retailers in violation fined up to $65,000.[138] Ardern said that intense public pressure was a key reason for enacting the ban, and pointed to a petition that garnered 65,000 signatures. "Every year in New Zealand we use **hundreds of millions of single-use plastic bags.**"

McCarthy added that in recent years, more than 60 countries have taken action on single-use plastics. Some countries like Kenya, Chile, Taiwan, and Morocco have **banned plastic bags**, while others have taken more gradual approaches. The movement is informed by the environmental impact of plastics. Plastic pollutes oceans at a rate of at least eight million tonnes a year, breaks down into harmful microplastics that contaminate food and water supplies, leaches toxins into the environment, and contributes to climate change. In spite of all the variants of good news, based on current projections, plastic production is expected to grow by 40 percent over the next decade.[139] In contrast, the amount of plastic bags on the seafloor surrounding the United Kingdom plummeted after grocery stores began banning the item.[140]

This decade is already revealing solutions we so desperately need, and showing signs that 2020 could actually be the year of the "Hundredth Monkey Effect." The true impact of EIGHT BILLION people is starting to be felt, forcing humanity to finally take, admittedly tentative, action.

---

135 Joe McCarthy, Global Citizen, "Pakistan's New Leader Vows to Plant 10 Billion Trees," August 6, 2018: https://www.globalcitizen.org/en/content/pakistan-plant-10-billion-trees/

136 Brandon Blackburn-Dwyer, Global Citizen, "India wants to spend over $6 billion on new forests," May 9, 2016: https://www.globalcitizen.org/en/content/india-forests-6-billion/

137 ABC News Radio, "Florida to purchase 20,000 acres of Everglades wetlands to prevent oil drilling," January 16, 2020, https://www.wbal.com/article/431267/109/florida-to-purchase-20000-acres-of-everglades-wetlands-to-prevent-oil-drilling

138 Joe McCarthy, Global Citizen, "New Zealand PM Jacinda Ardern Just Announced an Ambitious Ban on Plastic Bags," August 10, 2018, https://www.globalcitizen.org/en/content/new-zealand-bans-plastic-bags-six-months/

139 Daniele Selby, Global Citizen, "Plastic Production is Set to Increase by 40% Over the Next Decade, Experts Say," December 26, 2017: https://www.globalcitizen.org/en/content/plastic-production-increase-pollution-ocean-waste/

140 Joe McCarthy, Global Citizen, "There Are Fewer Plastic Bags on the Ocean Floor Because of New Laws," April 5, 2018: https://www.globalcitizen.org/en/content/fewer-plastic-bags-in-oceans-because-of-laws/

The World Wildlife Fund is also working in the UK to combat the crisis caused by a growing human population with their **seagrass project**. WWF explains that seagrass is the grass of the sea. Found along shallow coastlines, it absorbs carbon up to 35 times faster than tropical rainforests and is vital for marine life, which depends on the meadows for food and shelter. With public support, they worked with partners to collect one million seagrass seeds, which they'll plant next year to establish a pilot site in Wales. If successful, this could be scaled up around the UK to help support marine life and tackle the climate crisis.[141]

The Power of Social Media Groups

When we talk about extinction of wildlife, Extinction Rebellion will be one of the groups leading the pack on the social media scene in the future. According to Wikipedia, this group was established in the UK in 2018 with about 100 academics signing a call to action to compel government to avoid tipping points in the climate system, biodiversity loss, and the risk of social and ecological collapse.[142]

Extinction Rebellion (XR) is a global environmental movement with the stated aim of using non-violent civil disobedience to take action. In November 2018, five bridges across the River Thames in London were blockaded. In April 2019, Extinction Rebellion occupied five prominent sites in central London to rally support worldwide around a common sense of urgency to tackle climate breakdown and the sixth mass extinction. Their aims are as follows:

- Government must tell the truth by declaring a climate and ecological emergency, working with other institutions to communicate the urgency for change.
- Government must act now to halt biodiversity loss and reduce greenhouse gas emissions to net-zero by 2025.
- Government must create, and be led by the decisions of, a citizens' assembly on climate and ecological justice.[143]

A youth wing—XR Youth—of Extinction Rebellion had formed by July 2019. In contrast to the main XR, it is centred around consideration of the Global South and Indigenous peoples, and is more concerned with climate justice. By October 2019, there were 55 XR Youth groups in the UK and another 25 elsewhere. All XR Youth comprise people born after 1990, with an average age of 16, and some as young as 10. This dual generational approach is providing a model that could prove very successful for protecting wildlife.

Stopping Biodiversity Loss Through Dietary Choices

Meat consumption is considered one of the primary contributors of the sixth mass extinction. A 2017 study by the World Wildlife Fund found that **60 percent of global biodiversity loss is attributable to meat-based** diets, in particular from the vast scale of feed crop cultivation

141 World Wildlife Fund, "UK: Using seagrass to combat the climate crisis," 2019: https://www.wwf.org.uk/end-of-year-successes

142 Extinction Rebellion, Wikipedia: https://en.wikipedia.org/wiki/Extinction_Rebellion

143 Extinction Rebellion, "Our Demands": https://rebellion.earth/the-truth/demands/

needed to rear tens of billions of farm animals for human consumption. This puts an enormous strain on natural resources resulting in a wide-scale loss of lands and species. Currently, livestock make up 60 percent of the biomass of all mammals on Earth, followed by humans (36 percent) and wild mammals (4 percent).[144]

Solution: In November 2017, 15,364 world scientists signed a Warning to Humanity calling for, among other things, drastically diminishing our per capita consumption of meat and "dietary shifts towards mostly plant-based foods." The 2019 Global Assessment Report on Biodiversity and Ecosystem Services, released by IPBES, also recommended reductions in meat consumption in order to mitigate biodiversity loss. The report called on countries to begin focusing on "restoring habitats, growing food on less land, stopping illegal logging and fishing, protecting marine areas, and stopping the flow of heavy metals and wastewater into the environment." It also suggests that countries reduce their subsidies to industries that are harmful to nature, and increase subsidies and funding to environmentally beneficial programs. Restoring the sovereignty of Indigenous populations around the world is also suggested, as their lands have seen lower rates of biodiversity loss.[145]

A July 2018 study in *Science* says that meat consumption is set to rise as the human population increases along with affluence, which will increase greenhouse gas emissions and further reduce biodiversity. Here again, **population is undermining the progress** citizens have fought so hard for.

International Wildlife Protection

Meanwhile, the World Wildlife Fund has had their share of successes to announce in 2019. Rhino poaching was down by over 40 percent in Kenya from 2017–2018 thanks to their program to monitor and protect Rhinos to stop the illegal wildlife trade. WWF participated in Kenya's first national lion census, and results will be used to implement measures to protect them.[146]

Since 2017, WWF has been monitoring jaguars in the Amazon so that better protection can be provided. In the same area, Amazon River dolphins are being monitored. All of the tested dolphins have high levels of mercury, which leaches into the waterways from gold mining, polluting them. They are fighting for more protected areas in the region.

When we talk about cooperation, 2019 was an exciting year for wildlife protection. The Convention on International Trade in Endangered Species of Flora and Fauna (CITES) held their triennial World Wildlife Conference in Geneva. With 183 parties to the agreement, CITES remains one of the world's most powerful tools for wildlife conservation through the regulation of trade. Formally known as the CoP18, this conference revised the trade rules for dozens of wildlife species that are threatened by unsustainable trade linked to

> *Humanity needs to respond to the growing extinction crisis by transforming the way we manage the world's wild animals and plants. Business as usual is no longer an option.*
> **—CITES Secretary-General Ivonne Higuero**

144 "Meat consumption," Wikipedia: https://en.wikipedia.org/wiki/Meat

145 Global Assessment Report on Biodiversity and Ecosystem Services, 2019, Wikipedia: https://en.wikipedia.org/wiki/Global_Assessment_Report_on_Biodiversity_and_Ecosystem_Services

146 World Wildlife Fund, "End of Year Successes," 2019: https://www.wwf.org.uk/end-of-year-successes

overharvesting, overfishing or overhunting. These ranged from commercially valuable fish and trees to charismatic mammals such as giraffes to amphibians and reptiles sold as exotic pets.[147]

> *This IUCN Red List update offers a spark of hope in the midst of the biodiversity crisis. Though we have witnessed 73 genuine species declines, the stories behind the 10 genuine improvements prove that nature will recover if given half a chance. Climate change is adding to the multiple threats species face, and we need to act urgently and decisively to curb the crisis.*
> **—Dr. Grethel Aguilar, IUCN Acting Director General**

Thousands of species are internationally traded and used by people in their daily lives for food, health care, housing, tourist souvenirs, cosmetics, or fashion. CITES regulates international trade in over 35,000 species of plants and animals to ensure their survival in the wild. Responding to high and increasing demand for teak and rosewood from Africa, CITES increased protection for these rare trees. Noting that giraffes have declined by 36 to 40 percent over the past three decades due to habitat loss and other pressures, the conference increased protection for the world's tallest animal. The conference took decisions promoting capacity building and other activities aimed at strengthening wildlife management and compliance with and **enforcement of CITES trade rules**.

The latest **IUCN Red List** update reveals that despite the ever-increasing impacts of human activities, there have been **genuine improvements in the status of ten species**—eight bird species and two freshwater fishes. Captive breeding, combined with careful management of wild populations, has been key to these conservation successes.[148]

Among these improvements is the flightless, fast-running Guam Rail—the second bird in history to recover after being declared extinct in the wild, after the California Condor. However, the bird is still classified as critically endangered—one step away from extinction. In Mauritius, the echo parakeet continues its recovery, and two freshwater fish species—the Australian trout cod and Pedder galaxias—have likewise improved. Both species face threats from invasive species and habitat destruction and degradation.

> *We have never had a single unified statement from the world's governments that unambiguously makes clear the crisis we are facing for life on Earth. That is really the absolutely key novelty that we see here.*
> **—Thomas Brooks, chief scientist, International Union for Conservation of Nature**

Yes, it is true that we can stave off a species' extinction by taking extraordinary measures and protecting that specific habitat, but this has only saved a very select few. The long list of species facing extinction should be a good indication of how severe this crisis is, as wherever humans show up, ecological disaster follows. Every time human overpopulation is excluded from this conversation, a lie is being told and

147 CITES 2019 press release, "CITES conference responds to extinction crisis by strengthening international trade regime for wildlife": https://cites.org/eng/CITES_conference_responds_to_extinction_crisis_by_strengthening_international_trade_regime_for_wildlife_28082019#GlobalGoals

148 IUCN Red List, "Species recoveries bring hope amidst the biodiversity crisis," December 10, 2019: https://www.iucn.org/news/species/201912/species-recoveries-bring-hope-amidst-biodiversity-crisis-iucn-red-list

more species disappear. We are at a critical moment for the future of the planet and the species emergency ahead.

The Paradigm Shift to Real Solutions for Biodiversity Loss

It will take a global effort to achieve the UN 2030 Sustainable Development Goal designed to protect wildlife diversity.

Solutions: Every country needs to do the following:

- Compile a national wildlife inventory in conjunction with international agencies.
- Develop a fair and comprehensive population policy for humanity by doing a carrying capacity study and meaningful public consultation process.
- Implement a national family planning program to bring the human population into sync with our natural world and following international UN goals for 2030.
- Implement animal welfare legislation that reflects the humane treatment of our fellow beings in line with international law as well as societal expectations, and including a well-funded enforcement aspect with penalties that are appropriate to deter offenders.
- Monitor outcomes regularly and take measures to stay on course.

As voters and taxpayers, we have every right to demand dramatic changes to curb entrenched mismanagement of our land and wildlife. So where do we go from here? If we are to ebb the precipitous decline of wildlife and the remorseless destruction of habitat, we all have a lot more work to do. It is time to imagine the future we want to see, gather all of our collective wisdom, and demand that our leaders take action. Then we all have to define our own strategy for the 2020 decade and do our part, since we are all responsible for getting to this critical point. This all starts with YOU!

The good news in Canada is that in 2020, the Ontario government introduced new provincial animal welfare legislation, a move which Canada desperately needs to make into federal law as well, to update century-old Criminal Code laws that are sadly inadequate. Global News reporter Liam Casey pointed out that animal advocates and stakeholders have long spoken about the need for a revamped cruelty enforcement regime in Ontario.[149] Animal Justice's executive director Camille Labchuk stated, "We support a robust, well-funded public enforcement model because animal protection is a key societal value, and this is what Ontarians expect."

Lynn Perrier, the Canadian founder of Reform Advocates for Animal Welfare, applauded the move by Premier Doug Ford's government saying, "Of particular interest to me is the potential contribution Crown attorneys can make to enforce animal abuse laws. To date, animals have not been well represented in our courts and abusers get away with a slap on the hand. Hopefully that will end now." The legislation will also regulate the keeping of wild animals as pets and captive wildlife in roadside zoos. The solicitor general said there will be big fines for corporations found guilty of animal cruelty, including a $250,000 fine for a first-time offence and up to $1 million for repeat offenders. This is the kind of success we need to see globally in the coming decade!

149 Liam Casey, "Ontario government to introduce new animal welfare legislation," Canadian Press, Global News, October 29, 2019: https://globalnews.ca/news/6096521/ontario-government-animal-welfare-legislation/

CHAPTER 6

WOMEN'S MOVEMENT—UNFINISHED BUSINESS

She's just turned 16 and is already a world leader with more statesperson-like qualities, clear-eyed goals, plain speaking and sheer guts than almost any national head of today or recent history.
—Julian Cribb looks at the rise of Greta Thunberg

We know that since the 6th century BCE, women have wanted to postpone pregnancy and plan their families. Richard Cincotta and Robert Engelma tell us that a fennel-like plant in the carrot family, called sylphion, made this possible. This wild herb once grew in abundance along the coast of what is now Libya. Cyrenian women once valued this antifertility herb, and in fact oral contraception was its only recorded use.[150]

I'm sure women around the world would be growing sylphion in their gardens today, if it hadn't gone extinct around the height of the Roman Empire. Cincotta and Engelma explain that as knowledge of sylphion's properties spread throughout the Mediterranean basin, the herb was transformed from a home remedy, largely controlled by women themselves, into one of the principal commodities of Cyrene's foreign trade, controlled by men. Evidence of sylphion's economic importance is its silhouette on the face of Cyrenian four-drachma coins. The 4th-century BCE physician Hippocrates records Greek and Syrian attempts to cultivate the herb on home ground and cut into the Cyrenian market. The attempts failed, commercial traders over-harvested it, and around the 2nd or 3rd century CE, sylphion disappeared.

Whatever lessons might have been offered to today's three billion women seeking to manage their own fertility have been lost forever, but fortunately a variety of other contraceptives have been developed to take the place of sylphion. It is a crying shame that, according to PubMed, over **250 million** women in developing nations still had no access to these newer contraceptives in 1988.[151] In fact, Population Matters tells us that by 2019 this number had risen to **270 million women globally who still have an unmet need for contraceptives.**[152] Cincotta and Engelma emphasize, "The case of sylphion reminds us that women's interest in managing their own fertility is ancient. The herb's extinction speaks of humanity's

150 Richard P. Cincotta & Robert Engelma, Population Action International, *Sylphion: A Natural Contraceptive and Its Loss, Nature's Place,* https://www.biologicaldiversity.org/programs/population_and_sustainability/oceans/pdfs/Natures_Place.pdf

151 PubMed, *Access to birth control: a world estimate,* April 1988: https://pubmed.ncbi.nlm.nih.gov/12281360/

152 Population Matters website, "Family planning services still falling behind population growth," February 27, 2020: https://populationmatters.org/news/2020/02/27/family-planning-services-still-falling-behind-population-growth

longstanding but fragile economic association with biodiversity. It also demonstrates the importance of conserving critical biological resources for ourselves and for our posterity. As in the case of sylphion, we may not know the full price of our loss until it is too late."[153]

Ancient and Indigenous cultures throughout the centuries have celebrated the sacredness of nature and honoured Mother Earth's feminine presence. In the 20th century, many of these richly diverse spiritual ways have slipped away, and the life-givers—the feminine presence—have lost their rightful status. Now in the 21st century, it is time to bring back that feminine leadership to balance the masculine. We must work to equalize and unite all genders if we are to create a just and sustainable world past 2020.

That being said, working together we must also confront the spate of violence against humanity and our planet, and finally recognize the enormous contribution women can make in this decade of action. Both the World Health Organization and international human rights laws recognize that for women to be equal, they must be free from sexual violence *and their sex-specific health-care needs have to be accommodated.*[154]

The UN Sustainable Development **Goal #5** launched in 2015 aims to **"achieve gender equality and empower all women and girls" by 2030,** in large part through the access to contraception and education. The UN explains, *"In short, all the SDGs depend on the achievement of Goal #5. End all forms of discrimination and violence against all women everywhere by 2030."* That gives us only ten years to make some incredible changes in every country of the world.[155]

This goal officially declared that women and girls, everywhere, must have equal rights and opportunity, and be able to live free of violence and discrimination. Gender equality by 2030 requires urgent action to eliminate the many root causes of discrimination that still curtail women's rights in private and public spheres.

The UN has developed three scenarios for population growth, with various outcomes depending on our response (see Figure 6-1). They recommend moving toward the low-variant projection of **halting population growth at EIGHT BILLION** if we are to meet their SDGs.

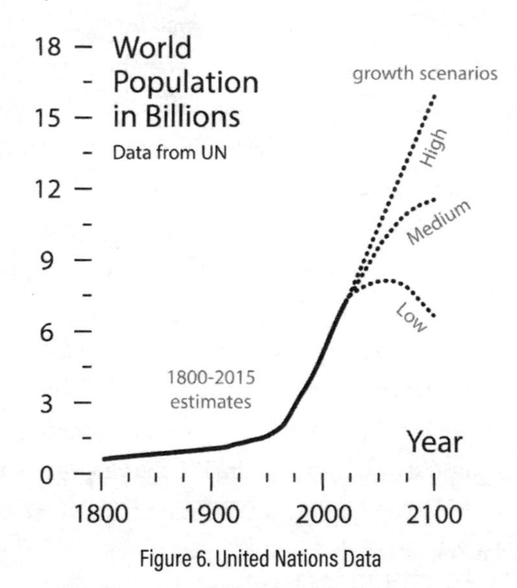

Figure 6. United Nations Data

153 Richard P. Cincotta & Robert Engelma, *Sylphion: A Natural Contraceptive and Its Loss, Nature's Place,* https://www.biologicaldiversity.org/programs/population_and_sustainability/oceans/pdfs/Natures_Place.pdf

154 World Health Organization online, *Sexual health, human rights and the law,* 2020: https://apps.who.int/iris/bitstream/handle/10665/175556/9789241564984_eng.pdf

155 UN Women, https://www.unwomen.org/en/news/in-focus/women-and-the-sdgs/sdg-5-gender-equality

Giving Women Control

Dr. Elizabeth Bagshaw certainly strived to give women more control when she became medical director of Canada's first and illegal birth control clinic in 1932. Stacy Gibson with University of Toronto magazine tells us that this kind of defiance was exceedingly rare for a woman in the early 20th century. Bagshaw spent more than three decades at the centre empowering women to control their reproductive health and plan the size of their own families.[156]

Gibson points out that the stakes for those involved in the clinic were extremely high: "It was an indictable offence—liable to two years in prison—to sell or advertise contraceptives or to instruct people on how to use them. One clause in the code offered hope: if one could prove '**public good**' was served by their actions, they could avoid conviction. But no one wanted to be put in the position of a long, expensive trial or risk being jailed."

It was the beginning of a life of defying expectations for Bagshaw, who witnessed the relentless poverty of the Great Depression, husbands on relief, and the great numbers of children that couldn't be fed or cared for. Many women had died trying to perform abortions on themselves, so she knew that family planning was the right thing to do. The number of women who sought her help was well beyond the clinic's predictions: In the first year, they expected about 60 women; almost 400 came.

Gibson adds, "In retrospect, this likely didn't shock them. The clinic provided the first opportunity most women ever had to learn about their own bodies and control their reproductive destinies. Prior to the clinic, men were the gatekeepers to women's health in Canada. Men headed the educational and medical systems that controlled knowledge about birth control. They helmed the pulpits that made the religious creeds against using contraceptives. They ran the courts that judged birth control as unlawful, and spoke loudest in the court of public opinion that deemed it immoral. The women who ran the clinic cut through this socially constructed shame, countering that birth control was 'about as immoral as a good day's washing is immoral.'"

One of the clinic's staunchest opposers was the Roman Catholic Church, with the bishop calling Bagshaw a heretic and incessantly attacking her and the clinic from the pulpit. Bagshaw remarked, "It was the best advertising we had. It was against the law to advertise, therefore we couldn't say anything, but he advertised it—then I'd tell the nurses, 'Be sure to be on time, don't be late at the next two or three clinics because we'll have a number of Roman Catholics there.' And we always did."

And then in September 1936, a trial brought the issue of the legality of distributing and advising on birth control to a head. Dorothea Palmer, a nurse with the Parents' Information Bureau in Ottawa, was charged with advertising birth control after visiting women in their homes to teach them about family planning. After a tense 20 days of trial, a verdict was reached: she was not guilty. The magistrate found that Palmer had, indeed, acted for the "**public good.**" The case was appealed. She won again. In 1969 the birth control law was finally officially changed.

156 Stacy Gibson, *University of Toronto Magazine*, "Doing the Devil's Work," June 25, 2019: https://magazine.utoronto.ca/people/alumni-donors/doing-the-devils-work-elizabeth-bagshaw-canadas-first-birth-control-clinic/

Gibson notes, "Bagshaw didn't retire from her own practice until the age of 95 in 1976, which made her the oldest physician practising in Canada. She died at the age of 100. For her pioneering work, she had been inducted into the Canadian Medical Hall of Fame invested into the Order of Canada and had an elementary school in Hamilton named in her honour. Bagshaw had greatly advanced the concept of health-care equality and her legacy continues to motivate doctors today."

The challenges that Bagshaw faced during the Depression are similar to challenges that doctors in many countries face today, whether it is in countries where bias against women prevails or in countries where women don't have access to safe and affordable family planning programs.

In 1968, during the time Bagshaw worked in the clinic, Paul Ehrlich wrote *The Population Bomb*, in an attempt to further empower women to make their own fertility decisions. At that time there were about 3.4 billion people on the planet, and now we are nearing EIGHT BILLION, with over 10 percent living in extreme poverty. In fact, in 2020 a child dies from hunger every ten seconds.[157] So it turns out that Ehrlich's predictions about food scarcity were correct, just a little premature in their timing.

The refusal to recognize the reality of the population factor by many activists in the women's movement is indeed a betrayal of those they claim to speak for, because poor women and children are the most adversely affected by the deterioration of the planet. It is a shame that sensible feminists are reluctant or afraid to challenge the false feminist orthodoxy that addressing the population issue violates the rights of women.

These rights include the right to a safe abortion, and morally we know that legalizing abortion has changed the lives of thousands of women for the better. A 2008 poll showed that 65 percent of Canadians supported the appointment of **Dr. Henry Morgentaler** to the prestigious **Order of Canada**; however, then Prime Minister Stephen Harper joined the minority by voicing his opposition.

Morgentaler was well known for leading a controversial movement to legalize abortion in Canada, opening his first clinic in Montreal in 1969. He risked his own life and that of his family for decades to do what he believed in. He spent ten months in jail because of it, despite being acquitted by successive juries of ordinary Canadians, including Catholics, when the Vatican opposes abortion. His clinic in Toronto was bombed by extremists. And Morgentaler's belief in a woman's right to choice is a position the overwhelming majority of Canadians agreed with and that Parliament finally supported.[158]

Countless pro-choice activists saw his induction as long overdue and believed that he should be recognized for his hard work and dedication to Canadian women's rights. Bestowing the highest honour in the country on Morgentaler showed recognition for an important milestone in women's rights.

---

157 The World Counts.com, "People who died from hunger," 2019: https://www.theworldcounts.com/challenges/people-and-poverty/hunger-and-obesity/how-many-people-die-from-hunger-each-year/story

158 Bill Tieleman, Blogspot, "Morgentaler deserves it," July 8, 2008: http://billtieleman.blogspot.com/2008/07/dr-henry-morgentaler-has-earned-order.html

If we want to prevent abortions, then we need to do more to reduce the high incidence of rape, eliminate the 50 percent of unintended pregnancies, and ensure that all women have access to safe and effective birth control. The "take home message" here is that we **must** take more aggressive action when urging our governments to make the interests of sexually abused citizens a priority.

So whether it is date rape, statutory rape, acquaintance rape, incest, or war rape, this is an alarmingly increasing epidemic in 2020, affecting far more women than COVID, yet garnering far less attention. If the amount spent on COVID-19 globally had been spent on eliminating rape and human trafficking, the planet and all beings would be far better off. However, these decisions on spending are made mainly by males who are unlikely to consider the rape epidemic a worthy cause.

Ultimately, we would all like to eliminate the need for abortion, and if every family used family planning and there were no rapes, there would be no need for abortions, but until that day comes, women should have a choice. Policies that forbid abortion not only demonstrate a shocking lack of compassion, they also directly contradict strong evidence that restricting abortion access does not eliminate abortion—it just results in more women dying or being injured by clandestine and unsafe procedures.

Are Women Our Best Hope for Surviving the 21st Century?

When we hear of all this violence, it is refreshing to hear author Julian Cribb speak of a flicker of hope when he describes climate activist Greta Thunberg: "Greta Thunberg is a small candle shining in a dark world. She symbolises new hope to young people, to women and to enlightened men alike.... Swedish climate campaigner and global school strike organiser Greta Thunberg is the personification of the emergent world order and its invigorating leadership. The one that aspires to rescue humanity from the consequences of colossal, mindless mistakes.... But Greta Thunberg is also another kind of new world leader. She's a woman."

In his enlightening new book *Surviving the 21st Century,* Cribb argued, "Universal female leadership is the only way that humanity and the Earth can be saved. If men are left in charge, we're doomed. Why? Because:

- Men start wars, women don't. Women mostly eschew violence.
- Men prefer simple solutions—machines, chemicals, weapons—that all cause unanticipated destruction.
- Women tend to think of the needs of future generations, children and grandchildren, to a greater degree
- The vast bulk of carbon emissions, toxic chemical emissions, wildlife extinctions, deforestation, desertification, land degradation, overfishing and mindless development is by men, not women. Men run all the dirty industries which ravage the planet. They run the governments and churches which support the rape.
- Women are already leading on overpopulation. Without consulting men, they have slashed birthrates from 4.4 to 2.4 since 1970. They are leaders in education, healthcare, sustainable food, regeneration of wilderness, social development....

This argument for female leadership has little or nothing to do with feminism or equal opportunity. It is, quite simply, a fundamental rule for human survival in the existential emergency we now face. Put women in charge and humanity, or a part of it, may pull through. Leave men in charge, and it's doubtful."[159]

Cribb pulls no punches when he talks about it being men that run the governments and churches, which support the status quo and the ongoing denial of women's rights and safety. On the other hand, in recent years we see a growing number of men stepping up to work with women to combat this problem, and I believe this trend will continue as international law becomes more stringent. Sexual assault is such a colossal hindrance to progress that overcoming it will take united cooperation and brutal honesty in acknowledging the scope of the problem.

World Influencers Leading the Way Through the 21st Century

Along with Jane Goodall and Greta Thunberg, another extraordinary world influencer is Chimamanda Ngozi Adichie, regarded as one of the most original writers of her generation. *Vogue*'s reporter Olivia Marks tells us that Adichie is an award-winning author, a Ted Talk sensation, and Beyonce's favourite feminist. But it was the 2013 TEDxEuston, a series of talks in London focusing on African affairs entitled "We Should All Be Feminists," which addressed a feminism beyond race or class, that created a viral explosion. Her speech there inspired Beyonce's "Flawless," which topped the charts in 104 countries and sold nearly 850,000 copies in three days. Adichie, 41 and living in Nigeria, has now cemented her transformation from bestselling author to feminist authority.[160]

"I was immediately drawn to her," says Beyoncé. "She was elegant and her words were powerful and honest. Her definition of a feminist described my own feeling: equality of the sexes as it pertains to human rights, equal pay and sexuality. She called the men in her family feminists, too, because they acknowledged the need for equality."

For all these reasons, when it came to compiling a dream list of women who should feature on *Vogue*'s Forces for Change cover, Ms Adichie was at the top. "Policy is important, laws are important but changing cultural norms and mindsets matter even more," came Adichie's response, when *Vogue* asked how we can bring about change.

Marks points out, "You don't have to spend long in Nigeria to witness the deeply patriarchal nature of the culture, where men are always greeted as 'sir' and women are lucky to be greeted at all." Adichie's dream is to change this mentality, and she is proving to be a major force in the development of gender equality with stories that highlight the oppression of women.

It is often assumed that women in developing countries choose to have large families, when in fact it is estimated that at present over 200 million women globally would like to limit the size of their families but cannot obtain access to family planning assistance. Even in areas where some

159 Julian Cribb, "JULIAN CRIBB. The Rise of Woman. Greta Thunberg," John Menadue, Pearls and Irritations, March 11, 2019: https://johnmenadue.com/julian-cribb-the-rise-of-woman-greta-thunberg/

160 146 Olivia Marks, *Vogue*, "Chimamanda Ngozi Adichie: 'Women Do Not Need To Be Extraordinary To Be Admirable,'" September 2019, https://www.vogue.co.uk/article/chimamanda-ngozi-adichie-september-2019-issue

form of family planning is available, most women have at best some input into deciding the size of their families. A woman's husband, or even her in-laws, are likely to have more say in the matter than she does. In many parts of the world, there is strong social and religious pressure on women to have more children than they personally would choose, and the pressure to produce a "sufficient" number of sons is often enormous.

How Family Planning and Planetary Responsibility Align

However, along with the human right to bear children comes a human's responsibilities to this planet and other living beings. People do not have the absolute right to implement personal decisions that affect others' lives without considering the common good. Parents do not pay all the costs of bringing up children—society as a whole picks up a large part of this burden. Overpopulation affects us all, and it threatens the life-support system that sustains us all. We are all forced to deal with its consequences.

If we do not deal with overpopulation now, through education and family planning, we may not be able to avoid less attractive methods down the road to attempt a late-stage rescue of the planet and the human species. The other option is to allow Mother Nature to deal with it through starvation, disease, increased human suffering, and loss of biodiversity. However, when making this decision, the public should be informed of the options, and everyone should be allowed to be involved in a fair and meaningful public consultation process.

John Guillebaud, co-chairman of Optimum Population and emeritus professor of family planning at University College in London, put it this way: "The effect on the planet of having one child less is an order of magnitude greater than all these other things we might do, such as switching off lights. An extra child is the equivalent of a lot of flights across the planet."[161]

Yet, it is shocking that population planning and management is still such a radioactive topic to the majority of people around the world. Since it intersects our public life, environmental crises, and changing personal situations, the very word confuses many citizens. After all, it doesn't get much more personal than suggesting each couple have fewer children—even if it is for the greater good, like reducing poverty, empowering women, reducing climate change, and returning some land to the other critters on the planet.

> *The greatest thing anyone in Britain could do to help the future of the planet would be to have one less child. When couples are planning a family they should think about the environmental consequences. As a general guideline, couples should produce no more than two offspring.*
> **—John Guillebaud, Optimum Population**

But just think of all the benefits! As women's rights increase, fertility rates begin to decrease. As men give up some of their power and control over women, we are seeing happier more dynamic families. As we let go of some of our human-controlled land to expand our forests, we are creating a

161 Sarah-Kate Templeton, "Children 'Bad For Planet,'" *Sunday Times* (London), May 7, 2007: https://www.amren.com/news/2007/05/children_bad_fo/

carbon sink to reduce climate change. As we turn towards a plant-based diet, we are returning more habitat to endangered wildlife and reducing animal suffering. For every excess we give up, we are getting many more real benefits in return. We are already beginning to see this happen in documented cases around the world as we restore peace, equality and environmental integrity where situations once seemed hopeless—as discussed later in this chapter and in Chapter 5.

> *More importantly, international population and family planning assistance is about giving women control, not taking it from them. A woman in the developing world is the center of her environment.... In order to effectively manage her life and her environment, every woman must have the right and the means to make informed decisions regarding the number and spacing of her children.*
> **—National Audubon Society U.S., 1993**

Universal Call to Action

Not only that, but national initiatives based on the UN Sustainable Development Goals are giving men, women, and children an opportunity to be part of a universal call to action never seen on this planet before. Movements for climate change like those led by Greta Thunberg and Leonardo DeCaprio, movements for sustainable population like those led by Thriving Together and UN Family Planning 2020, and movements for youth empowerment like Childfree, Fair Start Movement and One Planet - One Child are all changing the world. Social media is the catalyst behind this colossal shift in thinking, since mainstream media is reluctant to discuss the population issue. Also, the United Nations is finally beginning to fulfill their rightful duty of leading the world in this paradigm shift, now recommending smaller families and plant-based diets as effective solutions.

Of course, achieving a sustainable population level would mean that both individuals and governments would have to step up and start taking more responsibility for our environmental crisis. It means we would all have to start scaling back on our excesses, on consumption, greed, exploitation, violence, cruelty, and dare I say it—babies. It would mean acknowledging Earth's limitations to accommodate, and Earth's fragility in response to human abuse.

In fact, there are a growing number that now see it as an "environmental misdemeanour" to have a big family, and maintain that we must make family planning services available to all couples who would like to plan the size of their families. This is also true in situations where women presently have no choice but to bear unplanned or unwanted children, including in the case of human trafficking, rape, coerced parenthood, and child brides. Rape is a common and violent act that impacts not only the victim, but also the family of the victim, and ultimately society at large.

These brutal assaults can destroy trust, devastate families, and result in millions of unintended pregnancies.

Men Still Wield Most of the Power: Where Has It Gotten Us?

According to the Webster's Dictionary, *patriarchy* is defined as "social organization marked by the supremacy of the father in the clan or family." But we must understand that like women, men have been groomed by their parents and society to play a specific role. And like women, many men have been sexually abused as children by those in the education, sports, religious, and community sectors. In 2020, the news broke regarding the Boy Scouts of America filing for bankruptcy to survive hundreds of sexual abuse lawsuits. This deluge of sex-abuse claims follows Catholic dioceses and U.S.A. Gymnastics in seeking bankruptcy protection amid sex-abuse cases. Family abuse inflicted on boys by either parent or other relatives and friends is also common, so this is a systemic problem and boys are often the victims of sexual assault as well. Since this travesty has been occurring for decades, it might explain some of the sexual abuse men show towards women and children. This will be covered more in Chapter 12.

As women, we like to think that we have made a great deal of progress in gaining equality and improving women's rights. Yes, we have come a long way, but as Julian Cribb pointed out, this is still a world dominated by men. Men still wield most of the power, make most of the major decisions affecting women, and in many countries have the right to keep women in a submissive and obedient state. This is especially true when it comes to decisions regarding reproductive rights and population growth. The United Nations, the growth lobby, the Vatican, the war machine, organized crime, and the thriving slave trade industry are run predominantly by men. In most cases they need a steady surplus of the submissive and the poor to continue doing business as usual.

Since 1964, the United Nations has allowed the Vatican to have veto power at conferences, which has greatly interfered with decisions to promote family planning programs. The UN and most participating countries have also made little effort to end the slave trade, and this failing has greatly benefited organized crime and others invested in this travesty. Of course, the growth lobby benefits greatly from increased population, which results in increased demand for housing, food, and so on. All of these patriarchal dominated sectors of society have a vested interest in population growth. The Billionaires Club, which is also predominantly male, runs the world with little ethical or moral consideration.

The War Machine is another male-dominated entity that benefits from a steady supply of poor children, women, and men to fight their battles for them. War also perpetuates the sexual abuse of millions of women, resulting in untold numbers of pregnancies, and it takes funding away from more important causes, such as poverty and family planning programs. As outlined earlier, more of our leadership should be coming from women if we are to save the planet from today's crippling human population dilemma.

Male Influencers Champion Reproductive Rights

When acknowledging the male heroes that have campaigned for women's rights for decades, several extraordinary men come to mind. **Bill Ryerson**, founder and president of Population Media Center and chair of the board of the Population Institute, has fought tirelessly for family planning education and access to birth control worldwide. **John Meyer**, president of Zero Population Growth Canada and Canadians for Sustainable Society, has diligently supported efforts for a sustainable population. **Dave Gardner**, creator of Growth Busters and co-host for World Population Balance, has brilliantly used social media to promote family planning. It is men of this calibre that will lead us through the 21st century alongside our committed female activists.

The Global Rape Epidemic

> *We hope that this global report will open up a conversation to bring addressing sexual violence into the very centre of our collective thinking and action and to promote the right of women and girls everywhere to equality and to be free from violence.*
> **—Yasmeen Hassan, Global Executive Director, Equality Now 2019**

Equality Now emphasizes that sexual abuse and rape are everyday violent occurrences, affecting close to a billion women and girls globally over their lifetime. In the 21st century, a conservative estimate is that one in three women will experience physical or sexual violence in their lifetime, regardless of age, background, or country, and every country in the world has laws that treat women and girls as second-class citizens. Sexual exploitation, violence, harmful cultural practices, and systemic inequalities violate their human rights and prevent them from reaching their potential.[162]

Since Equality Now was founded in 1992, they have been working with the survivors of rape and sexual assault to get justice, strengthen laws, and increase enforcement. The International Bar Association has been their partner from the beginning, and they work closely to end this tragedy. They recently released a global report titled *The World's Shame: The Global Rape Epidemic.*[163]

The report emphasizes that by any measure, gender-based violence, including sexual violence, is being inflicted on women and girls in epidemic proportions. Fortunately in 2020 we have a powerful global team of champions for the cause of empowering women and girls. Global Citizen is a movement of engaged citizens from around the world who are using their collective voice to end extreme poverty by 2030 and empower women and girls. According to the Global Citizen website, their mission is to build a movement of 100 million action-taking Global Citizens to help achieve our vision of ending extreme poverty by 2030.[164]

On their platform, Global Citizens learn about the systemic causes of extreme poverty, how to take action on those issues, and earn rewards for their actions—as part of a global community committed to lasting change. This change includes empowering women and girls and eradicating sexual violence. Despite attempts to change laws to help prevent assaults and hold offenders

---

162 Yasmeen Hassan, Equality Now, "Why Gender Equality?" https://www.equalitynow.org/why_gender_equality1
163 Equality Now, "The World's Shame: The Global Rape Epidemic," https://www.equalitynow.org/the_global_rape_epidemic_campaign?locale=en
164 Global Citizen is a movement of engaged citizens who are using their collective voice to end extreme poverty by 2030 and empower women and girls. Global Citizen online: https://www.globalcitizen.org/en/about/who-we-are/

accountable, sexual violence continues to be a significant issue. UN Sustainable Global Goal #5 promotes expanding gender equality and eradicating gender-based violence worldwide.

According to Global Citizen reporter Catherine Caruso, **one rape is reported every minute in India**, and nearly 34,000 incidents of rape were reported in 2018. Activists and women's rights groups believe this is because crimes against women aren't always taken seriously, especially by the police.[165]

The 2020 annual crime report released by the Ministry of Home Affairs in India showed that while 85 percent of reported assaults resulted in charges, only 27 percent of offenders were actually convicted. This is not surprising because the country is still run by men and most judges are still men. The gang-rape and murder of a young woman on a bus in New Delhi seven years ago was the first major case to gain national and international attention.

Caruso points out, "The tragic attack sparked massive protests and civil unrest, which ultimately led to more severe sentences for convicted rapists and murderers. Ironically, since then the number of women and girls being raped and murdered has increased by 35%. The data in the report is not an accurate estimate of the total number of rapes in 2018, just those reported to the police. For many communities in India, it is considered taboo to report a rape."

Gender-based violence (GBV) is a profound problem in Africa as well where over 4,500 women reported missing in South Africa between 2016 and 2019. South Africa is one of the most dangerous places in the world to be a woman. This is a result of the gender-based violence levels that are so high that President Cyril Ramaphosa last year declared it a national crisis. Crime statistics released by Cele in September 2019 revealed that 2,771 women were killed between 2018 and 2019, while 24,387 sexual offences were reported in the same period.

Global Citizen reporter Lerato Mogoatlhe states,"GBV is rarely out of the national spotlight, and several high-profile cases have shocked the nation in recent years—including the murder of Uyinene Mrwetyana in August 2019—and sparked renewed calls for President Cyril Ramaphosa to prioritise creating a safer South Africa. In August 2018, gender activists staged national protests called the #TotalShutDown, which urged women and allies to stay away from work and instead join marches against GBV in various cities across South Africa."[166]

Mogoatlhe adds that up to 40 percent of women in South Africa have experienced sexual or physical violence. Millions of women still have no access to basic contraception or life-saving childbirth medicines. UNFPA Supplies supports access for women and girls worldwide but it needs international support.

Globally, governments have committed and recommitted to ending all forms of violence against women and girls, including sexual violence. However, existing laws are failing women and girls around the world. Let's hope that the updated UN 2030 goals are taken more seriously and that policymakers will strengthen and enforce laws on sexual violence.[167]

---

165 Catherine Caruso, "1 Rape Is Reported Every 15 Minutes in India," Global Citizen, January 13, 2020: https://www.globalcitizen.org/en/content/rape-reported-every-15-minutes-india/

166 *Over 4,500 Women Reported Missing in South Africa Between 2016 and 2019*, By Lerato Mogoatlhe, Jan. 22, 2020, Global Citizen, https://www.globalcitizen.org/en/content/4500-women-missing-south-africa-gender-violence/

167 United Nations, *Transforming Our World*, 2015, https://sustainabledevelopment.un.org/content/documents/21252030%20Agenda%20for%20Sustainable%20Development%20web.pdf

However, in countries where laws have not managed to progress as quickly and victims cannot afford to prosecute, rape is on the rise. In 2020, *Rape Population Review* commented, "South Africa has the highest rate of rape in the world of 132.4 incidents per 100,000 people. According to a survey conducted by the South African Medical Research Council, approximately one in four men surveyed admitted to committing rape, as it is accepted conduct there. Although the Parliament of South Africa enacted the Criminal Law (Sexual Offences and Related Matters) Amendment Act in 2007 attempting to amend and strengthen all laws dealing with sexual violence, the rates of reported rape, sexual abuse of children, and domestic violence have continued to rise.[168]

Madeline Holler with Microsoft News notes, "The United States has a rape rate of 27.3. As in many other countries, rape is grossly underreported in the United States due to victim shaming, fear of reprisal, fear of family knowing, cases not being taken seriously by law enforcement, and possible lack of prosecution for the perpetrator. Only 9% of rapists in the US get prosecuted and only 3% of rapists will spend a day in prison. **97% of rapists in the United States will walk free."**[169]

Not only that, but Holler adds, "The economic consequences of sexual abuse are costing Americans BILLIONS of dollars every year—some $450 billion annualy. Rape alone costs an estimated $127 billion per year (excluding child sex abuse), with each rape costing around $151,423. These costs are the result of funding sexual assault services, as well as lost wages over time due to lower educational attainment (which is common among adolescent assault survivors) and lost wages due to an inability to work, a frequent consequence of assault."

Many of the countries with high rape statistics need to look beyond just legislation to fix the problem. These countries need to look at the deep, systematic dysfunction of their cultures and social norms that have not prevented sexual violence. On the issue of rape, the Health Research Funding Organization points out that one of the biggest debates in the abortion arena is whether or not a woman who is raped should be legally allowed to have an abortion if the rape results in a pregnancy.[170]

Facts About Rape Victims Getting Pregnant in America:

1. The Centers for Disease Control and Prevention estimates that nearly 1.3 million American women are victims of rape or attempted rape every year.
2. A 1996 study found a national rape-related pregnancy rate of 5 percent per rape among victims between the ages of 12 and 45. That represents a staggering number of women in the U.S. alone: 32,000 in this one study. Of course, since 60 percent of rapes are not reported, this number could be well over a 10 percent pregnancy rate in the U.S. In countries like Saudi Arabia, the pregnancy rate is over 20 percent.

---

168 World Population Review, Rape Population Review, *Rape Statistics By Country 2020,* January 12, 2020 http://worldpopulationreview.com/countries/rape-statistics-by-country/

169 Madeline Holler, Microsoft News (MSN), *17 Staggering Sexual Assault Statistics Everyone Should Read,* 4/8/2019. https://www.msn.com/en-us/health/medical/17-staggering-sexual-assault-statistics-everyone-should-read/ar-BBVK6WB

170 Health Research Funding Organization, *18 Profound Statistics of Rape Victims Getting Pregnant,* , 2020, https://healthresearchfunding.org/18-profound-statistics-rape-victims-getting-pregnant/

A study of Ethiopian adolescents who reported being raped found that 17 percent subsequently became pregnant, and rape crisis centres in Mexico reported the rate of pregnancy from rape at 15 to 18 percent. Estimates of rape-related pregnancy rates may be inaccurate since the crime is greatly underreported.

In Nicaragua, between 2000 and 2010, around 172,500 births were recorded for girls under 14, representing around 13 percent of the 10.3 million births during that period. These were attributed to poverty, laws forbidding abortion for rape and incest, lack of access to justice, and beliefs held in the culture and legal system. A 1992 study in Peru found that **90 percent of babies delivered** to mothers aged 12 to 16 were conceived through **rape**, typically by a father, stepfather, or other close relative. In 1991 in Costa Rica, the figure was similar, with **95 percent of adolescent mothers under 15 having become pregnant through rape.**

Is there a link between pornography and violence against women? The amount of pornographic material available on the web is staggering. Women's rights groups warn that it is pervasive and a destructive threat to society, causing vast suffering, ruined lives, and millions of unplanned pregnancies due to rapists or partners acting out the sexual violence portrayed in porn and not using protection.

The conundrum—Why the apathy? Where is the outrage? With the high incidence of sexual assault, overpopulation, poverty, suffering, loss of biodiversity, injustice, corruption, and child cruelty, where is the outrage? Gender inequality and lack of family planning services are recognized by the UN as having one of the greatest impacts on driving climate change. On a planet of EIGHT BILLION, there is little hope of achieving either the 1948 Universal Declaration of Human Rights goals or the 2020 United Nations Development Goals. Gender-based violence will only begin to decline when human numbers decline and our attitude towards sexual violence changes in a monumental way.

Slave Trade (aka Human Trafficking)

The point of population stabilization is to reduce or minimize misery.
—Roger Bengston, World Population Balance

When we talk about the rape epidemic plaguing our planet, one predominating aspect—the slave trade—is perhaps the most insidious evil that is decaying our society. There are millions of slaves in the world today, more than ever before in human history. The slave trade is thriving, and now more lucrative than the drug trade, partly because drugs can only be sold once. When one person exerts control over another person in order to exploit them for economic gain, it is called human trafficking, and it's a form of

> *Slavery is the illegal trade of human beings. It's the recruitment, control, and use of people for their bodies and for their labor. Through force, fraud, and coercion, people everywhere are being bought and sold against their will–right now in the 21st century. Every 30 seconds, another person is sexually exploited.*
>
> **—A21**

modern slavery that is happening at this very moment, in almost every country across the surface of Earth. Yet it is dramatically unreported.

As the population increases, there are more unwanted women and children being enticed or forced into the human trafficking business, which continues to grow because of lack of regulations and enforcement, not to mention the rampant corruption of government and those in the law enforcement and justice systems. Plus, with the MeToo and ChurchToo movements taking hold in many countries, the **risks of committing rape** are getting much higher. **Sex traffickers are offering a far more attractive option in almost every major city of the world, as this hidden form of rape is less risky.**

There are four means of exploitation through human trafficking—for forced labour, organ harvesting, sexual exploitation, and child soldiers.

The anti-slavery group A21 explains that sex trafficking is complex and deeply rooted in global social issues, such as poverty, unemployment, globalization, gender discrimination across different cultures, lack of education, corrupt law enforcement agencies or other organizations, as well as a commercial demand for sex. As many families in developing countries face a lack of economic opportunities and/or gender discrimination, they often migrate under the pretense of a promised job and better life, but end up being forced into sex trafficking. Sometimes family members are sold to traffickers. This is not always the case. Young boys and girls struggling through life can be befriended by traffickers who show them false love and friendship, slowly gaining their trust, until they begin prostituting them for huge profits.[171]

Citing a 2014 report by the International Labour Organisation, A21 **reveals that 40 million people are being trafficked worldwide every year, delivering $150 billion a year to their captors—that's more revenue than Nike, Google, and Starbucks combined**. The A21 video, ***Stopping Traffic***, explores cases throughout the Philippines, Mexico, Thailand, Iraq, and major U.S. cities to provide raw images and first-hand documentation of human trafficking crime across the globe. It explores the practices within families and in the streets, explaining how a child or young adult is turned into being trafficked, at an average cost to a trafficker of $90, but with a potential to yield thousands. The creators establish the U.S. as the biggest source of sex traffickers' customers, with the Super Bowl being the most popular event for scoring a trafficking arrangement.[172]

In the United States, the average age of girls and boys forced to enter the sex trade is between 12 and 14 years, similar to global statistics. A trafficker exploits an average of four to six children at a time, and can make $150,000 to $200,000 per child each year. Children are trafficked at such a young age because they are easily manipulated, physically weak, and can be taught to believe what they are doing is normal and acceptable. (Children in slavery will be discussed in Chapter 14.)

They are often forced to participate in child pornography for Internet viewing, which is commonly a gateway into prostitution. So when we see prostitutes on the street anywhere in the world, there is a good chance they are not there willingly. Also, when we see pornography on the Internet, there is a good chance the participants are not there willingly either. Fortunately, the

171 A21, Stopping Traffic, *The Movement to End Sex Trafficking*, February 27, 2018, https://www.kickstarter.com/projects/siddhalishree/stoppingtrafficfilm

172 Stopping Traffic, Wikipedia, 2017 , https://en.wikipedia.org/wiki/Stopping_Traffic

anti-slavery group A21 created a video, *Stopping Traffic*, in order to end this travesty, as discussed in the Solutions section.

The Rise of Technology in Trafficking and Sexual Exploitation

Technology, in particular the Internet, has enabled sex trafficking and sexual exploitation to become the fastest growing criminal enterprise in the world. The Internet has completely changed the face of trafficking, with an increase in access **for both buyers and sellers** and rapid increase in activity without regulation. Technology and the Internet provide the tools that traffickers can use. Online resources such as **open and classified advertisement sites, adult websites, social media platforms, chatrooms, and extending into the dark web** enable traffickers to interact with an increasing number of potential victims.[173]

Equality Now stresses, "Solutions must be global, multi-dimensional and supported by actors including governments, tech companies, civil society and UN agencies. They must be informed by the experiences and perspectives of survivors." Equality Now is exploring the role of technology in sexual exploitation in order to advocate for the best approach and most effective solutions for adult women and adolescent girls. Citizens are also invited to join their campaign to address the misuse of technology in trafficking and online sexual exploitation.

Harvard law graduate Yasmeen Muhsam became the global executive director of Equality Now in 2011. She was previously with the United Nations Division for the Advancement of Women where she worked on the implementation of the Convention on the Elimination of all Forms of Discrimination Against Women (CEDAW) and the Secretary-General's study on violence against women. Muhsam has dedicated most of her life to eliminating gender-based violence, and her success story will be featured later in this chapter.

Solutions and Successes

> *Family planning could bring more benefits to more people at less cost than any other single technology available to the human race.*
> —United Nations

World Influencers Stepping Up

The Conversation reporters Niki Rust and Laura Kehoe ask, "**Could the Pill save the polar bear**?" They explain, "Conservationists tend to spend their time worrying about protecting forests, catching poachers or keeping carbon out of the atmosphere. But all these things (and more) are driven by humans. Given that it's easier and cheaper to reduce the human birth rate than it is to address these other issues, why aren't conservationists more concerned about keeping our population down?"[174]

173 Equality Now, *The Rise of Technology in Trafficking and Sexual Exploitation*, 2020, https://www.equalitynow.org/stop_tech_exploitation

174 Niki Rust and Laura Kehoe, The Conversation , Could the Pill save the polar bear?, July 13, 2015 https://theconversation.com/could-the-pill-save-the-polar-bear-44401

This question has been foremost on my mind for decades as well, especially since nothing could empower women more than allowing them reproductive choices and reducing population to a sustainable level. The UN predicts that by 2100, we could be at as many as 12 billion people. Shockingly, with each child a woman has, her carbon emissions legacy is increased six-fold. We also know that women will find it almost impossible to feed their children on a planet of 12 billion people. On the other hand, if we can stop population growth at EIGHT BILLION and start reducing it to a manageable number, food will be much easier to come by. Although regrettably, population doesn't seem to be a concern to many women's groups, clearly it should be.

> *Equality under the law is the first essential step to gender equality This is the struggle of men and women working together to fix humanity so that we can all thrive.*
> **—Yasmeen Hassan, Equality Now**

We are not talking about coercive measures in any form, as the things women want the most, education and contraception, would also be the biggest solutions for overpopulation. Rust and Kehow explain, "Fertility rates decline the longer a girl spends in school, so by simply providing better female education, the overall population in 2050 could be 1 billion less than current projections. This is because women who are empowered through education have fewer children, as well as having them later in life and therefore have the resources to provide them with better care. The benefits can be seen relatively quickly: between 1960-2000, contraceptive use by married women in developing nations increased from 10% to 60%, reducing the average number of children per woman from six to three."

In 2020, the women's movement started to get help from an unexpected source—conservation groups—as wildlife conservation is being integrated with family planning. The US Center for Biological Diversity distributed 40,000 condoms in 2018 wrapped in packaging with catchy slogans like *In the sack?... Save the leatherback*, and *Don't go bare... ... Panthers are rare.*[175] A more holistic approach combines family planning and other health care services with alternative livelihood options; this has been implemented in some key high biodiversity areas that have an unmet need for contraception and health care. One program in Nepal led to an increased use of condoms and **reduced wood fuel usage** equivalent to saving nearly 9,000 trees annually—due to reduced population requiring less wood for fuel and providing alternative employment. This same tactic would work to **reduce demand for fossil fuels** by reducing population worldwide.

One challenge the planet has yet to overcome is the increasing gap between funding and demand for contraception. Filling the unmet need for family planning across developing countries would cost US$8.1 billion annually. However, we do know that initiatives like Thriving Together and Fair Start Movement are making huge progress in this area. Also, by reallocating funds from our overspending military and pricey space missions, we could easily meet these needs. It is the political will that is sadly missing here and priorities gone awry. In fact, if governments just fulfilled

---

175 Sarah Baillie, US Center for Biological Diversity, *February 12, 2019, 40,000 Endangered Species Condoms to Be Handed Out in Top Wedding Cities*, February 12, 2019, *https://www.biologicaldiversity.org/news/press_releases/2019/endangered-species-condoms-02-12-2019.php*

their promised funding commitments to UN family planning programs, we would be halfway there. Fortunately in 2020, with Jo Biden committed to reinstate universal funding for family planning and reverse the global gag rule, this will set a positive example for the world.

Addressing human population growth may be a relatively fast and cheap remedy for wildlife loss, which also brings us closer to achieving gender equality. This is clearly a win-win. The Center for Biological Diversity urges, "The sooner we start to pull the brakes, the easier it will be to eventually come to a stop. So what are we waiting for?"

When it comes to meeting the UN Global Goals for 2030, this is where the decade of action starts—**with women**. We all have a role to play, and this is where women are starting to step up side by side with men and tackle the world's problems together. As the UN stated, "With just 10 years to go, an ambitious global effort is underway to deliver the 2030 promise—by mobilizing more governments, civil society, businesses and calling on all people to make the Global Goals their own."[176]

One of the heavy hitters of population growth and women's empowerment would concur when it comes to addressing violence as a major public health issue. Dr. Gro Harlem Brundtland was Norway's first female prime minister and director-general of the World Health Organization (WHO) from 1998 to 2003. Suzanne York, reporting on the 1992 Rio+20 Earth Summit, noted that Brundtland had been around big international and environmental negotiations for decades talking about how our rapid population growth was not sustainable and was still a challenge we face.[177]

Brundtland, affectionately known as "Mother of Sustainable Development," specified that we can only succeed by enabling young women to avoid early pregnancy, by breaking the intergenerational cycle of poverty, and by providing education. Mary Robinson, former president of Ireland, agreed and stressed the importance of bringing the message of supporting family planning and reproductive health to a much wider audience and linking it with women's issues, reassuring people that this is a mainstream human development issue.

In regards to the Rio+20 negotiations, Robinson said it was very worrying that these issues were not on the agenda. She challenged, "**Who is trying to prevent this?** This is a fight that still has to be fought and we have to be prepared to fight it. We need leadership on this issue. We need to connect the dots and integrate issues, and the essential element is the economic empowerment of women. And the removal of discrimination, especially of the girl child, has to be front and center."[178] Of course it is no secret that it is the **Vatican** that continues to derail progress on reproductive health through its veto power in the UN.

Dr. Brundtland asked, "What is our key mission? I see WHO's role as being the moral voice and the technical leader in improving health of the people of the world." Since 2003, Brundtland has been a UN special envoy on climate change, and member and deputy chair of The Elders—a group

176 UN Sustainable Development Goals – Decade of Action, UN online Jan. 2020 https://www.un.org/sustainabledevelopment/decade-of-action/

177 Suzanne York, Population Growth, The Heavy Hitters of Population Growth & Women's Rights, Jun 19th, 2012http://populationgrowth.org/the-heavy-hitters-of-population-growth-womens-rights/

178 Norway in the UN, *Mother of Sustainable Development*, Jun 14, 2017, https://www.norway.no/en/missions/UN/norway-and-the-un/norways-rich-history-at-the-un/important-norwegians-in-un-history/gro/

of former leaders convened by Nelson Mandela, Graça Machel, and Desmond Tutu to contribute their wisdom, leadership, and integrity to tackling world problems.

When it comes to fighting for women's rights, attorney Yasmeen Hassan was awarded the National Public Interest Award from Stanford Law School in 2019 for her work with Equality Now. The award recognized her international impact on social justice.[179]

Yasmeen points out the many successes, "In the last twenty years or so we have seen a sea of change in laws on violence against women. We have come a long way from domestic violence being seen as just 'life' (in the words of Gloria Steinem), to over 119 countries having laws against such violence and many have criminalized marital rape. More than 125 countries have enacted laws against sexual harassment. We also have positive changes in governments recognizing their responsibility to end harmful traditional practices: 23 African countries that have the practice of female genital mutilation ban it by law and most recently, there is a great impetus in Pakistan to change the laws that allow perpetrators of honour crimes to go free. Enactment of such laws is the first step; implementation is critical and needs political will."[180]

Protests for Justice

In 2019, women around the world joined the ongoing protests that began in Chile outside the country's Supreme Court on November 25, 2019, to mark the International Day for the Elimination of Violence Against Women. The Latin American feminist group, Las Tesis, sang their anthem—"The Rapist Is You"—which went viral and reached London in early December. Their campaign was in protest of the vilest forms of gender violence: femicides, disappearances, rapes, and lack of legal abortion.

Athanasia Francis, reporter with *The Conversation*, states, "The song rapidly crossed national borders. From Mexico to Greece, and from the Basque country to Kenya, it has become a global vehicle of feminist protest with local adaptations built around a solid lyrical core:

> *Patriarchy is a judge who judges us for being born*
> *and our punishment is the violence you don't see.*
> *It's femicide, impunity for my murderer,*
> *it's disappearance, it's rape.*
> *And it wasn't my fault, where I was, or how I dressed.*
> *The rapist is you, the rapist is you.*
> *It's the police, the judges, the state, the president.*
> *"The oppressive state is a macho rapist."*[181]

179 Equality Now, *Equality Now's Yasmeen Hassan Awarded the National Public Interest Award from Stanford Law Schoo,l* Nov 01, 2019 https://www.equalitynow.org/yasmeen_hassan_national_public_service_award

180 Janvi Patel, Equality Now, *Interview With Yasmeen Hassan*, August, 2016, https://www.equalitynow.org/interview_with_yasmeen_hassan?locale=es

181 Athanasia Francis, The Conversation, *'The rapist is you': why a viral Latin American feminist anthem spread around the world*, December 17, 2019 https://theconversation.com/the-rapist-is-you-why-a-viral-latin-american-feminist-anthem-spread-around-the-world-128488

Francis reports that amid ongoing protests in Chile demanding social reforms, there has been widespread criticism by international human rights organizations over the extremely violent way in which protestors are being treated. Protesters have reported the use of kidnapping, torture, and sexual abuse against women in Chile. The use of sexual violence concerned almost 15 percent of all complaints made within the first 40 days of the protests, according to a report by the campaign organization Human Rights Watch.

For protesters, the lyrics and choreography directly reflect their experiences. The pillars of institutional power—police, judges, political leaders—either turn a blind eye to sexual violence or become its perpetrators. The Chilean state's response became a major theme in the song. Gendered violence is an international common denominator affecting our communities.

"This is a war." Women in both the UK and Spain reported having to carry weapons to avoid being raped when going home alone, or submitting themselves to a form of curfew after dark for fear of gendered violence.

Justice systems in most countries seem to be equally complacent when dealing with this type of violence. This transnational feminist response in the form of a song proposes solidarity and empathy in turbulent political times. It's calling for an acknowledgement of sexual violence as a systemic and global problem in democratic institutions that, ironically, were created to prevent it.

One man who shares these same goals is author Ziauddin Yousafzai, father of Nobel Peace Prize winner **Malala**, claiming he has been fighting for gender equality for decades. He declared, "I'm not sure why I chose to start that journey, while other men accept the values passed down to them for centuries. Maybe it's because I was bullied as a child, for my dark skin and my stammering problem, so I was angry about any kind of discrimination against someone for the way they are born. I am sure of one thing: patriarchy is sheer stupidity. Fathers have a great interest in dismantling it. And we as campaigners need to communicate that to them."[182]

In Pakistan in 2012, the Taliban shot his daughter Malala in the head in an attempt to assassinate this 15-year-old girl for disobeying their law prohibiting girls from attending school. She lived—and began fiercely campaigning for women's and human rights, for which she won the Nobel Peace Prize. In his attempt to dismantle patriarchy, Ziauddin Yousafzai has since founded the Malala Fund charity with his daughter to help the 130 million girls currently not in the education system.

But it is his work as a parent that makes Ziauddin most proud. "When I married my wife, Toor Pekai, we chose to build an egalitarian family, respecting each other as equal partners and raising our daughter Malala the same way we raised our sons." It is because of activists like this father–daughter team that girls are being given the power to speak out, make communities safer, and create jobs. In these jobs, women can **talk to local leaders, teach equality in schools, and invest in female talent.**

> *In patriarchal societies, fathers are known by their sons. But I'm one of the few fathers in the world who is known by his daughter. And I'm so proud of it.*
> **—Ziauddin Yousafzai**

182 Sophie Gorman, France 24, June 22, 2019, *Malala's father on becoming a feminist and the battle for gender equality,* https://www.france24.com/en/20190622-let-dismantle-patriarchy-malala-yousafzai-feminist-father-pakistan

Demanding Social Reform

Nothing is as powerful as an idea whose time has come.
—Victor Hugo

When it comes to the media, women are starting to demand that their voices be heard. When it comes to the women's movement, it is rare to find any mainstream media sources that are willing to cover the fundamental issues of overpopulation, family planning, and violence against women. So it is an exceptional treat to find *The Guardian*'s award-winning Global Development website and news source, funded in part by the Bill and Melinda Gates Foundation. The women's movement could never have made the progress it has without social media sites like this, and this is a sad testament to who controls mainstream media on the planet.

In 2010 *The Guardian* began to focus on the millennium development goals—the eight targets set in 2000 by the United Nations Millennium Declaration with the aim of improving the lives of the world's poorest people by 2015. The journalism and other content is editorially independent. *The Guardian* assures that they are committed to open journalism, recognizing that the best understanding of the world is achieved when we collaborate, share knowledge, encourage debate, welcome challenge, and harness the expertise of specialists and their communities. Many articles used in this book come from *The Guardian*'s site, and I am thankful to them for this rich resource.

Eliminating Slavery In Our Lifetime

The secret of change is to focus all of your energy, not on fighting the old, but on building the new.
—Socrates

Foremost, to build a better future, we must ensure that all children are planned and wanted so that they are not sold or enticed into slavery. This would require that every woman has access to contraception, has the freedom to utilize it, and fully understands all of the impacts of her choices on the environment and future generations. Since presently half of children born globally are unintended, we must support family planning initiatives like Thriving Together and Fair Start Movement so that every child is planned for. We must also create stringent international laws to protect women and children against sexual violence, and these laws must be strictly enforced. We must build a better justice system globally, so that those dealing in the slave trade are held accountable and brought to justice. We also must build a better working relationship with the media to encourage them to give modern slavery the attention it deserves, so that people can recognize and report these cases to the authorities and understand the magnitude of this problem. We must create a world where there is no place for sexual violence to exist, by stopping population growth at EIGHT BILLION, then reducing it to a sustainable level.

The A21 documentary *Stopping Traffic* is a 2017 American film directed by Sadhvi Siddhali Shree and produced by the team of monks at Siddhayatan Tirth in the U.S. **This production is significant because it shows how women and men working together can inspire the world to**

take action. The documentary was inspired and is based on the teaching of non-violence. *Stopping Traffic* made its world premiere at the Global Cinema Film Festival in Boston, a human rights festival, on March 11, 2017, where it won Best Picture. The film is a volunteer endeavour funded by a Kickstarter campaign.

A21 is the non-profit organization that has made this documentary happen, and it is their mission to end slavery. They have outlined 21 ways that we can abolish slavery, and have declared that they are the abolitionists of the 21st century, **working with the public to free slaves and disrupt the demand.**[183]

A21 states, "We believe in a world without slavery. For 10 years now, we've rolled up our sleeves and set our feet to action. Why? Because in a single moment a number can turn into a name, a tragedy into a victory, and a belief into an action. But phrases like 'slavery' and 'human trafficking' can still feel ambiguous. This is the reality: slavery is violence. It's physical, verbal, and sexual abuse. It's forced prostitution. It's barbaric working conditions.... Traffickers use different methods to recruit victims. When we know how they are trafficked, we are given the power to stop slavery before it starts."

Law and Order

In an effort to bring about this equality and justice, a remarkable group of global influencers has launched a life-changing mission. Since being founded in December 2013, **Operation Underground Railroad** (O.U.R.) has gathered the world's experts in extraction operations and in anti-child trafficking efforts to bring an end to child slavery. O.U.R. explains that their operations team consists of former CIA, past and current law enforcement, and highly skilled operatives that lead coordinated identification and extraction efforts. These operations are always in conjunction with law enforcement throughout the world.[184]

O.U.R is committed to enhancing law enforcement efforts by providing resources where budget shortfalls prohibit a child pornography, child exploitation, or human trafficking operation from going forward. Once victims are rescued, a comprehensive process involving justice for the perpetrators and recovery and rehabilitation for the survivors begins. O.U.R. advocates, "It is time for private citizens and organizations to rise up and help. It is our duty as a free and blessed people."

In one such high-risk mission, Operation Whale Watch, 24 young women were rescued in Latin America. Only one was a minor. But that one minor is key in the potential punishment of the five traffickers. Because of that 16-year-old girl, they are now facing 15 to 30 years in prison. That's a good day's work. The Latin American authorities of this country were involved every step of the way in this undercover operation. In 2020, a revealing movie, *Sound of Freedom*, was released to highlight their cause. Because their main focus is rescuing children being exploited for child slavery, I will be discussing this encouraging success story more in Chapter 13.

---

183 A21, *We Are The New Abolitionists*, 2020, https://www.a21.org/content/who-we-are/gnihwo

184 Operation Underground Railroad, Operation Whale Watch, https://ourrescue.org/stories, https://ourrescue.org/about

In another effort to bring about law and order in 2020, models posed at a Toronto storefront window for the Shoppable Girls Campaign to illustrate how victims are seen as profitable items for sale in Canada's multibillion-dollar slave trade industry. However, most Canadians have a misconceived belief that this human trafficking is only something that goes on in other countries where victims are physically detained and sold through large international trafficking rings.

CTV News reports that most recent data from Statistics Canada show that 90 percent of sex trafficking victims are females between the ages of 18 and 24—and sometimes as young as 13. "In reality sex trafficking is a Canadian issue. Ninety-three per cent of sex trafficking victims are Canadian citizens—this is happening all around us," said Julie Neubauer, manager of anti-human trafficking services at Covenant House Toronto. Many victims are exploited by trusted people such as boyfriends, and they are often manipulated into thinking they are not being trafficked.[185]

In 2018, the Canadian government announced a national strategy to provide $75 million in funding initiatives to aid in victim support and resources, and starting in 2020, a further $1.4 million will be invested per year over the next five years for new anti-trafficking initiatives. Increased reporting in recent years confirms the global trend showing human trafficking is getting far worse, but also that the **government is starting to take it seriously.**

Neubauer advised, "We need to encourage people to really understand what sex trafficking is and to have empathy and compassion for what these people have endured; most poignantly to recognize that no one chooses to be sex trafficked, it's not about choice here." She added that while Canada's Criminal Code recognizes sex trafficking as a crime, there need to be stricter laws and harsher penalties for convicting sex traffickers.

In 2019, Canadian police dealt a serious blow when they arrested 31 people in a human trafficking ring; they faced more than 300 charges between them. Dozens of women were victimized in this one operation that operated throughout the country. They were controlled by means including physical violence, emotional manipulation, drugs, and alcohol.

"To the naked eye it may appear that these females that are involved in the sex trade are willing participants," Inspector Thai Truong of the York Regional Police said at a news conference. "They may smile at you, they may not even appear to be controlled or victimized." But, he added, police "have seen the horrific things that are happening to these women. They're controlled in every way imaginable." He said the women were forced to work constantly, allegedly bringing in what he called a conservative average of $1,000 a day for the organization.[186]

This small sample of amazing women and men from around the world make it clear just how much progress we have made in setting the stage for this 2020 decade of action. These stories have a compelling energy that can't help but give one hope. When it comes to being motivated, it is **social activism that has put meaning in my life**. Previously I was very self-involved, into personal

185 Melissa Lopez-Martinez, CTVNews.ca, *Sex trafficking still a prevalent issue across Canada, advocates and police say*, February 20, 2020, https://www.ctvnews.ca/canada/sex-trafficking-still-a-prevalent-issue-across-canada-advocates-and-police-say-1.4820944

186 Michelle McQuigge, The Canadian Press, *More than 300 charges laid in multi-province human trafficking investigation*, October 16, 2019, https://windsor.ctvnews.ca/more-than-300-charges-laid-in-multi-province human-trafficking-investigation-1.4641479

achievement, how I looked, and how others perceived me. Now I feel like I'm part of a meaningful mission, working for the "greater good."

The Failings of Traditional Media and Eco Groups

It is not only **new words**—like eco anxiety, compassion fatigue, degrowth, covidiots, quarantini, megadrought, insectgedon, and existential—that are needed to deal with today's calamities, but **new eco groups and media sources** as well. The old sources have appallingly failed us.

What mighty time-honoured systems we would have at our fingertips if only traditional media and environmental groups were to get on board. Working together for the greater good, these two entities truly have the power to change the world as no other mechanism could. Yet, why is it that they are both ignoring the most critical issues on the planet? It is little wonder that the youth today feel betrayed and have to create all new groups and use online media to tackle the important issues causing climate change— overpopulation, depletion of natural resources, and inhumane animal agriculture.

It is the media to which we look for balanced reporting on world events and the state of our planet. Yet, by making these topics taboo, most have managed to undermine any efforts to have open and meaningful dialogue regarding these matters. It is especially irresponsible for public broadcasting networks like the CBC and BBC to refuse to deal with these urgent issues when so many citizens have expressed a growing anxiety about them. The solution is to encourage the few that are brave enough to tackle these issues, and support the many new groups and media sites that are now emerging to fill the void. Some of these are Global Citizen, Fair Start Movement, Center for Biological Diversity, The Guardian, LivingKindly, and One Planet – One Child.

There are numerous programs on TV, like the *Nature of Things*, which vividly show the devastation our growing population is causing. However, they stop short of actually pointing out that it is caused by overpopulation or mentioning that reducing population would be a viable solution. TV talk shows are eager to interview celebrities who are pregnant, glorifying parenthood and promoting big families. Yet they seldom mention the realities of how difficult parenting is in these trying times, and the many neglected and abused children resulting from the lack of family planning. They don't mention the burden that overpopulation puts on society or the environment.

Plus, if the media does deal with the issue of population, the material almost always suggests that there are benefits from increased population, reflecting the biased interests of the growth industry and our capitalist societies. To keep the wheels of industry turning out excessive profits, they almost always promote population growth and increased immigration—despite scientific evidence to the contrary.

Have you noticed that news reports daily mention the effects of growth and the need for spending billions of dollars on expanding, building, replacing, or updating our infrastructure and increasing our energy supply, housing, schools, prisons, and so on? Yet seldom do they give the other side of the story, including suggestions that we reduce the demand by reducing population. The news is constantly quoting the Vatican on the virtues of being celibate or avoiding the use of birth control or family planning. Yet never do they balance this out with views from family planning or population groups.

On the other hand, the population movement online reflects the rising tide of discontent and is growing by leaps and bounds. This underground movement is using the increasing number of blogs, alternate news sites, discussion groups, and Twitter feeds to let their voice be heard.

In this new decade of change, we are seeing citizens ramp up pressure on governments, businesses, and social entities demanding social reform and justice. It is clear that old habits and senseless traditions that lead to violence are no longer going to be tolerated, especially when it comes to sexual abuse and torture. This was made evident in Malawi, Africa, in 2019 when protesters called for a government investigation into allegations of rape by police officers during ongoing post-election violence.[187]

In 2020, *Guardian* reporter Charles Pinsulo stated that the EU ambassador to Malawi had condemned the alleged sexual violence and the British high commisioner had called for a thorough investigation. Following the May 2019 election, protesters took to the streets leading to clashes with the police. The NGO Gender Coordination Network (NGO-GCN) documented accounts from women and girls who said they had been sexually assaulted by police officers. The network is disturbed with reports that some of the police officers dispatched in the area raped women, defiled self-boarding girl students, tortured people, and looted private property.

The report demanded that the president and other authorities ensure the allegations were thoroughly investigated and perpetrators punished. Emma Kaliya, a women's rights campaigner who led the protests, described the conduct as the "worst kind of gender-based violence." The ministry of gender in Malawi earlier issued a press statement saying, "The ministry is deeply concerned by these allegations that border on disregard for women's rights and exploitation of their vulnerability."

Another blot on our civilization is the number of sexual assaults taking place in ride-sharing cars in the U.S. alone. *Business Insider* reporter Yusuf Khan stated, "Uber is set to lose $1 billion in market value after it reported 6,000 sex assault cases. The number of cases were 2,936 in 2017 and 3,045 in 2018, or 5,981 over the two years. Uber's stock fell 2.2% in premarket trading. That translates to about $1.1. billion off a market cap of $48.9 billion."[188]

Uber CEO Dara Khosrowshahi said, "Doing the right thing means counting, confronting, and taking action to end sexual assault. My heart is with every survivor of this all-too-pervasive crime. Our work will never be done, but we take an important step forward today. In the long run, we will be a better company for taking this step today—because I firmly believe that companies who are open, accountable, and unafraid are ultimately the companies that succeed."

---

187 Charles Pensulo, The Guardian, *Malawi protesters demand inquiry into allegations of rape by police officers,* Jan 27, 2020, https://www.theguardian.com/global-development/2019/oct/25/malawi-protesters-demand-inquiry-into-allegations-of-by-police-officers

188 Yusuf Khan, Business Insider, *Uber is set to lose $1 billion in market value after it reported 6,000 sex assault cases,* Dec. 6, 2019, https://markets.businessinsider.com/news/stocks/uber-stock-price-1-billion-to-get-wiped-from-market-cap-on-sex-assaults-2019-12-1028743506

188 Angelique Chrisafis, The Guardian, *G7 leaders told to scrap discriminatory gender laws from statute books,* Aug 25 2019, https://www.theguardian.com/world/2019/aug/25/g7-leaders-told-to-scrap-discriminatory-gender-laws-from-statute-books

Government Leadership

> *Leaders were advised to implement at least one progressive law and told they would be held accountable at future summits.*
> **—G7 Summit, France, 2019**

In this 2020 decade of change, it isn't only companies that are being held accountable, but countries as well. From the G7 summit in France in 2019, *Guardian* reporter Angelique Chrisafis stated, "All G7 countries, including the UK, still have discriminatory laws on their statute books or substantial loopholes that allow discrimination, the G7's gender equality advisory council said at a key summit session attended by all leaders, including former US president, Donald Trump. Discriminatory laws range from the UK's legal position on child marriage, with children as young as 16 allowed to marry if their parents give consent, to limits on free contraception in Germany, or regressive laws on abortion in certain US states."[188]

Chrisafis noted that the leaders were addressed by the Nobel laureates Nadia Murad—a Yazidi woman from Iraq who was kidnapped and raped by the Islamic State before she escaped, going on to become a human rights campaigner—and Denis Mukwege, a Congolese gynecologist who has treated tens of thousands of rape survivors. With them, Phumzile Mlambo-Ngcuka, the head of UN Women, urged the leaders, "We are not only asking you to be brave, we are asking you to be the great generation of leaders who will take bold steps. We call upon you to pronounce 2030 as the expiry date of gender inequality."

Chrisafis emphasized, "Trump attended the key meeting on gender equality despite senior US administration officials complaining that the French president, Emmanuel Macron, had moved the focus of the G7 summit too far towards what they called "niche issues" such as equal rights and the climate emergency."

Leaders were given a long list of progressive laws that countries could take inspiration from. These ranged from Denmark's new law on cyber-harassment and Iceland's equal pay laws, to France's new law on street harassment and its law on banning misinformation on abortion online. France pledged to increase funding for help for survivors of sexual violence, as campaigners also called for more financing for women's rights groups.

"We encouraged leaders to go home and look at their laws," said Katja Iversen, the president of the Women Deliver campaign group. "Many said they had already identified laws they would improve, that was very encouraging."

Identifying and improving laws that are failing women is also the goal of the UN Sustainable Development Program. Their goals to end poverty, rescue the planet, and build a peaceful world are gaining momentum. Today, progress is being made in many places, but, overall, action to meet the goals is not yet advancing at the speed or scale required, especially with COVID-19 hampering efforts. Yet the 2020s need to produce a decade of ambitious action to deliver the goals by 2030.[189]

189 UN online, UN Sustainable Development Goals – Decade of Action, Jan. 2020 https://www.un.org/sustainabledevelopment/decade-of-action/

In September 2019, the UN Secretary-General called on all sectors of society to mobilize for a decade of action on three levels: **global action** to secure greater leadership, more resources, and smarter solutions for the Sustainable Development Goals; **local action** embedding the needed transitions in the policies, budgets, institutions, and regulatory frameworks of governments, cities, and local authorities; and **people action**, including by youth, civil society, the media, the private sector, unions, academia, and other stakeholders, to generate an unstoppable movement pushing for the required transformations. In January 2020, the UN Secretary-General outlined his priorities for the year, including the Decade of Action:

> *The 2030 Agenda is our roadmap for the world we all want. The Global Goals are our best hope—for people, for planet, for prosperity, for peace and for partnerships.*
> **—United Nations**

> We will mobilize everyone, everywhere to create an unstoppable force linked to the Global Goals. This requires each of us to take action—individually and collectively, locally and globally. We must be the generation to end extreme poverty, win the race against climate change and conquer injustice and gender inequality. We will hold leaders to account and point to what is possible when action delivers results. We will shine a light on solutions that expand access and demonstrate the possibilities of ideas. We will drive sustainable innovation, financial investments and technology—while making space in our communities and cities for young people to lead. The spirit of human endeavor has demonstrated our shared ability to deliver the extraordinary.[190]

UN Women focuses on Sustainable Development Goal #5 to achieve gender equality and empower all women and girls. Targets include the following:

- End all forms of discrimination against all women and girls everywhere.
- Eliminate all forms of violence against all women and girls in the public and private spheres, including trafficking and sexual and other types of exploitation.[191]

UN Women advocates that gender equality by 2030 requires urgent action to eliminate the many root causes of discrimination that still curtail women's rights in private and public spheres. For example, 49 countries still lack laws protecting women from domestic violence, while 39 countries bar equal inheritance rights for daughters and sons. Eliminating gender-based violence is a priority, given that this is one of the most pervasive human rights violations in the world today. Based on data from 87 countries, one in five women and girls under the age of 50 will have experienced physical and/or sexual violence by an intimate partner within the last 12 months. Harmful practices, such as child marriage, steal the childhood of 15 million girls under age 18 every year.

Women do 2.6 times more unpaid care and domestic work than men. While families, societies, and economies depend on this work, for women, it leads to lower earnings and less time to engage

190 ibid

191 **UN Women,** *SDG 5: Achieve gender equality and empower all women and girls,* 2019 https://www.unwomen.org/en/news/in-focus/women-and-the-sdgs/sdg-5-gender-equality

in non-work activities. In addition to equal distribution of economic resources, which is not only a right, but accelerates development in multiple areas, there needs to be a fair balance of responsibility for unpaid care work between men and women.

UN Women adds that sexual and reproductive rights are critical in their own right. Shortfalls in these multiply other forms of discrimination, depriving women of education and decent work, for example. Yet only 52 percent of women married or in a union freely make their own decisions about sexual relations, contraceptive use, and health care.

While more women have entered political positions in recent years, including through the use of special quotas, they still hold a mere 23.7 percent of parliamentary seats, far short of parity. The situation is not much better in the private sector, where women globally occupy less than a third of senior and middle management positions.

UN Women acts to empower women and girls across all its programs and advocacy. With stepped up action on gender equality, every part of the world can make progress towards sustainable development by 2030, leaving no one behind.

This is an ideology that the Jane Goodall Institute of Canada strongly supports with her Delivering Healthy Futures project. This is a project based in the Democratic Republic of Congo (DRC) that is funded in partnership with the Government of Canada, through Global Affairs Canada. It is based on the understanding of a strong and interactive relationship between the health and sustainability of human communities and the integrity of the local ecosystem.[192]

In the Congo, 78 percent of Grauer's gorilla populations have disappeared from eastern Congo, declining from 17,000 in 1995 to 3,800 in 2020. Goodall points out, "The places where we work in the DRC are politically volatile. People and animals alike have suffered from the ongoing conflict there. JGI is committed to implementing multi-year projects despite uncertainty and danger. The eastern DRC includes enormous rainforests—the 'green heart' of Africa—which is critically important habitat for chimpanzees and the highly endangered Grauer's gorilla, also known as the eastern lowland gorilla. Protecting the forests is essential to the long-term survival of great apes." Solutions:

- Conduct reproductive health and family planning outreach.
- Train local community members on reproductive health and prevention of childhood diseases.
- Support the implementation of a region-wide vaccination program.
- Build new maternal health clinics.

The Delivering Healthy Futures project to eliminate poverty, empower people, and save chimps is the kind of project that can truly create healthy and sustainable communities where the people and wildlife can both benefit. It represents the essence of the UN Sustainable Development Goals in action. Kudos to Jane Goodall, her leadership, and great wisdom!

192 Jane Goodall Institute of Canada, *Delivering Healthy Futures Program*, 2020, https://janegoodall.ca/our-work/community-conservation/delivering-healthy-futures/

Musicians and Birdwatchers Champion Change

When it comes to changing the world, music is a universal language that has the potential to inspire and uplift the next generation of activists. Through Global Citizen, musicians can champion change and encourage their fans around the world to join the fight against extreme poverty. Musicians and activists from John Legend to Beyoncé were recognized for their musical accomplishments in 2020. The 2020 Grammy Award nominations feature a wide variety of artists who have worked and performed with Global Citizen.[193]

Among the list of Grammy Award nominees are breakout artists Lizzo, who volunteered at a food bank in Australia to help victims of the bushfires in 2019, and Billie Eilish, who is working to combat climate change. They were both nominated for Best New Artist and were set to perform at Global Goal Live: The Possible Dream in 2020. Other artists that prove that activism and music go hand in hand are Shawn Mendes, John Legend, Ed Sheeran, Cardi B, and H.E.R. The activists also include Beyoncé, who co-founded Chime for Change, an organization promoting the rights and education of women and girls around the world.

When it comes to global celebrities and forces for change, naturalists have a role to play as well. Teenage birdwatcher Mya-Rose Craig became a celebrity when she received Bristol University honorary doctorate at the age of 17. Known as Birdgirl, Craig earned this accolade for creating Black2Nature to help engage more children from minority ethnic backgrounds to tackle the environmental crisis threatening our planet, says BBC News.[194]

Now more than ever, it is important to recognise that inequality of engagement creates inequality of opportunity and an unequal world is not a sustainable one.
—Mya-Rose Craig, "Birdgirl"

193 Catherine Caruso, Global Citizen, *8 Global Citizens Were Nominated for Grammy Awards This Year*, Jan. 23, 2020, https://www.globalcitizen.org/en/content/global-citizens-grammy-nominations-2020/?utm_source=twitter&utm_ medium=social&utm_content=global&utm_campaign=general-content&linkId=81210616

194 BBC News, *'Birdgirl' Mya-Rose Craig receives Bristol University honorary doctorate,* February 20, 2020, https://www.bbc.com/news/uk-england-bristol-51561747

CHAPTER 7

CLIMATE DISRUPTION—REAL SOLUTIONS

If any fraction of the observed global warming can be attributed to the action of humans, then this by itself is positive proof that the human population has exceeded the carrying capacity of the earth. As a result, it is an inconvenient truth that any actions or programs to reduce global warming that do not center on population reductions are what Mark Twain called "Silent Lies."
—Professor Al Bartlett

Climate change is the emergency of the decade that has brought the world together and unified a movement to finally deal with some of the planet's woes at the eleventh hour. But like all of the other critical issues, climate change is just a symptom of the ultimate problem. We only have one problem—human overpopulation—with many symptoms attached that all urgently need to be dealt with. This is the chilling truth and the basis we must use to tackle climate change and all of the other world issues if we are to survive. The UN has also made this critical connection (see Figure 6).

POPULATION AND CO_2 EMISSIONS, 1750-2015

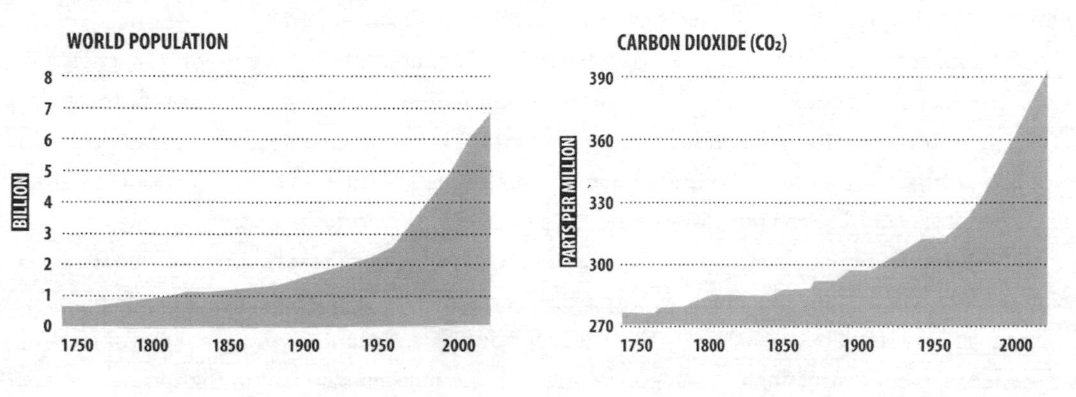

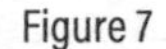
Figure 7

In this chapter, I intend to establish how OVERPOPULATION is the driving force behind climate change, and how together they have been—and increasingly will be—the cause of immense human suffering and environmental destruction if we refuse to acknowledge their significance. Therefore, I would prioritize the causes (and the solutions) of climate change in the following order:

1. Population
2. Meat and dairy consumption
3. Deforestation and loss of plant life
4. Energy waste in all its forms

NewSecurityBeat reporter Robert Engelman tells us that long in the background, population is becoming a bigger issue at climate change discussions. He notes that this growing awareness matches one of the emerging conclusions of the Worldwatch Institute's Family Planning and Environmental Assessment project: **Researchers around the world are increasingly recognizing the strength of the population–climate change link.**

In 2015, Engelman stated, "As most of the world's governments are puzzling out what they can offer to combat global climate change, a sensitive but critical aspect of the problem is coming into clearer focus: population. The word appears 20 times in a new 66-page synthesis of country pledges to cut greenhouse-gas emissions by the UN Framework Convention on Climate Change's Secretariat. And those are the mentions of population in the context of size or growth, not the word's more frequent use as a synonym for 'people.'"[195]

Finally, in 2019, the Intergovernmental Panel on Climate Change also issued a strong statement on the contributions of population growth to rising carbon emissions and promoted the benefits of family planning programs. We know that **almost half of population growth today results from unintended pregnancies rather than self-determination**. Therefore, assuring that all pregnancies are welcomed would in itself greatly reduce population growth. In crafting policies in the future, population should be viewed as a significant factor in climate change. Sustainable population as a major solution to climate disruption will be discussed in detail in the Solutions section below, along with tree planting projects and a move towards a plant-based diet.

Unfortunately, most of humanity has decided to focus on alternative energy—the issue that will bring the least benefit of the four, instead of the most fundamental one that could provide the greatest benefit. Consequently, in 2019, science indicated that emissions are still going up, not down. In turn, this is triggering a spate of billion-dollar climate-fuelled weather events that are devastating our planet. This could be the folly for which future generations least forgive us.

In the 2020 documentary *Planet of the Humans*, Michael Moore emphasizes the threat of too many people demanding too much energy from one small planet. *Bennington Banner* reporter Suzanna Jones states, "Is 'green' energy the savior for our climatic and environmental problems or is it a false prophet distracting us from confronting the gargantuan elephant in the room? Frankly,

195 Robert Engelman, NewSecurity Beat, *Long in the Background, Population Becoming a Bigger Issue at Climate Change Discussions,* November 10, 2015, https://www.newsecuritybeat.org/2015/11/population-bigger-issue-climate-change-discussions/

the green energy 'movement' is really about sustaining our way of life and the economic system that it depends upon, not the health of the biosphere. Capitalism is brilliant at co-opting anything that resists it. **Green energy**—like much of the broader environmental movement—is no exception. **It's business-as-usual in camouflage.** The film has struck a nerve. The criticisms reveal how much power and money lie behind the renewables-as-savior myth. With so much at stake, the industry and big environmental organizations have little appetite for discussing or even acknowledging the unsavory side of the technologies. And ultimately, the core issues remain unaddressed; the most important things remain unspoken."[196]

Planet of the Humans reveals the diabolical truth about the green energy fiasco, and points us towards the root of our environmental crisis—human numbers out of control.

There is no denying that climate change is causing havoc, and because this is in front of our faces right now, we are busy putting out fires. However, if we are truly to make any progress towards limiting to a 1.5 degree increase in global temperature, we will need to focus on the most pressing issue at the same time—overpopulation. I am not going to get caught up in the ongoing debates taking place in the realm of the skeptics. I don't need to dwell on the overwhelming scientific evidence, charts, graphs, or other convincing data. For me, the most compelling evidence that humankind has managed to put our planet in peril can be highlighted in just a few images. It is the heartbreaking image of a polar bear stranded on an ice flow in the middle of the Arctic. That is real. That is evidence. And that is testament to our disturbing disconnection to nature. It is the devastating fires burning koalas alive in the trees of Australia. It is the one million dead sea birds washing ashore along the Pacific coast of North America.

These images have touched the hearts of millions of people around the world, causing much of the eco-anxiety troubling most of our youth today. To deny that humankind played a role in these tragedies is extremely irresponsible and delusional—when we have condemned most of our endangered animals to certain death. Do we honestly have the right to do this?

Putting Out Fires

So, has a new poster child for climate change emerged out of the ashes of Australia's fire to join the polar bear of the Arctic? The iconic koalas represent the victims of an ancient dry continent now in ruin, suffering from the worst drought and bushfire season in history. Wildlife carers estimate 50 percent of the koala population has been killed by fires, heat, and drought. MAHB Stanford reporter Sue Arnold agonizes, "With no sign of the intense catastrophic fires subsiding, or the drought ending, the future for Australia's unique biodiversity is grim indeed.... As we watch, in horror, apocalyptic scenes play out across the continent."[197]

Arnold points out that adding to the tsunami of threats is a Pentacostal Prime Minister Morrison dedicated to ensuring the end of times for Australia's wildlife and environment. Like former U.S.

---

196 Suzanna Jones, Bennington Banner, *What 'Planet of the Humans' reveals about us,* May 19, 2020, https://www.benningtonbanner.com/stories/suzanna-jones-what-planet-of-the-humans-reveals-about-us,604994

197 Sue Arnold, Millennium Alliance for Humanity and the Biosphere, *Australia can no longer call itself the lucky country,* January 16, 2020, https://mahb.stanford.edu/blog/australia-can-no-longer-call-lucky-country/

President Trump, he continues to reject the existence of climate change and to insist that the disasters devastating our planet are normal. Arnold asserts, "He has distinguished himself as the worst Prime Minister in Australia's history. Morrison has ripped up environmental protections, ensuring his government *removes the green tape holding up developments.*" Unfortunately, this insanity is supported by many of the premiers of the country.

The impacts of the fire holocaust on wildlife are almost impossible to grasp. Professor Chris Dickman, an ecologist with the University of Sydney, makes a conservative estimate that 800 million mammals, birds, and reptiles have perished. He says, "If you add bats, frogs, and invertebrates in the death toll, over a billion would be a very conservative figure."

Australia's bushfires are believed to have spewed as much ***as two-thirds of the nation's annual carbon dioxide emissions*** *in just the past three months.*
—CSIRO Climate Science Center

In Adelaide Hills in South Australia, carers report 3,000 bats died in one day because of heat stress. Kangaroos and wallabies that have survived the fires are dying of thirst and starvation. Months ago, kangaroos and emus in outback Australia were dying in similar agony. They say that the Queensland government refused to send aerial water bombers—the only way to get to the fires—because it was "too expensive." National media has reported that 23 former fire and emergency leaders say they tried for months to warn Prime Minister Scott Morrison that Australia needed more water-bombers to tackle bigger, faster, and hotter bushfires. According to firefighters, the fires have been so intense that the earth has burned down as far as two metres, ensuring that insects, beetles, worms, grubs, and millipedes have been cremated. The concept of forests having any role in rain-making is completely foreign to the current dismal array of state and federal politicians.

Arnold stated, "Australia's mammal extinction rate is the highest in the world according to the country's Threatened Species Hub. Rampant clearing of native vegetation in NSW alone has jumped 800% in three years. An estimated half of Australia's forest cover in the last 200 years has been logged and cleared. The intensity of the current fires has resulted in the loss of entire ecosystems. According to a senior research CSIRO scientist Australia's bushfires are believed to have spewed as much as two-thirds of the nation's annual carbon dioxide emissions in just the past three months, with experts warning forests may take more than 100 years to absorb what's been released so far this season. The bush will be silent. And the heavy cloud of mourning for what has been lost is a stone in the heart of all caring Australians. However, grief is transforming into massive public anger and the extraordinary flow of dollars to shelters from all over the world has put our wildlife on the global front page."

Will Australia's forests and biodiversity ever be able to recover? Not according to *New York Times* reporter Richard Flanagan, who asserts that Australia is committing climate suicide. He indicated, "Australia today is ground zero for the climate catastrophe. Its glorious Great Barrier Reef is dying, its world-heritage rain forests are burning, its giant kelp forests have largely vanished, numerous towns have run out of water or are about to, and now the vast continent is burning on a scale never before seen."[198]

Flanagan describes the thousands, "driven onto beaches in a dull orange haze, a crowded tableaux of people and animals almost medieval in their strange muteness—half-Bruegel, half-Bosch, ringed by fire, survivors' faces hidden behind masks and swimming goggles. Day turns to night as smoke extinguishes all light in the horrifying minutes before the red glow announces the imminence of the inferno. Flames leaping 200 feet into the air. Fire tornadoes. Terrified children at the helm of dinghies, piloting away from the flames, refugees in their own country. All this and it was only just the beginning of the peak fire season."

Flanagan says, "As I write, a state of emergency has been declared in New South Wales and a state of disaster in Victoria, mass evacuations are taking place, a humanitarian catastrophe is feared, and towns up and down the east coast are surrounded by fires, all transport and most communication links cut, their fate unknown. The bookstore in the fire-ravaged village of Cobargo, New South Wales, has a new sign outside: '*Post-Apocalyptic Fiction has been moved to Current Affairs*.'"

And yet, notes Arnold, the response of Australia's leaders to this unprecedented national crisis has been not to defend their country but to defend the fossil fuel industry, a big donor to both major parties—as if they were willing the country to its doom. While the fires were exploding in mid-December, the leader of the opposition Labour Party went on a tour of coal-mining communities expressing his unequivocal support for coal exports. Prime Minister Morrison went on vacation to Hawaii, and he announced plans to build the biggest coal mine in Australia. Since 1996, successive conservative Australian governments have successfully fought to subvert international agreements on climate change in defence of the country's fossil fuel industries.

> *Australia is the world's largest exporter of both coal and gas. It recently was ranked 57th out of 57 countries on climate-change action.*
> **—Richard Flanagan,** *New York Times*

Flanagan remarks, "Mr. Morrison has tried to present the fires as catastrophe-as-usual, nothing out of the ordinary. This posture seems to be a chilling political calculation: With no effective opposition from a Labour Party reeling from its election loss and with media dominated by Rupert Murdoch—58 percent of daily newspaper circulation—firmly behind his climate denialism, Mr. Morrison appears to hope that he will prevail as long as he doesn't acknowledge the magnitude of the disaster engulfing Australia. More than one-third of Australians are estimated to be affected by the fires. Mr. Morrison may have a massive propaganda machine in the Murdoch press and no opposition, but his moral authority is bleeding away by the hour. As Mikhail Gorbachev, the last Soviet leader, once observed, the collapse of the Soviet Union began with the nuclear disaster at Chernobyl in 1986. In the wake of that catastrophe, 'the system as we knew it became untenable,' he wrote in 2006. Could it be that the immense, still-unfolding tragedy of the Australian fires may yet prove to be the Chernobyl of climate crisis?"

198 Richard Flanagan, NY Times, *Australia Is Committing Climate Suicide,* January 3, 2020, https://www.nytimes.com/2020/01/03/opinion/australia-fires-climate-change.html

Many countries are willing to throw their defenceless wildlife into the black hole of climate change, as long as it doesn't affect the bloated human population. While less than a hundred people died in Australia's devastating fire, over a billion animals perished. Not only are these extreme weather events a catastrophe for the animals, but for the economy as well because extreme weather climate risks are increasingly hard to predict. Consequently, a significant and increasing majority of Australians are demanding action on climate change.

Extreme Weather and Economic Recession

According to Iran News, a shocking new study says extreme weather events caused by climate change could result in an economic recession "the likes of which we've never seen before."[199]

This study was conducted by Paul Griffin, an accounting professor at the University of California. He cautions, "Physical climate risk from extreme weather events remains unaccounted for in financial markets. Without better knowledge of the risk, the average energy investor can only hope that the next extreme event won't trigger a sudden correction."[200]

Iran News notes that the central message in this latest research is that there is too much "unpriced risk" in the energy market. "Unpriced risk was the main cause of the Great Recession in 2007-2008," Griffin said. "Right now, energy companies shoulder much of that risk. The market needs to better assess risk, and factor a risk of extreme weather into securities prices. Loss of property is what grabs all the headlines, but how are businesses coping? Threats to businesses could disrupt the entire economic system," he said.

For example, excessive-high temperatures, like those experienced in the United States and Europe last summer, can be deadly. Not only do they disrupt agriculture, harm human health, and stunt economic growth, they also can overwhelm and shut down vast parts of energy delivery, as they did in Northern California when Pacific Gas & Electric shut down a delivery during fires and weather that could trigger a fire. Extreme weather can also threaten other services such as water delivery and transportation, which in turn affects businesses, families, and entire cities and regions, sometimes permanently. All of these strain local and broader economies.

"Despite these obvious risks, investors and asset managers have been conspicuously slow to connect physical climate risk to company market valuations," Griffin said in his article.

The study warns that climate-vulnerable locations also factor into risk for energy markets. In the United States, oil refining is located on the Gulf Coast, an area exposed to sea-level rise and intense storms. Oil refining in Benicia and Richmond, in Northern California, can be exposed to coastal flooding. Energy companies' transmission infrastructure is located in arid areas, increasing the risk of damage, such as the destruction from recent wildfires in California. In addition, it is not clear insurance will be available to cover such risks.

---

199 Tasnim, Iran News, *Extreme Weather May Result in Economic Recession,* February 18,2020, https://irannewsdaily.com/2020/02/extreme-weather-may-result-in-economic-recession/

200 Paul Griffin, Nature Energy, Feb 17 2020, *Energy Finance Must Account for Extreme Weather Risk,* https://www.nature.com/articles/s41560-020-0548 2?utm_campaign=Hot%20News&utm_source=hs_email&utm_medium=email&utm_content=83639122&_hsenc=p2ANqtz-

Griffin concluded, "Add to those risks litigation, sanctions and even loss of business from the property destroyed. While proprietary climate risk models may help some firms and organizations better understand future conditions attributable to climate change, extreme weather risk is still highly problematic from a risk estimation standpoint. This is because with climate change, the patterns of the past are no guide to the future, whether it be one year, five years or 20 years out. Investors may also normalize extreme weather impacts over time, discounting their future importance."

What Does Reason Tell Us?

For way too long, the politicians and the people in power have gotten away with not doing anything to fight the climate crisis, but we will make sure that they will not get away with it any longer. We are striking because we have done our homework and they have not."
—Greta Thunberg

These extreme weather events will become increasingly important in the future as they become more and more difficult to predict. In 2020, the world is already 1.1 degrees (Celsius) warmer than it was at the onset of the industrial revolution, and it is predicted to reach up to 3.5 degrees warmer by 2100 at our present rate of consumption. There is a growing chorus from around the world from youth, Indigenous peoples, women, and science demanding action as we face this relentless spite of disasters. The rising toll it is taking on both humans and wildlife is a relentless onslaught on the world's most vulnerable.

UN Call to Action

According to a 2019 UN Climate Change report, despite efforts over the last ten years to move towards sustainable energy, we are failing miserably at reducing emissions. **The science shows that emissions are still going up, not down.** This continuing long-term trend means that future generations will be confronted with increasingly severe impacts of climate change, including rising temperatures, more extreme weather, water stress, sea level rise, mega droughts, and disruption to marine and land ecosystems.

Even if countries meet commitments made under the 2015 Paris Agreement, the world is heading for a 3.2 degrees Celsius global temperature rise over pre-industrial levels, leading to even wider-ranging and more destructive climate impacts.
—United Nations

They quantified that the annual *Emissions Gap Report,* which compares where greenhouse gas emissions are heading versus where they need to be, shows that emissions need to fall by 7.6 percent each year over the next decade, if the world is to get back on track towards the goal of limiting temperature rises to close to 1.5 degrees Celsius.[201]

201 UN News, UN emissions report: *World on course for more than 3 degree spike, even if climate commitments are met,* November 26, 2019, https://news.un.org/en/story/2019/11/1052171

If the world warms by more than 1.5 degrees, we will see more frequent, and intense, climate impacts—as the Intergovernmental Panel on Climate Change (IPCC) has demonstrated in several hard-hitting reports—such as the heat waves and storms witnessed in recent years. The report stresses the urgency of the situation, and urges every country, city, business, and individual to take action now to reduce our impact.

The UNEP (United Nations Environment Program) report calls on all countries to reduce their emissions and substantially increase their "nationally determined contributions" (the commitments made under the Paris Agreement) in 2020, and put into place the policies to implement them. **The lead, however, must be taken by the world's most developed economies (the G20), which contribute some 78 percent of all emissions**: currently, only five of these countries have committed to a long-term zero emissions target (as indicated in figure below). They also report that the summit served as a springboard ahead of crucial 2020 deadlines established by the Paris Agreement, focusing global attention on the climate emergency. And leaders from many countries and sectors stepped up, showing that transition is possible.[202]

The good news coming from this report is that more than 70 countries committed to net zero carbon emissions by 2050, even if major emitters have not yet done so. More than 100 cities did the same, including several of the world's largest. Small island-states together committed to achieve carbon neutrality and to move to 100 percent renewable energy by 2030. And countries from Pakistan to Guatemala, Colombia to Nigeria, New Zealand to Barbados vowed to plant more than 11 billion trees. More than 100 leaders in the private sector committed to accelerating the green economy. A group of the world's largest asset-owners, controlling $2 trillion, pledged to move to carbon-neutral investment portfolios by 2050. This is in addition to a recent call by asset managers representing nearly half the world's invested capital, some $34 trillion, for global leaders to put a meaningful price on carbon and phase out fossil fuel subsidies and thermal coal power globally.

The UN reports that among the many elements that need to be ironed out is the financing of climate action worldwide. Currently, not enough is being done to meet the three climate goals: **reducing emissions 45 percent by 2030, achieving climate neutrality by 2050 (which means a net zero carbon footprint), and stabilizing global temperature rise at 1.5 degrees Celsius by the end of the century.**

Because the clock is ticking on climate change, the world cannot afford to waste more time, and a bold, decisive, ambitious way forward needs to be determined. Unfortunately, population sustainability is the missing link here, which is sadly not being taken seriously at these negotiations. With a population nearing EIGHT BILLION, and adding a quarter of a million more energy users each day, these goals may be out of reach.

Washington Post reporter Brady Dennis points out that world leaders agreed to hold warming to "well below" 2 degrees Celsius compared with preindustrial levels; the current trajectory is nearly twice that. If that rate continues, the result could be widespread, catastrophic effects. Coral reefs, already dying in some places, would probably dissolve in increasingly acidic oceans. Some

202 Mauritania/Freya Morales, UNDP, COP25: UN climate change conference, 5 things you need to know, December 1, 2019, https://news.un.org/en/story/2019/12/105225

coastal cities, already wrestling with flooding, would be constantly inundated by rising seas. In much of the world, severe heat, already intense, could become unbearable. The findings are rather grim, since world leaders have never demonstrated the ability to cut greenhouse gas emissions on such a scale.

> *The world has squandered so much time mustering the action necessary to combat climate change that rapid, unprecedented cuts in greenhouse gas emissions offer the only hope of averting an ever-intensifying cascade of consequences.*
> **—Brady Dennis,** *Washington Post*

Dennis adds, "The past year has brought devastating hurricanes, relentless wildfires and crippling heat waves, prompting millions of protesters to take to the streets to demand more attention to a problem that seems increasingly urgent."[203]

A 2019 *Washington Post* analysis found that roughly 20 percent of the world has already warmed to troubling levels. For example, examination of the fastest-warming places around the world has found that Canada's Magdalen Islands have warmed 2.3 degrees Celsius (4.2 degrees Fahrenheit) since the late 19th century, twice the global average—and they are crumbling into the sea. They assert that slowing this kind of warming in the future will require monumental changes, such as phasing out gas-powered cars, halting the construction of coal-fired power plants, and overhauling how humans grow food and manage land. Sadly, there is no mention of reducing the human population here.

Dennis outlines a number of alarming trends: The world's carbon emissions have moved in the opposite direction. The United States' energy-related CO_2 emissions **rose 2.7 percent in 2018,** after a gradual decline. That increase came as the Trump administration continued to roll back Obama-era climate regulations and made clear that the United States, once a leader in pushing for climate action, would withdraw from the Paris Agreement in 2020. Of course, with Biden as U.S. president beginning in 2021, the situation is likely to reverse once again. Investment in renewable energy in the developing world also dropped significantly in 2018, according to an analysis released by BloombergNEF, which tracks worldwide energy trends.

"Last year [2018], **developing countries added an extra Texas worth of coal generation,**" said Ethan Zindler, head of Americas for BloombergNEF. "And that's obviously scary. At the same time, China's investment in clean energy projects dropped from $122 billion in 2017 to $86 billion in 2018. In 2019 the intergovernmental World Meteorological Organization reported that levels of greenhouse gases in the atmosphere had hit a record high and that the trend means that future generations will be confronted with increasingly severe impacts of climate change."

According to the UN, the 2019 IPCC report indicates that the ocean and the cryosphere—the frozen parts of the planet—play a critical role for life on Earth. A total of 670 million people in high mountain regions and 680 million people in low-lying coastal zones depend directly on these

203 Brady Dennis, Washington Post, November 26, 2019, https://www.washingtonpost.com/climate-environment/2019/11/26/bleak-report-un-says-drastic-action-is-only-way-avoid-worst-impacts-climate-change/, https://www.washingtonpost.com/podcasts/post-reports/the-canadian-islands-crumbling-into-the-sea/

systems. Four million people live permanently in the Arctic region, and small developing island-states are home to 65 million people.[204]

The report reveals the benefits of ambitious and effective action and, conversely, the escalating costs and risks of delayed action. Increasingly, the ocean is becoming warmer, more acidic, and less productive. Melting glaciers and ice sheets are causing sea level rise, and coastal extreme events are becoming more severe. Urgently reducing greenhouse gas emissions limits the scale of ocean and cryosphere changes. Ecosystems and the livelihoods that depend on them can be preserved.

"The open sea, the Arctic, the Antarctic and the high mountains may seem far away to many people," said Hoesung Lee, chair of the IPCC. "But we depend on them and are influenced by them directly and indirectly in many ways—for weather and climate, for food and water, for energy, trade, transport, recreation and tourism, for health and wellbeing, for culture and identity. If we reduce emissions sharply, consequences for people and their livelihoods will still be challenging, but potentially more manageable for those who are most vulnerable. We increase our ability to build resilience and there will be more benefits for sustainable development."

A key theme of the report was that we must strengthen our ability to adapt. "The world's ocean and cryosphere have been 'taking the heat' from climate change for decades, and consequences for nature and humanity are sweeping and severe," said Ko Barrett, vice-chair of the IPCC.

Of course, nothing from this report should come as a surprise, nor is it telling us anything new. In the *State of the World 2002* report, the Worldwatch Institute warned that there was already evidence that regional climate changes had affected a wide range of physical and biological systems. These changes include glacier shrinkage, permafrost thawing, later freezing and earlier buildup of ice on rivers and lakes, lengthening of mid- to high-latitude growing seasons, shifts of plant and animal ranges, declines of plant and animal populations, and earlier flowering of trees, emergence of insects, and egg-laying by birds.

They pointed out that several natural systems were recognized as especially at risk of irreversible damage, including glaciers, coral reefs and atolls, mangroves, boreal and tropical forests, polar and alpine ecosystems, prairie wetlands, and remnant native grasslands. Climate change will increase existing risk of extinction of the more vulnerable species and the loss of biodiversity, with the extent of the damage increasing with the rate and magnitude of change. The 2020 projected adverse impacts to human systems include the following:

- A reduction in potential crop yields in most tropical and subtropical regions for most temperature increases.
- Decreased water availability for populations in many water-scarce regions, notably in the subtropics.
- An increase in the number of people exposed to vector-borne and water-borne diseases (such as COVID-19, malaria, and cholera) and an increase in heat stress mortality.
- A widespread increase in the risk of flooding for tens of millions of people, due to both increased heavy precipitation events and sea level rise.

204 United Nations IPCC website, *Choices made now are critical for the future of our ocean and cryosphere*, Sept. 25, 2019, https://www.ipcc.ch/2019/09/25/srocc-press-release/

Why Doesn't Reason Reach Us?

> *Our collective failure to act early and hard on climate change means we now must deliver deep cuts to emissions. We need to catch up on the years in which we procrastinated.*
>
> —Inger Andersen, United Nations

Even in 2007, the oil giants were not happy with these scientific reports regarding climate change. According to a Greenpeace USA report, ExxonMobil channeled about $30 million to researchers and activist groups promoting disinformation about global warming over the years. But the oil company pledged to stop such funding in 2007, in response to pressure from shareholder activists. ExxonMobil gave more than $2.3 million to members of Congress and corporate lobbying group that deny climate change and block efforts to fight climate change—eight years after pledging to stop its funding of climate denial, the *Guardian* learned.[205]

In 2018, *Guardian* reporter Suzanne Goldberg noted, "Climate denial— from Republicans in Congress and lobby groups operating at the state level—is seen as a major obstacle to US and global efforts to fight climate change, closing off the possibility of federal and state regulations cutting greenhouse gas emissions and the ability to plan for a future of sea-level rise and extreme weather." In a statement to the *Guardian* in 2018, Exxon spokesman Richard Keil reiterated, "ExxonMobil does not fund climate denial."

We have known for some time that Earth's checks and balances have been disrupted. In 2007, the *Independent*'s Chris Rapley asked, "What do the following have in common: the carbon dioxide content of the atmosphere, Earth's average temperature and the size of the human population? Answer: each was, for a long period of Earth's history, held in a state of equilibrium. Whether it's the burning of fossil fuels versus the rate at which plants absorb carbon, or the heat absorbed from sunshine versus the heat reflected back into space, or global birth rates versus death rates—each is governed by the difference between an inflow and an outflow, and even small imbalances can have large effects. At present, all of these three are out of balance as a result of human actions. **And each of these imbalances is creating a major problem.**"[206] This is especially evident when we consider in the exponential growth rate, which has been highlighted with the COVID-19 pandemic exponential growth rate.

These imbalances were discussed again at the 2009 UN Climate Conference in Copenhagen, where governments of the world united at the biggest summit on climate change ever to take place. The question on everyone's mind was, "Will this event prove to be more successful than those of the past, which produced nothing more than business as usual?" Unfortunately, the honest answer is NO, as the Growth Lobby [corporate, political, and economic forces supporting growth] and the Vatican managed to derail efforts to deal with climate change effectively or to stabilize human

205 Suzanne Goldenberg, The Guardian, *ExxonMobil gave millions to climate-denying lawmakers despite pledge,* Feb 14, 2018, https://www.theguardian.com/environment/2015/jul/15/exxon-mobil-gave-millions-climate-denying-lawmakers

206 Chris Rapley, The Independent, *This Planet Ain't Big Enough for the 6,500,000,000,* June 27,2007, https://www.resilience.org/stories/2007-06-30/population-environment-june-30/

population. It appears that history has yet again repeated itself in 2019 as greenhouse gas emissions continue to rise, even though the stakes are much higher and the urgency is much greater.

So why does this keep happening? Protecting the health of our planet and its people—our global commons—requires all our best efforts. So does the task of creating gender equality, caring for the most vulnerable on the planet, ensuring social justice, and securing economic well-being. This requires international action beyond what nations or markets can provide by themselves. This is where we look to the United Nations for leadership because only the United Nations can take on the issues that affect us all, that shape the fate of the entire Earth and its peoples. These are big responsibilities, and the UN needs international laws and an international justice system prepared to help carry out these responsibilities. These are powerful concepts that seem to be lacking here, especially when it comes to protecting our planet.

At present, there is no law that encompasses a crime against nature of the magnitude of the climate disruption, extinction of species, or the many other disasters occurring due to overpopulation and overconsumption. Until we have a law to prosecute those who destroy the planet, corporations and governments will never be held accountable for their crimes. Polly Higgins, international environmental lawyer, states, "We do not currently have a legal crime in place that fits this description but there is one fast looming on the horizon and that crime is **ecocide.**"[207]

> *Currently there is no law to prosecute those who are destroying the planet. Instead, climate campaigners do not have the support of the judiciary in preventing the corporate ecocide that is daily occurring under our very noses.*
> **—Polly Higgins, International Environmental Lawyer**

Ecocide is the extensive destruction of, damage to, or loss of ecosystems. In 2010, Higgins submitted a proposal to the United Nations to accept "ecocide" as a fifth "crime against peace," which could be tried at the International Criminal Court (ICC). Adoption of this law by the UN and enforcement by the UN International Court would be the two basic buiding blocks of a modern globalized society that are sadly missing here. As of 2021, the UN still had implemented no such law. Reforming the United Nations must be a priority if we hope to respond effectively to the challenges of our time and make any progress on climate change.

Higgins explains, "Ecocide is permitted (as genocide was in Nazi Germany) by the government and, by dint of the global reach of modern-day transnational business, every government in the world. Corporate ecocide has now reached a point where we stand on the brink of collapse of our ecosystems, triggering the death of many millions in the face of human-aggravated cataclysmic tragedies."

Over a billion of the world's poorest are living in poverty now, with the threat of starvation and disease looming. Over a billion animals have perished in Australia's 2019 forest fires alone, with billions more the victims of widespread human-caused extinctions. Consequently, it is abundantly clear that ecocide is responsible for much of this tragedy, and those responsible must be held

207 Polly Higgins, Why we need a law on ecocide, Jan. 5, 2011, https://www.theguardian.com/environment/cif-green/2011/jan/05/ecocide-law-ratcliffe

accountable. If not, tyranny—the cruel, unacceptable, or arbitrary use of power that is oblivious to consequence—will continue to dictate the laws we live by.

One-Dimensional Solutions Cannot Address Multidimensional Problems

On Social Europe's website, Gro Harlem Brundtland weighs in on the "changing climate on climate change" and how she sees everything as being interconnected. She emphasizes the complexity of nature and recognizes that one-dimensional solutions cannot address multidimensional problems like those we currently face. She is hopeful that with China now working to reverse the growth in its greenhouse-gas emissions and the European Union setting a high bar for action, other developing countries will find it increasingly hard to argue against controlling their own emissions.

Brundtland points out, "Indeed, today's challenges are seldom simply environmental, social, or economic, and their solutions do not lie within the area of competence of a single government ministry. Without broad, multidisciplinary impact analysis, such narrow thinking can lead to new problems. This is particularly true of climate change. Fortunately, a growing realization that rising global temperatures are not simply an environmental concern provides reason to hope that world leaders are finally ready to address the problem in an effective way."[208]

It is her contention that the **Sustainable Development Goals (SDGs) represent a quantum leap forward** from their predecessor, the Millennium Development Goals, in that they set global targets and embed sustainability in every aspect of policy and practice. However, she warns that there is no guarantee that the agreement reached in Paris will work. As former UN Secretary-General Kofi Annan stated, "If we fail to tackle climate change, the worst effects will be suffered by future generations and by poor countries far from global power centers."

> *All too frequently, leaders will concern themselves with matters that are closest at hand, while the most serious issues are often more distant—geographically or in time.*
> **—Kofi Annan, former UN Secretary-General**

But the SDG targets are unlikely to be met if world leaders are unable to forge a credible accord to limit the rise in global temperatures to 2 degrees Celsius. A stable climate provides the underpinnings for poverty reduction, prosperity, and the rule of law—in short, human development. The UN has emphasized that of the 17 SDGs, Goal #5 to achieve gender equality and empower all women and girls is pivotal, for **all the SDGs depend on the achievement of Goal 5**. Gender equality by 2030 requires urgent action to eliminate the many root causes of discrimination that still curtail women's rights in private and public spheres. This would include tackling sexual abuse and family planning services if any real progress is to be made.

Of course, some countries don't feel that these goals apply to them. *The Hill* reports that U.S. carbon emissions are expected to fall just 4 percent by 2050, according to data released in 2020 by

208 Gro Harlem Brundtland, Social Europe, *The Changing Climate On Climate Change*, September 18, 2015, Social Europehttps://www.socialeurope.eu/the-changing-climate-on-climate-change

the U.S. Energy Information Administration (EIA), falling far short of the changes scientists say are necessary to avoid the worst effects of climate change.[209]

Many environmental groups have rallied around reports from the United Nations, as well as the U.S.'s Fourth National Climate Assessment, calling for the country to take dramatic action to reduce carbon emissions before 2030. "This projection of relentless climate pollution is nothing short of terrifying," the Center for Biological Diversity said in a statement. "With Trump officials crippling emissions rules, climate-friendly lawmakers must build support for truly bold policies that avert the bleak future predicted by the EIA. We need much stronger measures."

Former U.S. Secretary of Energy Steven Chu points out the flaw in Trump's thinking in a *Forbes* article by Jeff McMahon. Chu stated, "Greenhouse-gas emitting is a lot like cigarette smoking, you can do it for decades before the cancer shows up, and then it's very difficult to recover. No one explains the delay between emissions and the climate changes they produce, nor the enduring nature of those changes. Once it's carbon dioxide, some of the models are saying that circulates with a half-life of about 10,000 years. So don't think 2100, think 12100. Let that sink in. We never talk about beyond 2100. So the longevity of this is going to be much longer than the next century or the next millennium."[210]

Like cigarette smoking, climate change is likely to pose many health issues in the future. Global Citizen reporter Catherine Caruso tells us that with a new decade in full swing, the World Health Organization (WHO) has released a list of pressing health challenges for the next ten years, developed by public health and policy experts around the world. She states, "The United Nations General Assembly has labeled the 2020s as 'the decade of action.' Addressing global health concerns and crisis head-on and finding ways to support nations that require aid will be vital in ensuring significant action is taken."[211]

Caruso points out that nearly seven million people die from air pollution every year. Natural disasters and catastrophic weather events caused or fuelled by climate change worsen the spread of disease and can lead to or inflame malnutrition. Outbreaks of diseases are very difficult to treat or contain in countries experiencing conflict or crisis. Health care, medication, reproductive health services, and vaccines should be made accessible to people of all socio-economic backgrounds. Nearly one-third of the world's population doesn't have access to quality medicine and vaccines, which puts millions of lives in danger, as we have seen with COVID-19.

---

209 Rebecca Beitsch, The Hill News, *Carbon emissions will fall just 4 percent by 2050, according to government projections,* Jan. 29, 2020, https://thehill.com/policy/energy-environment/480548-carbon-emissions-will-fall-just-4-percent-by-2050-according-to

210 Jeff McMahon, Forbes, *Recent Carbon Emissions Will Affect The Atmosphere For 10,000 Years,* Steven Chu Says, Apr 3, 2019, https://www.forbes.com/sites/jeffmcmahon/2019/04/03/recent-carbon-emissions-will-affect-the-atmosphere-for-10000-years-steven-chu-says/#27d7761f272d

211 Catherine Carso, Global Citizen, *The World Health Organization released a list of health issues to be addressed in the next 10 years,* Jan. 16, 2020, https://www.globalcitizen.org/en/content/most-urgent-health-challenges-for-the-2020s/

Arctic Ice Loss Is the Canary in the Coal Mine

Washington Post reporter Chris Mooney tells us that the Arctic Ocean has lost 95 percent of its oldest and thickest floating sea ice, according the U.S. National Oceanic and Atmospheric Administration's annual Arctic Report Card—a startling sign of what's to come. He warns that if the thinning trend continues, scientists fear an added boost to global warming.[212]

The finding suggests that over the past three decades of global warming, the sea at the top of the world has already morphed into a new and very different state, with major implications not only for creatures such as walruses and polar bears but, in the long term, perhaps for the pace of global warming itself. Mooney explains that the oldest ice can be thought of as a kind of glue that holds the Arctic together and, through its relative permanence, helps keep the Arctic cold even in long summers.

"The younger the ice, the thinner the ice, the easier it is to go away," said Don Perovich, a scientist at Dartmouth College who coordinated the sea ice section of the 2018 yearly report. If the Arctic begins to experience entirely ice-free summers, scientists say, the planet will warm even more, as the dark ocean water absorbs large amounts of solar heating that used to be deflected by the cover of ice. The new findings were published as climate negotiators in Poland are trying to reach a global consensus on how to address climate change.[213]

> *The Arctic is often cited as the canary in the coal mine for climate warming ... and now as a sign of climate warming, the canary has died.*
> **—Dr. Jay Zwally, NASA climate scientist**

Malte Humpert reports, "As the Arctic Ocean has experienced a rapid loss of sea ice over the past decade or so, numerous studies have aimed at predicting when the region could experience its first 'ice-free' summer season. Especially after the dramatic drops in ice extent during the summer of 2007 and again in 2012, many scientists forecast a continued rapid decline of ice coverage. Now, researchers at UCLA have made an attempt to narrow down the prediction window to a 25-year period in a new study published by *Nature Climate Change*. The new study predicts an 'ice-free' Arctic, meaning ice-extent of just one million square kilometers, between 2044 and 2067. Only the most resilient ice to the north of Greenland and in the Canadian archipelago would survive, with the rest of the Arctic Ocean practically ice-free for part of the year."

This **information is extremely important** because the amount of the Arctic Ocean covered in ice is crucial not only to weather patterns in the lower latitudes, including Western Europe, but also because it serves as a **global cooling mechanism**. "Arctic sea ice is a key component of the earth system because of its highly reflective nature, which keeps the global climate relatively cool," explains Chad Thackeray, lead author at UCLA.

---

212 Chris Mooney, Washington Post, *The Arctic Ocean has lost 95 percent of its oldest ice — a startling sign of what's to come,* Dec. 11, 2018, https://www.washingtonpost.com/energy-environment/2018/12/11/arctic-is-even-worse-shape-than-you-realize/

213 Malte Humpert, High North News, *New Study Narrows Window for Ice Free Arctic To As Early As 2044,* Nov 28, 2019, https://www.highnorthnews.com/en/new-study-narrows-window-ice-free-arctic-early-2044

Global environment editor for the *Guardian*, Jonathan Watts, warns that climate change will also affect melting permafrost in the Arctic and will have a $70 trillion climate impact, according to the most advanced study yet of the economic consequences of a melting Arctic.[214]

Watts notes, "Greenhouse gases, which have been frozen below the soil for centuries, have already begun to escape. So far the impact is small. Ten gigatonnes of carbon have been released from the permafrost but this source of emissions will grow rapidly once temperatures rise beyond 1.5C. If countries fail to improve on their Paris Agreement commitments, this feedback mechanism, combined with a loss of heat-deflecting white ice, **will cause a near 5% amplification of global warming and its associated costs.**" The authors say their study is the first to calculate the economic impact of permafrost melt, in a paper published in the March 2020 edition of *Nature*.

On the current trajectory of at least 3 degrees Celsius of warming by the end of the century, melting permafrost is expected to discharge up to 280 gigatonnes of carbon dioxide and 3 gigatonnes of methane, which has a climate effect that is 10 to 20 times stronger than CO_2.

"It's disheartening that we have this in front of us," said Dmitry Yumashev of Lancaster University. "Even at 1.5C to 2C, there are impacts and costs due to thawing permafrost. But they are considerably lower for these scenarios compared to business as usual. **We have the technology and policy instruments to limit the warming but we are not moving fast enough.**" Scientists are starting to pay tribute to some of our tragic losses as the damages accumulate. Gizmodo reporter Brian Kahn tells us that in 2019, scientists wrote a eulogy for Iceland's first glacier lost to climate change. He states, "That may sound like an Onion headline, but alas, it is not. We've reached the point in our wild planetary experiment where humans are memorializing the things we're knowingly wiping out."[215]

Kahn commiserates that **Iceland lost its first glacier to rising temperatures**. Now, scientists from Rice University and Iceland are planning to install a plaque near the sad pile of ice and snow formerly known as Ok Glacier. The researchers say it's the first memorial to a disappearing glacier, but climate change ensures it almost certainly will not be the last. The plaque contains a melancholy message for future generations.

Ok is the first Icelandic glacier to lose its status as a glacier. In the next 200 years all our glaciers are expected to follow the same path. This monument is to acknowledge that we know what is happening and know what needs to be done. Only you know if we did it.

—Rice University and Iceland, August 2019, 415 ppm CO_2

Climate and Deadly World Weather Events

We do know that the last decade has seen the most extreme weather conditions within recorded history, corresponding with the highest population levels in recorded history. This scenario has led

214 Jonathan Watts, The Guardian, *Melting permafrost in Arctic will have $70tn climate impact – study*, Apr 23, 2019, https://www.theguardian.com/environment/2019/apr/23/melting-permafrost-in-arctic-will-have-70tn-climate-impact-study

215 Brian Kahn, Gizmodo, *Scientists Wrote a Eulogy for Icelands First Glacier Lost to Climate Change*, July 20, 2019 https://gizmodo.com/scientists-wrote-a-eulogy-for-icelands-first-glacier-lo-1836542745

to increased species extinction, the spread of arid regions, and stronger floods, droughts, and forest fires. Scientists have acknowledged that we now need to adapt to the new reality. **Extreme weather is the new normal in this century**. They suggest that the sooner that is recognized, the sooner we can set about creating the tools and systems that can make extreme weather events less deadly.

With the terrible human and wildlife toll from forest fires in 2019 behind us, people might be thinking that the worst must be over. Surely we could not see such a devastating year repeated? A year when the flooding, typhoons, and earthquakes that usually make headlines became sideshows to the devastating forest fires in Australia. In March, Typhoon Idai claimed more than 900 lives in Africa, the heat wave that struck Japan in July killed more than 160 people, and successive droughts and floods in Europe and Asia also took their toll. Together these disasters have killed an unprecedented number of people and wildlife.

But what if this is just the beginning? We already have enough data from the last decade to know that natural disasters kill and displace more and more people each year. The United Nations has repeatedly predicted more drought and reduced agriculture yields in Africa, widespread heavy flooding in Asia, more wildfires in Australia, and stronger and more frequent storms in the Caribbean, and these predictions are playing out right now before our eyes.

The number of people affected by drought alone is increasing by around five million every year. Liam Gilliver with *Plant Based News* tells us, "Residents of Fairbourne, a seaside village in Wales, have been branded as the **UK's 'first climate change refugees'** after they were told to leave their homes due to the threat of coastal flooding. All 450 houses, amongst shops, pub, and post office, will be 'decommissioned' by 2054 as sea-levels continue to rise. It really is getting to a point where coastal towns like ours are getting into severe danger." [216]

Business as Usual

How did we allow things to get this bad? Kate Yoder from Grist explains that one reason is that the oil industry has pumped Americans full of fake news. She questions, "The world has known about the dangers of climate change for decades—so why are oil and gas companies still drilling for crude as if there's no tomorrow? There's no simple answer. But any explanation would have to give some credit to the wizards of public relations. For more than a century, these spinmasters downplayed misdeeds, twisted facts, and cajoled the media into mimicking their talking points."[217]

Yoder points out that creating a cloud of confusion around established science is one of their well-known tactics. Exxon and the coal industry knew about global warming as early as the 1960s; instead of telling the public, they spread doubt about the science behind it. That's just one facet of the **fossil fuel industry's propaganda machine**. But it appears that recently that has all started to

216 Liam Gilliver, Plant Based News, *UK's 'First Climate Change Refugees' Told To Leave Their Homes Due To Threat Of Flooding*, Mar 2, 2020, https://www.plantbasednews.org/news/uk-first-climate-change-refugees
217 Kate Yoder, Grist, How the oil industry pumped Americans full of fake news, Feb 7, 2020, https://grist.org/climate/how-the-oil-industry-pumped-americans-full-of-fake news/?utm_source=newsletter&utm_medium=email&utm_campaign=beacon

change. Yoder notes that many in the media have decided to stop playing along. And there are other signs that the tide is turning against the oil industry. Once the world's most valuable company, Exxon's stock has dropped by a third over the last five years, wiping away nearly $200 billion in market value. Jim Cramer, the loudmouth host of CNBC's *Mad Money*, recently said that it's time to ditch oil stocks. Even public relations companies are now taking their services elsewhere.

Unfortunately, a similar scenario has played out with the population issue, as the growth industry and the Vatican continue to use fake news to deny the connection between population and climate change—or any of the critical world issues (discussed further in Chapter 11). A lot of the PR that exists today was specifically designed to promote economic and population growth. Playing on established myths and taboos and strategically creating doubt about the facts and science surrounding population has proven to be very effective. This **highlights the need for media and governments to demystify the causes and consequences of our population crisis**, and expose the myths and taboos for the deceptions they are.

In this situation as well, the public is largely unaware of the counter-rebellion under way, designed to create distrust for those brave enough to tackle the overpopulation pandemic plaguing the planet. Those PR magicians have been quite successful in influencing the media, politicians, corporations, and especially the working class. This fake news lingers on the Internet for decades, and is promoted through cleverly designed TV ads and shows promoting a false utopian-like lifestyle for those choosing the traditional family with many children, like *The Brady Bunch* or *The Waltons*.

The world's corporate elite know that population is the missing link preventing the planet from achieving a sustainable lifestyle, but they also are confident that it is future generations and the poor that will have to deal with the worst of it. And they are very willing to incinerate their great-grandchildren along with all the dwindling wildlife. So instead of dealing with population, they choose to focus on turning off lights and buying a Prius.

According to Dan Gearino with *Inside Climate News*, even a renewable energy leader like Xcel, one of the first to pledge net zero emissions by mid-century, is finding it hard to end coal without adding natural gas. This raises questions about any utility's ability to break from coal without adding new carbon energy in other forms, mainly natural gas. Gearino explains that Xcel, which serves 3.6 million customers in eight states, laid out a detailed proposal in 2019 for the Upper Midwest part of its territory. **It is proposing to retire coal plants early, extend the life of a nuclear plant, and dramatically expand solar and wind energy.**

"They are putting some very tangible flesh on the bones to get to their 2050 goal," said Kevin Lee, director of the climate and energy program for the Minnesota Center for Environmental Advocacy. It's a "pretty monumental thing," he said.

But Gearino warns, "Xcel's plan would also **expand its use of natural gas** by purchasing a gas plant and converting one of its coal plants to run on gas. The utility says it needs the fossil fuel to maintain reliability.... Four other large utilities issued plans in 2019 to get to net-zero emissions, or close to it, but most of them also have plans to **expand natural gas** in the coming years... ...That decision is dividing local activists and is a microcosm of the broader debate around gas's future that has become pervasive in the fight over the nation's energy economy, as scientists warn about locking in decades of planet-warming greenhouse gas emissions."[218]

So betting all of our money on alternate energy to reduce the use of fossil fuels, when we only have a ten-year window to show results, is a real longshot. However, if we put this same amount of funding and effort into moving towards a sustainable population, we would improve all the critical world issues at the same time. This would be a priority that could pay off sooner and be much easier to accomplish—if we start immediately. If achieving a sustainable population were in the works at the same time as moving towards reducing energy consumption, there would be a much greater chance of success by 2030. **This is a plan I'd be willing to bet my boots on.**

The High Cost of Green Energy

One reason I'm not betting on renewable energy is that most forms of renewables have their own host of drawbacks, if you consider wind, hydro dams, and nuclear disasters, for a start. Take windmills, for example—we know that they kill millions of birds, bats, and insects every year. According to Seth Slabaugh with *Star Press*, migratory bats are dropping like flies, with some species almost impossible to find in 2020. He cites university biologist Tim Carter: "People call this green energy," he said recently to a crowd of bird lovers at the Kennedy Library. "I call it red energy.... A single wind farm can kill 4,000 bats in a single season," he said. "The 150-foot-long blade of a wind turbine might not look like it's moving fast, but on a windy day, it can complete one revolution in four seconds, which equates to the tip of the blade traveling more than 200 mph."[219]

Figure 8

The plight of the iconic bald eagle is just as perilous. *The Blade* tells us that in 2020, property owners in Ohio have legally challenged a proposed wind turbine fearing degradation of the landscape, noise, health risks, reduced property value, and dangerous turbine breakdowns. But the most compelling reason, and their strongest ally in this battle, is the bald eagle—the U.S. national symbol revered for its strength, freedom, majestic appearance, and long life.[220]

The numerous wind projects planned by Apex Clean Energy could be enough to undo all the progress that has been made. In fact, according to Save The Eagles International, wind farms kill 10

218 Dan Gearino, Inside Climate News, *Utilities Have Big Plans to Cut Emissions, But They're Struggling to Shed Fossil Fuels,* Jan 6, 2020, https://insideclimatenews.org/news/06012020/xcel-utility-renewable-energy-100-percent-roadmap-2019-year-review-duke-dte-southern-pseg-natural-gas-coal

219 Seth Slabaugh, Star Press, Professor: Many of Indiana's bats are 'dropping like flies', January 26, 2020, https://www.wind-watch.org/news/2020/01/26/professor-many-of-indianas-bats-are-dropping-like-flies/

220 Matt Markey, The Blade, Outdoors: Bald eagles fly into turbulent wind farm debate, January 25, 2020,www.toledoblade.com

to 20 times more birds than previously thought. Contrary to what we are told, wind farms will cause the extinction of many bird and bat species as they are slaughtering millions annually. Meanwhile, governments are issuing 30-year permits for the legal right of "taking" (killing) bald and golden eagles as part of the wind industry's doing business as usual. This is happening all over the world.[221]

In 2012, the Spanish Ornithological Society (SEO/Birdlife) reviewed actual carcass counts from 136 monitoring studies. They concluded that Spain's 18,000 **wind turbines** are killing 6 to 18 million birds and bats yearly. Extrapolating from these little-publicized studies, it is estimated that U.S. wind farms are killing 13 to 39 million birds and bats every year. The number of wind projects has doubled since then, so it is little wonder that many of our feathered friends could face extinction by 2050. However, Save the Eagles emphasizes, "This carnage is being covered up by self-serving and/or politically motivated government agencies, wind industry lobbyists, environmental groups and ornithologists, under a pile of misleading studies paid for with more taxpayer money."

Studies done in 2020 are naturally showing even more disturbing findings. Helena Horton with the *Telegraph* asserts that the wind farms built to tackle climate change could be the final nail in the coffin for seabirds, which have faced a 70 percent drop worldwide since the 1970s, and numbers continue to fall.[222]

Wind energy facilities kill a significant number of bats far exceeding any documented natural or human-caused sources of mortality in the affected species.
Biologist Paul Cryan, US Geological Survey

In 2019, *Forbes* contributor Michael Shellenberger pointed out that in many countries, wind turbines pose the single greatest threat to endangered bird and bat species. He adds, "In 2017, a team of scientists warned that the hoary bat, a migratory species, could go extinct if the expansion of wind farms continues. *Come on*, you might be thinking. *Surely there are greater threats to migratory bats than wind turbines?* **There aren't**."[223]

For decades, the wind industry has put out a steady stream of grossly misleading information about its wildlife impacts, claiming they are relatively low.

The decline of insect populations may be worsening the threat to endangered bird and bat species. "There is strong evidence that many insect populations are under serious threat and are declining in many places across the globe," notes Extinction Rebellion. "A 27-year long population monitoring study in Germany revealed a dramatic 76% decline in flying insect biomass."[224]

Shellenberger emphasizes, "What Extinction Rebellion does not mention is that scientists in Germany say wind turbines appear to be contributing significantly to what it calls the "insect die-off." Dr. Franz Trieb of the Institute of Engineering Thermodynamics concludes that a "rough but

221 Save The Eagles International, *Wind turbines are actually slaughtering millions of birds and bats annually,* 2020, http://savetheeaglesinternational.org/new/us-windfarms-kill-10-20-times-more-than-previously-thought.html

222 Helena Horton, The Telegraph, *Wind farms built to tackle climate change could be final nail in coffin for seabirds, RSPB warns,* 15 January 2020, https://www.wind-watch.org/news/2020/01/16/wind-farms-built-to-tackle-climate-change-could-be-final-nail-in-coffin-for-seabirds-rspb-warns

223 Michael Shellenberger, Forbes Contributor, *Why Wind Turbines Threaten Endangered Species with Extinction,* Jun 26, 2019, https://www.forbes.com/sites/michaelshellenberger/2019/06/26/why-wind-turbines-threaten-endangered-species-with-extinction/#5306ba5464b4

224 ibid

conservative estimate of the impact of wind farms on flying insects in Germany" is a "loss of **about 1.2 trillion insects of different species per year**," which "could be relevant for population stability." This is an astonishing *one-third* of the total annual insect migration in southern England.

Insects, birds, bats, and wind farm developers are all attracted to the same thing, Trieb notes: high winds. "Wind-rich migration trails used by insects for millions of years are increasingly seamed by wind farms."

Dead insects on wind turbine blades are as visible as dead insects on a car windshield. Scientists have reported the significant build-up of dead insects on wind turbine blades for three decades and in different regions around the world. In the 1990s, the wind industry claimed its turbine blades were too high to threaten flying insects. It also claimed insects flew too slowly to be impacted. Both claims turned out to be completely wrong. Researchers calculated in 2001 that the **build-up of dead insects on wind turbine blades can reduce the electricity they generate by 50 percent**.

So who is hiding the truth about the carnage of wind farms? The American Bird Conservancy says, "The wind industry and its proponents have contributed to this situation themselves, downplaying its impacts on wildlife while simultaneously overselling the industry's ability to mitigate associated problems."

Shellenberger emphasizes, "Environmental journalists deserve a significant amount of blame for suggesting the problem is either small or has been solved. 'Wind farm works to reduce eagle deaths from old turbines,' reads the headline of a *PBS Newshour* story that typifies journalistic bias. **But greater responsibility for the threatened extinction of birds and bats lies with environmentalists who promote wind energy as good for the environment.**"

Other animals, like bears, desert tortoises, and migratory mammals, are also being significantly impacted by wind farms and solar projects, with opponents challenging some projects all the way to federal court. As increasing numbers of renewable energy generation facilities are installed around the world, we are starting to realize the harsh realities of their disturbing operational impacts.[225]

Also, Dusty Miller with Black & Veatch tells us that large concentrating **solar plants** use "power towers" that consist of hundreds of thousands of computer-controlled mirrors to track the sun throughout the day, reflecting the sunlight to boilers at the tops of two or three approximately 450-foot-tall towers. Birds, insects, and bats that fly through the highly concentrated, high-temperature (800 to 1,000 degrees F) solar beams—sometimes called solar flux—at concentrating solar plants have been given the name "streamers" by operators of these facilities. When the insects, birds, and bats fly through these beams, they are ignited in midair, creating a plume of smoke, or streamer. The animals may be killed by the heat, by the force of falling to the ground, or by a waiting predator.[226]

This apparently unforeseen impact is alarming, Miller notes, because of the frequency of the streamer phenomenon that has been observed—**an estimated one every two minutes**—by U.S.

225 Michael Bielawski, True North Reports, Years later, Deerfield Wind impact on bear habitat in question , January 31, 2020, https://www.wind-watch.org/news/2020/02/01/years-later-deerfield-wind-impact-on-bear-habitat-in-question/

226 Dusty Miller, Black & Veatch, *Impact of Solar Energy on Wildlife Is an Emerging Environmental Issue,* January 1, 2017, https://www.bv.com/perspectives/impact-solar-energy-wildlife-emerging-environmental-issue

Fish and Wildlife Service (USFWS) law enforcement personnel at a large concentrated solar project in California. Then there is the "lake effect" of large fields of thousands of solar panels that convert sunlight into electricity that have been built in numerous countries, including China and the U.S. They can fool birds into changing flight direction, sometimes during migration, to approach them because they appear to be lakes from a distance.

Miller adds, "Many of the birds that have been killed at these large solar sites are waterbirds, which indicates that these birds fly to solar fields and realize too late in their descent that the solar panels are not water. The water birds then collide with the solar panels and are critically wounded or killed. Some water birds also have great difficulty taking off from non-water surfaces, which could leave them stranded in desert areas without food, water or shelter."

Although there are several mitigating measures being explored, they may be too little too late to prevent major species decline, especially with many more solar projects still being planned. Creation of habitat areas located outside large solar project areas, to draw birds away from the solar projects, has also been proposed. However, these present migratory routes have been established over hundreds of years. So **governments are considering selling solar companies permits, similar to the wind farm permits that allow companies to kill endangered California condors**, which would allow them to legally injure or kill a specified number of birds, bats, and insects. Alarmingly, none of these mitigation measures for wind or solar projects are designed to prevent the insect apocalypse that is occurring around the world.

Yet, renewables advocates would like to see wind and solar technologies grow exponentially — from today's 5 percent globally to somewhere between 30 and 100 percent of our electricity supply. What might the wildlife impacts of a 6- to 20-fold increase in solar and wind be?

The issue has caused division among environmentalists, and is all the more reason that we should be looking at achieving a sustainable human population, drastically reducing energy consumption, moving towards a plant-based diet, and planting trees and industrial hemp as win-win solutions to our climate crisis instead—they have no down side, are less expensive, and are easier to accomplish without destroying our wildlife. They are also fundamental to achieving the UN Sustainable Development Goals.

Many environmental groups and other proponents of the solar and wind industry have contributed to this unfortunate situation. While eco groups claim to put wildlife first, this doesn't seem to apply when **it comes to creating jobs in the renewable energy sector that they have made commitments to**—the options that are an easy sell to **misinformed citizens**. This is a **betrayal of the public trust.**

Financial Times reporter Simon Kuper adds, "Electric vehicles won't save us: their lifetime emissions are unacceptably high. (Mining lithium, making car batteries, shipping cars and generating most electricity isn't clean.)"[227]

227 Simon Kuper, Financial Times, *The myth of green growth,* October 23 2019, https://www.ft.com/content/47b0917c-f523-11e9-a79c-bc9acae3b654

Solutions and Successes

There are few times in human history where voices are amplified at such pivotal moments and in such transformational ways—but Greta Thunberg has become a leader of our time.... History will judge us for what we do today to help guarantee that future generations can enjoy the same livable planet that we have so clearly taken for granted.
—Leonardo DiCaprio

When it comes to climate successes in the last decade, one image dominates the landscape—that is the image of a furious young Swedish girl confronting the UN head-on, and tackling the climate emergency threatening our planet. *LiveKindly* reporter Jill Ettinger states, "She warned them that young people of today will be watching and holding them accountable for climate offenses."

That image is what has fueled the Fridays for Future global movement, empowering youth to rally behind a climate revolution. Ettinger notes that 16-year-old vegan Greta Thunberg and 44-year-old activist Leonardo DiCaprio have committed to each other to fight climate change and deliver the next generation a safe and brighter world.

At the UN in 2019, DiCaprio asserted, "Climate change is not hysteria—it's a fact. Climate change is real, it is happening right now. It is the most urgent threat facing our entire species, and we need to work collectively together and stop procrastinating."[228]

Time Editor-in-Chief Edward Felsenthal writes, "Meaningful change rarely happens without the galvanizing force of influential individuals, and in 2019, the earth's existential crisis found one in Greta Thunberg.... Thunberg has become the biggest voice on the biggest issue facing the planet—and the avatar of a broader generational shift in our culture that is playing out everywhere from the campuses of Hong Kong to the halls of Congress in Washington.... this was the year the climate crisis went from behind the curtain to center stage, from ambient political noise to squarely on the world's agenda, and no one did more to make that happen than Thunberg."[229]

PRNewswire explains how this perfect storm created a legend. Felsenthal writes, "That Thunberg is the youngest individual ever named TIME's Person of the Year says as much about the moment as it does about her.... in this moment when so many traditional institutions seem to be failing us, amid staggering inequality and social upheaval and political paralysis, we are seeing new kinds of influence take hold. It is wielded by people like **Thunberg, leaders with a cause and a phone who don't fit the old rubrics but who connect with us in ways that institutions can't and perhaps never could.**"

What we are seeing here is that common sense and verve far outweigh PhDs and the power elite. Understandably, many lost faith in politics and religion long ago, as these systems have failed the world and proven to lack any reliable answers in a time of crisis. So we are now searching for solutions based on science rather than outdated traditions, unjustified taboos, and self-defeating

228 Jill Ettinger, LIVEKINDLY, *Leonardo DiCaprio and Greta Thunberg Team Up to Stop the Climate Crisis*, Nov 1/19, https://www.livekindly.co/greta-thunberg-leonardo-dicaprio-climate-crisis/

229 Edward Felsenthat, PRNewswire, *TIME names Greta Thunberg the 2019 Person of the Year*, Dec. 11, 2019, https://www.prnewswire.com/news-releases/time-names-the-2019-person-of-the-year-greta-thunberg-300973562.html

myths. It is time we consider the distasteful and inconvenient truth. It is time we include population in the climate equation in order to ensure the best chance of success.

"Our house is on fire," Ms. Thunberg said at a rally. "We will not just stand aside and watch.... Don't invite us here to just tell us how inspiring we are without actually doing anything about it ... the empty promises are the same, the lies are the same and the inaction is the same." It is little wonder that Greta Thunberg was nominated for the Nobel Peace Prize and her speeches have gone viral around the world.

For sounding the alarm about humanity's predatory relationship with the only home we have, for bringing to a fragmented world a voice that transcends backgrounds and borders, for showing us all what it might look like when a new generation leads, Greta Thunberg is TIME's 2019 Person of the Year.
—Time Magazine

In the cover story, *Time*'s Charlotte Alter, Suyin Haynes, and Justin Worland write, "For decades, researchers and activists have struggled to get world leaders to take the climate threat seriously. But this year, an unlikely teenager somehow got the world's attention.... Thunberg is not a leader of any political party or advocacy group. She is neither the first to sound the alarm about the climate crisis nor the most qualified to fix it. She is not a scientist or a politician. She has no access to traditional levers of influence: she's not a billionaire or a princess, a pop star or even an adult. She is an ordinary teenage girl who, in summoning the courage to speak truth to power, became the icon of a generation."

A UK-based climate activist group that holds this same creedo is Extinction Rebellion. In 2019, Grist reporter Jane Hu stated, "Extinction Rebellion carried out a day of international civil disobedience in 60 countries around the world, from Australia to Sri Lanka to the United States, shutting down roads and bridges and splattering fake blood over the cobblestones of Wall Street. The goal: to **call on elected officials and world leaders to 'tell the truth' about the climate emergency** and make concrete plans to address it."[230]

On the other end of the spectrum, we are also hearing encouraging words from the United Nations that the solution exists if we all work together. UN News points out that their emission report sets the measures necessary to reach the 1.5 degree goal by 2030. They specify, "The technology exists, and there is increased understanding of the additional benefits of climate action, in terms of health and the economy. Many governments, cities, businesses and investors are engaged in ambitious **initiatives to lower emissions.**"[231]

Developing countries, which suffer disproportionately from climate change, can learn from successful efforts in developed countries, notes UNEP, and they can even leapfrog them, adopting cleaner technologies at a faster rate.

---

230 Jane C. Hu, Grist.org, *The story behind the Extinction Rebellion symbol at climate protests,* Oct 7, 2019, https://grist.org/article/the-story-behind-the-extinction-rebellion-symbol-at-climate-protests/?utm_source=newsletter&utm_medium=email&utm_campaign=beacon

231 UN News, UN emissions report: World on course for more than 3 degree spike, even if climate commitments are met, November 26, 2019, https://news.un.org/en/story/2019/11/1052171

The UNEP chief said that despite the figures, it was possible to avert disaster: "Because of climate procrastination which we have essentially had during these (past) 10 years, we are looking at a 7.6 per cent reduction every year" in emissions. "**Is that possible? Absolutely**. Will it take political will? Yes. Will we need to have the private sector lean in? Yes. But the science tells us that we can do this."

Alternative Energy is not the Superhero We've Been Sold

One method of moving towards "sustainable" energy has been to harness the wind to produce enough energy to meet the needs of EIGHT BILLION people. However, this strategy has led to the death of millions of bats, birds, and insects every year—what is being called an insect apocalypse by many.

WindWatch.org reports that a wind farm in the Netherlands is the first to fit a system to switch off turbines automatically to guarantee the safety of nearby birds and bats. At a wind farm near the Krammer protected nature reserve, 34 turbines have been fitted with cameras, microphones, and loudspeakers, and mainly cover the area where the rare sea eagle hunts and breeds.

Dutch News notes, "If the camera spots a sea eagle at a distance of 600 metres the turbine is stopped automatically. The detection system also cuts the power when other birds venture too near, such as cranes, spoonbills and great egrets. The wind farm loses between €120,000 and €180,000 a year because of the stoppages. 'It's quite a lot of money. But we really want the birds to stay in the area, particularly the sea eagle,' project manager Gijs van Hout told the broadcaster."[232]

Bats are detected by recording the ultrasonic sounds the animals make. Three such indications of the bat's presence will stop the turbine until no ultrasonic sound is heard for the duration of 15 minutes. While these measures are not protecting the majority of the birds or bats, it is a move in the right direction to mitigate the devastating loss of life these monstrous machines claim every year. That is 34 windmills killing fewer birds, bats, and insects—now **is the will and funding there to equip the thousands of remaining windmills around the world, and the thousands more yet to be built?**

Because alternative energies like wind, solar, nuclear, and hydro present so many of their own negative impacts that are seldom mentioned, I will not be promoting them in this chapter as successes. There are already numerous books out there that explore the many technologies being developed to mitigate the effects of climate change, but few reveal the negative impacts they have on wildlife and the environment. To date, none of them compare to the mostly ignored and already established technology of birth control. While it is true that small-scale or household sustainable energy solutions are preferable to fossil fuels, reducing the demand for energy through reducing the number of consumers would be by far the more logical solution and has no down side. This is not an either/or solution, but both must be implemented simultaneously.

---

232 WindWatch, *Wind farm shuts down turbines to protect birds and bats,* February 7, 2020, https://www.wind-watch.org/news/2020/02/07/wind-farm-shuts-down-turbines-to-protect-birds-and-bats/

Climate and Population

All our environmental problems become easier to solve with fewer people, and harder—and ultimately impossible—to solve with ever more people.
—Sir David Attenborough, Population Matters

Carter Dillard with the Independent Media Institute maintains that humanity's survival on earth starts with having smaller families. Carter Dillard is the founder of HavingKids.org. He served as an Honors Program attorney at the United States Department of Justice, and served with a national security law agency before developing a comprehensive account of reforming family planning for the *Yale Human Rights and Development Journal*.

He has begun to implement the transition to child-centric "Fair Start" family planning, both as a member of the steering committee of the Population Ethics and Policy Research Project, and as a visiting scholar of the Uehiro Center, both at the University of Oxford.[233]

Dillard stated, "Drive less. Fly less. Eat less meat. These are common prescriptions offered by scientists and environmentalists for reducing our individual carbon footprint to help combat climate change. But there's one recommendation frequently left off that list that would have the greatest impact on mitigating the climate crisis and moving toward the United Nations' sustainable development goals: have fewer kids."

> *Living car-free would reduce your annual carbon emissions by 2.4 metric tons, avoiding airplane travel would reduce those emissions by 1.6 metric tons, and eating a plant-based diet would reduce them by 0.8 metric tons. Having one fewer child, however, outpaces them all—reducing an individual's yearly carbon emissions by nearly 60 metric tons.*
> **—Carter Dillard, Fair Start Movement**

According to the Population Media Center, **contraception is not only the greenest technology, but also the cheapest.** They cite an Optimum Population Trust news release saying, "Contraception is almost five times cheaper than conventional green technologies as a means of combating climate change. Each **seven dollars** spent on basic family planning over the next four decades would reduce global CO_2 emissions by more than a ton. To achieve the same result with low-carbon technologies would cost a minimum of **thirty-two dollars.**"[234]

The Hill reporter Ashley Berke agrees, advising that we can fix the climate with smaller families. She notes, "Carbon dioxide and other greenhouse gas emissions are at an all-time high. That growth tracks explosive world population growth, which is the greatest driver of climate change and a threat multiplier. The Green New Deal has justifiably generated excitement and enthusiasm among people who believe its goal of cutting greenhouse-gas emissions to net zero over 10 years

233 Carter Dillard / Independent Media Institute, *Humanity's survival on Earth starts with having smaller families,* February 20, 2020, https://www.alternet.org/2020/02/humanitys-survival-on-earth-starts-with-having-smaller-families/
234 Population Media Center *CONTRACEPTION IS "GREENEST" TECHNOLOGY,* September 16, 2009, https://www.populationmedia.org/2009/09/16/contraception-is-greenest-technology/

is a crucial goal for humankind. However, it could be far more effective in reaching its goal if it addressed family planning and the empowerment of girls and women."[235]

Regardless of how many solar panels or wind turbines we build, and how many people buy a Prius, no green policy will do more to reverse global greenhouse emissions than reducing population growth. Even the most environmentally conscious person creates a significant carbon footprint simply by consuming food, building adequate shelter, and using energy for heating, to name just a few obvious impacts.

Berke advises, "Given the existential threat our planet faces, let's break the taboo on family planning policies. One way to break the taboo is to **think about what future kids need rather than what current parents want**."

Population Matters' Alistair Currie asserts, "We can't eradicate poverty and avert our environmental crisis unless we address population. Having fewer children makes an enormous difference to our carbon emissions and to our environmental footprint. And within the UK, it makes a difference to our quality of life, our public services and in many other respects."[236]

The **good news** according to Currie is that we know what the solutions are: "Ending and reversing population growth is achieved by positive measures: ending poverty, empowering women and girls, providing high quality education, ensuring everyone can use modern contraception and promoting small families. These make people's lives better and **we've seen that formula work time and time again** in the past in individual countries. Now, we have to apply ourselves to it globally, strategically and in a just and equitable way. This isn't just about countries in Africa with high fertility rates—in the UK we produce 60 times the amount of greenhouse gas emissions as someone in Niger. We have to foster the recognition here that choosing fewer children means a better life for the children we have."

BBC News reporter Ted Scheinman has also noticed that couples are rethinking having kids because of climate change. He points out, "There is an emerging generational movement linking climate change to its impact on personal reproductive choices. As people factor climate concerns into future plans, the decision to have a family is becoming increasingly fraught for some: with the way things are going on Earth, how many children is it responsible to have? *Should* you even have kids at all?"[237]

Many girls have nurtured a vision of having a large family throughout their childhood, but as they become aware of the impacts of climate change, the reality sets in that having a child in America would be the single most carbon-intensive choice they could make. They often feel guilty about their high-consumption way of life and want to help on a personal level. So it is becoming increasingly common that as young people marry and decide to revise their family plans, they are giving their parents warning not to expect grandchildren.

We don't need smaller carbon footprints, we need less feet.
—Bill Maher, HBO

235 Ashley Berke, The Hill, *Fix the climate with smaller families*, May 20, 2019, https://thehill.com/opinion/energy-environment/444640-fix-the-climate-with-smaller-families

236 The Standard, World Population Day: How smaller families will reduce our carbon footprint, says Population Matters' Alistair Currie, July 11, 2019, https://www.standard.co.uk/futurelondon/cleanair/world-population-day-smaller-families-population-matters-alistair-currie-a4188056.html

237 Ted Scheinman, BBC News, The couples rethinking kids because of climate change, October 1, 2019, https://www.bbc.com/worklife/article/20190920-the-couples-reconsidering-kids-because-of-climate-change

Scheinman adds, "In the developed world the carbon footprint of a child is roughly 58.6 metric tonnes annually, whereas that of a Malawian child has consistently been estimated between 0.07 and 0.1 metric tonnes annually. A 2019 poll by *Business Insider* reported that almost 38% of Americans aged 18 to 29 believe that couples should consider climate change when deciding to have children."

This wide variety of perspectives would indicate quite simply that we can't add more people without adding more misery, suffering, and stress to an already crippled planet. It is a certainty that unless we get our numbers in line with what Earth can support, all the other good things we do are ultimately doomed to fail. A Population Matters poll in 2018 showed 73 percent of Londoners favoured the introduction of a population strategy in the UK. In most countries, policymakers haven't caught up with that public mood yet—so we must let them know that they need to. We do have a small window of time to take transformative actions to turn things around, and that time starts NOW.

Climate and Politics

People are suffering; people are dying; entire ecosystems are collapsing. We are at the beginning of a mass extinction and all you can talk about is money and fairytales of eternal economic growth. How dare you.

—Greta Thunberg

Why is this tiny country of Bhutan—with a low population and enormous political will—setting an astounding example for the world? CNN reporters Mark Tutton and Katy Scott explain, "High up in the Eastern Himalayas is one of the greenest countries in the world. Sandwiched between China and India, Bhutan spans approximately 14,800 square miles—roughly the size of Maryland. While many nations are struggling to reduce their carbon emissions, the Kingdom of Bhutan is already carbon negative: it takes more greenhouse gasses from the atmosphere than it emits."[238]

> *For the past 46 years the Bhutanese government has opted to measure progress not through its Gross Domestic Product, but through* ***"Gross National Happiness,"*** *which places great emphasis on the protection of the country's rich natural environment.*
> **—Mark Tutton and Katy Scott, CNN News**

Carpeted in forest that covers about 70 percent of the country, Bhutan acts as a natural carbon sink and absorbs three times more CO_2 than it emits. Luxembourg, for example, with a smaller population, emits four times as much. This isolated and relatively undeveloped Buddhist kingdom controls its population and tourism by charging visitors a daily fee of $250 per person to ensure the environment is not spoiled by mass tourism. This nation of around 750,000 people also strives to manage foreign influence so they won't lose their cultural identity.

"Bhutan is the only country in the world that by its own constitution protects its forests," explains Juergen Nagler, of the UN Development Program in Bhutan. Environmental protection is enshrined in the constitution, which states that a minimum of 60 percent of Bhutan's total land

238 226 Mark Tutton and Katy Scott, CNN, *What tiny Bhutan can teach the world about being carbon negative,* October 11, 2018, https://www.cnn.com/2018/10/11/asia/bhutan-carbon-negative/index.html

should be maintained under forest cover for all time. The country even banned logging exports in 1999. What's more, almost all the country's electricity comes from hydropower generated from its mountain streams.

Nagler says that remaining carbon negative is of utmost importance to the Bhutanese as they have a "very high environmental awareness" and "appreciate harmony with the natural environment." He explains that Bhutan is on a "green and low-carbon development pathway" with government initiatives to make the country's agriculture 100 percent organic by 2020 and waste-free by 2030. Clearly, other countries can learn from Bhutan's decision to put the environment first.

> *Individual citizens and businesses are reducing their emissions, but these amount to drops in the ocean when compared to a collective effort supported by a legislated price signal that levels the playing field.*
> **—Rod Mitchell, Citizens Climate Lobby - Australia**

On the opposite end of the scale, one country that desperately needs to learn from Bhutan is environmentally devastated Australia. Rod Mitchell, national chair for Citizens Climate Lobby Australia commiserates, "Tragically, we now have a disaster serious enough to wake up almost all of us to the catastrophe that is unfolding in large parts of **Australia** and advertising itself to the rest of the world. The government now has the opportunity to break with the narrow sectional interests that have been constraining its climate policies and begin to lead in the national interest. It can now move to 'climate rescue' mode."[239]

Mitchell is hopeful that the government will be urgently discussing how best to **respond to the current disaster** and how best to definitively **shift to a zero-carbon economy**. The people of Australia need to know that their government is responding appropriately to both the immediate emergency and to the climate and ecological crisis that is escalating before the planet's eyes. He emphasizes that both are vitally important, the second one more so because it will affect humanity and the environment for generations to come.

In the form of solutions Mitchell offers, "There are a range of policy options available to them. Hopefully they will take the courageous step of seriously considering the long-excluded and oft-demonised option that happens to be one of the best—**putting a price on carbon.** A well-designed carbon price can drive an economy-wide decarbonisation process that will be much smoother than disastrous disruptions forced on us by escalating natural disasters. Business, industry, scientists, economists and citizens are asking for a price signal to guide their daily decisions towards zero carbon."

We know that globally fossil fuel industries are planning to continue exploration and expand production over the next decade. Shareholder expectations would dictate this pathway as long as it is a legal option. However, if government imposed a carbon price system calling for a fade-out of production over a decade, this would compel industry to smoothly transition to zero carbon.

Mitchell points out, "They also need the **true costs of fossil fuels to be revealed and implicit subsidies removed** so that the advantages of zero-carbon alternatives become even clearer. The

---

239 Rod Mitchell, Citizens Climate Lobby Australia, *SPECIAL BLOG FROM AUSTRALIA: Carbon pricing to the rescue – for climate and government,* January 7, 2020, https://canada.citizensclimatelobby.org/special-blog-from-australia-carbon-pricing-to-the-rescue-for-climate-and government/

IMF estimates that Australian fossil fuels are subsidised by US$29 billion a year. And they estimate that 'efficient pricing' from 2015 to 2019 would have reduced global emissions by 28%, almost halved air pollution deaths and increased government revenue by 3.8% of GDP. An efficient carbon price ends these subsidies, slashes emissions and enables the industries of the future to flourish. But we are faced with an industry that naturally does not want to give up its advantage and the legacy of decades of effective lobbying and politicking to keep carbon pricing off the table. Today's bushfires require that it goes back onto the table and that courageous leadership is taken to keep it there."

Citizens Climate Lobby has a policy option ready for consideration and, hopefully, timely adoption. Carbon Fee and Dividend puts a steadily increasing fee on the carbon content of fossil fuels (at the point where they enter the economy) and immediately returns the revenue to Australian households. And it ensures new growth will be mostly in sustainable, low-carbon activities. It gives the government a voter-friendly policy option that enables it to rescue the climate and, in doing so, rescue its severely tarnished credentials as protector of its citizens. A win-win for everyone, but especially for the environment.

Grist reporter Nathanael Johnson also asserts that California could meet its 2030 climate goals, but it would mean **saving money**. He explains, "California is not on track to meet its 2030 benchmark. To reduce state-wide emissions to 40 percent below the 1990 level, as enshrined in law, California needs to double the rate at which it is cutting carbon. How much would all this cost? **Nothing.** *Energy Innovation* calculates that these changes would **save** Californians $7 billion over the next decade, mostly by reducing the gas, oil, and other fuels they have to buy. If you include savings from averted social costs—like pollution-triggered asthma and climate-related disasters—the windfall increases to $21 billion."[240]

That's according to a new report from Energy Innovation, an environmental think tank. Despite a discouraging start, the report suggests that California could still meet its goals by making a few policy changes: things like more renewable energy, more electric cars, and a higher carbon price. Is this possible? Though some skeptics doubted California could reach its 2020 emission goals, they did it four years early and still had one of the strongest economies in the industrial world.

Johnson adds, "California is one giant experiment in combating climate change. The state is trying all sorts of policies to choke off emissions in time to hit a series of ambitious, self-imposed goals. If it is successful, the other 49 states will have a model as they find their way."

The Fixers Get Fired Up!

The Grist's new "Meet the Fixers - Grist 50! 2020" initiative highlights two promising projects certain to make a difference: Jerome Foster, founder and executive director of OneMillionOfUs, promises to get a million young people to the polls.

Grist explains, "As a child, Jerome Foster II loved exploring the Maryland forest. Then he saw *Avatar*—and understood that the natural world he loved was in trouble. His middle-school classmates scoffed, dubbing him 'the climate kid.' Foster didn't care. 'It's an impending crisis and no one's talking about it!' he says. Then Trump was elected, and people began listening. In 10th grade, Foster started

240 Nathanael Johnson, The Beacon, *California could meet its 2030 climate goals — but it would mean saving money,* Jan 16, 2020, https://grist.org/climate/california-could-meet-its-2030-climate-goals-but-it-would-mean-saving-money/?utm_source=newsletter&utm_medium=email&utm_campaign=beacon

The Climate Reporter, a blog that connected him with other young environmental leaders. He helped organize the 2018 Youth Climate marches and started striking in front of the White House every Friday morning before dashing to his internship with U.S. Representative John Lewis."[241]

In 2019, Foster started the advocacy organization OneMillionOfUs with the goal of mobilizing one million young people to vote in 2020. The campaign asked for a "vote pledge" and planned to offer free rides to the polls. "We need politicians to talk to young people," he says, "and see us not just as victims of gun violence and climate change, but as voters who can sway an election."

Other youth-centred climate groups are also weighing in on what the government's role in climate action should be. Sunrise Movement is one group that is building an army of young people to stop climate change and create millions of jobs in the process. Their strategy in the U.S. is to elect candidates down the ballot who can help usher in the movement's vision for the Green New Deal.

When describing this new climate-driven movement, *Vox* reporter Ella Nilsen states, "The new face of climate activism is young, angry—and effective. A growing sea of crusaders known as the Sunrise Movement has helped put climate change on the national agenda. Most aren't even 30.... The 15,000 supporters who have turned out in person have spent the past year occupying the offices and hallways of the US Capitol, state houses, and Democratic National Committee meetings across the country, yelling at the top of their lungs."

> *A third of American men and women aged 20 to 45 cited climate change as a factor in their decision to have fewer children.*
> *—Business Insider* **report - 2019**

Most of these young activists don't dare think too far into the future, as many feel that the uncertainty and bleak prospects of climate change are a weight they must carry with them. Their mission is twofold: trying to force politicians to act on one of the most dire issues facing humankind and building an army of young people to send the message.[242]

These mostly vegetarian activists estimate that 80,000 young people from across the country participated in their movement in 2019—a sharp rise from 2018 when there were just 11. Nilsen notes, "Their methods are straight out of the playbook of the civil rights movement of the 1960s: Frequently, they sing protest songs. They stand quietly as police officers zip-tie their hands behind their backs and lead them into vans for civil disobedience. The new face of climate resistance is young and diverse. It is scared, and it is loud."

In a short amount of time, the Sunrise Movement's assertive tactics have brought about a profound change, forcing climate change and the Green New Deal—their vision to solve it—to become defining issues of the 2020 election. These would-be young voters pushed Democrats running for president to release serious, detailed plans to drastically cut America's fossil fuel emissions.

There are numerous other groups working on the issue, like Zero Hour, Power Shift, Sierra Student Coalition, Youth v. Gov, and Earth Guardians. But the parents aren't being left out as groups like Mothers Out Front and Climate Parents are showing support for their children out on the front lines.

---

241 The Grist, *Meet the Fixers, Grist 50! 2020*, https://grist.org/grist-50/2020/

242 Ella Nilsen, Vox.com, *The new face of climate activism is young, angry – and effective*, Sep 17, 2019, https://www.vox.com/the-highlight/2019/9/10/20847401/sunrise-movement-climate-change-activist-millennials-global-warming

Meet Alexandria Villaseñor, also a youth activist, and she wants the world to go on stike. As founder and executive director of Earth Uprising International in New York, she worked with Greta Thunberg and 14 other young people to file a human-rights violation complaint to the UN against five countries for failing to uphold Paris Agreement pledges.

In 2018, at the age of 14, this young girl had her life turned around when in one week, she attended a leadership summit, hopped on a red-eye, protested in front of the U.N., and starred as a keynote speaker at a Model U.N. conference. In between, she coordinated with fellow activists for Earth Uprising, the climate-education nonprofit she founded in 2019, which now has members in 30 countries. Meanwhile, she'll have another big challenge to face in 2020: starting high school.

The United Nations Feeling the Heat

Phil McKenna with InsideClimate News points us to a little different approach to addressing climate change. He states, "A new United Nations proposal calls for national parks, marine sanctuaries and other **protected areas to cover nearly one-third or more of the planet by 2030** as part of an effort to stop a sixth mass extinction and slow global warming."[243]

> *Tuvaijuittuq, meaning "the ice never melts" in Inuktitut, is now an interim Marine Protected Area. Almost the size of Germany, it will provide a future climate refuge for sea ice-dependent species like narwhal, polar bear and walrus.*
> **—World Wildlife Fund**

The UN Convention on Biological Diversity explains that this proposal aims to halt species extinctions and also limit climate change by protecting critical wildlife habitat and conserving forests, grasslands, and other carbon sinks. This will not be easy, since only 15 percent of all land and 7 percent of oceans is currently protected, according to the United Nations Environment World Conservation Monitoring Centre.

Ecologists hailed the plan as a good starting point, while simultaneously urging that more needs to be done if we are to meet the goals of the Paris Agreement—a further 20 percent as a buffer zone would be required where trees and grasslands woud be conserved as a carbon sink. This will take some bold actions on the part of all countries around the globe. It will also require a reduction in human population if we are to reduce human encroachment on wildlife habitat and return some of this land to nature.

McKenna notes that approximately 190 countries have ratified the Convention on Biological Diversity since it was drafted in 1992. One major exception is the United States, which signed but has not ratified the agreement. The 2030 protected area targets, which could increase or decrease in ambition before being finalized, are anticipated to be adopted by governments at the UN's Convention on Biological Diversity in Kunming, China, in 2021. It is expected that funding to achieve these targets will come from countries, corporations, and private donors.

243 Phil McKenna, Inside Climate News, *UN Proposes Protecting 30% of Earth to Slow Extinctions and Climate Change*, Jan 14, 2020 , https://insideclimatenews.org/news/14012020/biodiversity-treaty-climate-change-marine-sanctuary-conservation-protected-areas-wildlife-habitat-forests

Media Attention

You can be sure that all of these protests have grabbed the attention of the media in a big way. Covering Climate Now is a global collaboration of more than 170 news outlets in 2018 (230 in 2019) from around the world working to strengthen coverage of the climate crisis. With a combined audience of hundreds of millions of people, this ranks as one of the most ambitious efforts ever to organize the world's media around a single coverage topic.

According to *Columbia Journalism Review* (CJR) reporter Mark Hertsgaard, "In addition to *The Guardian*—the lead media partner in Covering Climate Now—*CJR* and *The Nation* are joined by major newspapers, magazines, television and radio broadcasters, and global news and photo agencies in North and South America, Europe, Africa, and Asia."[244]

Among the outlets represented are Bloomberg; CBS News; *El País*; the *Asahi Shimbun*; *La Repubblica*; *The Times of India*; Getty Images; Agence France-Presse; national public TV broadcasters in Italy, Sweden, and the United States; most of the biggest public radio stations in the U.S.; scholarly journals such as *Nature, Science,* and the *Harvard Business Review*; and publications such as *Vanity Fair, HuffPost, BuzzFeed News,* and *The Daily Beast.* Covering Climate Now also includes a wide array of local news outlets and non-profit websites reporting from Rhode Island, Nevada, Turkey, Togo, and dozens of places in between.

"The need for solid climate coverage has never been greater," said Kyle Pope, CJR's editor and publisher. "We're proud that so many organizations from across the US and around the world have **joined with Covering Climate Now to do our duty as journalists**—to report this hugely important story."

Climate Mobilization reported that the list of U.S. cities declaring climate emergency grew by two in February 2020. The U.S. population represented by municipalities that have declared a climate emergency is now at 29,242,934 people, or 8.9 percent of the country. Across the Atlantic, Spain declared a climate emergency last week, bringing the worldwide declaration total to over 1,330 governments within 26 countries.[245]

According to a new poll reported by *Vice News,* **80 percent of Gen Z and Millennials** think that "global warming is a major threat to life as we know it." The poll reveals that young people are waiting eagerly for governments to act and feel prepared for "bold action" to address the crisis.

Grist reporter Shannon Osaka noted that during U.S. climate change talks in 2020, Senator Bernie Sanders weighed in suggesting that the money spent on militaries internationally could be used to stop warming. "This is not an American issue, it's a global issue," he said.[246]

Osaka added that the comparison of climate change spending to military spending is apt. Not only did the U.S. military receive approximately $700 billion in funding in 2019, it also emits huge

---

244 Mark Hertsgaard, Columbia Journalism Review, Covering Climate Now signs on over 170 news outlets August 28, 2019, https://www.cjr.org/covering_climate_now/covering-climate-now-170-outlets.php

245 Climate Mobilization, February 5, 2020, *Losing the Frame on global Warming*, https://www.theclimatemobilization.org/blog/climate-emergency-movement-news-in-brief

246 Shannon Osaka, Grist, *It was climate action vs. military spending at the 8th Democratic debate,* Feb 8, 2020, https://grist.org/politics/it-was-climate-action-vs-military-spending-at-the-8th-democratic-debate/?utm_source=newsletter&utm_medium=email&utm_campaign=beacon

amounts of greenhouse gases, more than some small countries. It's unclear exactly how much the U.S. government spends on addressing climate change, but best estimates have it somewhere in the $13 billion range. (And let's not forget America's continued fossil fuel subsidies.)

Other key strategies being offered were the U.S. getting back into the Paris Agreement, introducing a carbon tax, promoting clean power, and implementing gas mile standards.

When it comes to addressing the climate crisis, Indigenous nations have a lot to offer. *Guardian* reporter Jenni Monet states, "As presidents, prime ministers and corporate executives gathered at the 2019 UN climate action summit on Monday, for the first time, an indigenous representative joined the event in a formal capacity."[247]

> *We are more than 400 million indigenous peoples in the world and we protect 80% of the world's biodiversity.*
> **—Tuntiak Katan, Ecuadorian Shuar People**

Tuntiak Katan of the Ecuadorian Shuar People spoke on behalf of the International Indigenous People's Forum on Climate Change (IIPFCC), a caucus of Indigenous rights advocates who, for years, have been working towards more robust participation and inclusion at the UN level. Katan emphasized the need for Indigenous inclusion—even more so after the importance of traditional knowledge was mentioned in the 2015 Paris Agreement.

Monet noted, "Formal indigenous representation has chronically been absent at UN climate talks even though indigenous peoples have been voicing concerns about a warming planet for decades." "The Inuit have been bringing forth warnings about global warming to the international community since the first Earth summit in Rio de Janeiro in 1992," said Kuupik Kleist, former prime minister of Greenland, in a statement.

For all the promising gains, though, it was Kayapó tribal leader and 2020 candidate for the Nobel Peace Prize, Raoni Metuktire of the Brazilian Amazon, who perhaps demonstrated best the need for more Indigenous inclusion at the UN level. Raoni was denied access to Monday's climate summit despite explaining to organizers that his own ancestral homelands were still going up in flames—fires blamed on the multinational development logging plans of Brazil's president, Jair Bolsonaro.

In 2019, UN Secretary-General António Guterres conveyed a message of hope as he called on world governments to provide "accountability, responsibility and leadership" to end the global climate crisis. He pointed out that **scientists have provided a roadmap** to limit global temperature rise to just 1.5 degrees Celsius above pre-industrial levels, reach carbon neutrality by 2050, and cut greenhouse gas emissions by 45 percent from 2010 levels by 2030.[248]

"The technologies that are necessary to make this possible are already available," he added, "The signals of hope are multiplying. Public opinion is waking up everywhere. Young people are showing remarkable leadership and mobilization." The key missing ingredient is a lack of political will, he

247 Jenni Monet, The Guardian, *Indigenous representative joins UN climate summit: 'They need us,'* Sep 26 2019, https://www.theguardian.com/world/2019/sep/26/tuntiak-katan-indigenous-representative-un-climate-summit

248 United Nations, COP25: *'Signals of hope' multiplying in face of global climate crisis, insists UN's Guterre,* Dec 1 2019, https://news.un.org/en/story/2019/12/1052491

said: "Political will to put a price on carbon. Political will to stop subsidies on fossil fuels," or to shift taxation from income to carbon, "taxing pollution instead of people."

He emphasized that digging and drilling had to stop, replaced by renewable energy and **nature-based solutions** to drastically slow climate change. "In the crucial 12 months ahead, it is essential that we secure more ambitious national commitments—particularly from the main emitters—to immediately start reducing greenhouse gas emissions at a pace consistent to reaching carbon neutrality by 2050."

To assist in reaching these goals, the Secretary-General announced that a former head of the Bank of England, Mark Carney, is to become the UN's new special envoy for climate action and finance. Describing the Canadian as "a remarkable pioneer in pushing the financial sector to act on climate," Mr. Guterres said the new envoy would be focusing on ambitious implementation of action, especially shifting markets and mobilizing private finance, towards limiting global warming to the key 1.5 degrees mark.

> *We're often told about these negative tipping points, things that we can't change - There can also be positive tipping points, like when people decide they have had enough.*
> **—Greta Thunberg**

Meanwhile, Grist reporter Zoya Teirstein shares four more encouraging ways that climate politics went mainstream in 2019. She said, "In a video produced by Time magazine, Greta Thunberg—whose rapid transformation from lone climate striker to Person of the Year is proof on its own that 2019 was a momentous year for climate politics—says not all is lost.[249]

Thunberg is right. This was the year, a moment in time decades in the making, that the scales finally tipped in favour of political action to address climate change. How do we know?

Teirstein adds, "**The kids took matters into their own hands**. In 2019, millions of youth activists took to the streets to make it known that they vehemently disapprove of the job adults are doing to curtail rising emissions. Following in the footsteps of generation Z, teachers, parents, and employees of some of the world's biggest companies walked out in protest of government inaction on the crisis. These climate strikes took place on multiple Fridays throughout the fall of 2019, in cities around the world like Islamabad, Seoul, Berlin, New York, and Seattle."

Politicians Stepping Up

Many of the 2020 U.S. election candidates unveiled ambitious climate proposals that aim to protect everything from national parks, to coastal communities, to the planet's oceans. And it hasn't stopped them from tying other issues like health care and agriculture to the climate crisis during debates and town halls. And all of the Democrats that qualified for the December 2019 primary debate endorsed the Green New Deal.

---

249 Zoya Teirstein, The Grist, *4 encouraging ways climate politics went mainstream in 2019*, Dec 26, 2019, https://grist.org/article/4-encouraging-ways-climate-politics-went-mainstream-in-2019/

Teirstein concluded by saying, "**The Green New Deal burst onto the national stage.** In February (2019), freshman New York Representative Alexandria Ocasio-Cortez and junior Massachusetts Senator Ed Markey co-introduced a resolution 'recognizing the duty of the Federal Government to create a Green New Deal.' Such a deal, as its authors see it, would transition the U.S. economy to renewable energy and create millions of jobs in the process... It's prompted the Democratic establishment *and* some members of the GOP to start thinking about their own climate plans. That means that, after decades of relative silence, climate chatter has finally returned to Congress. When President Biden finally takes office in 2021, we will be seeing a powerful move on climate change, starting with renewed support of the Paris Agreement."

Although "renewable energy," as in mega solar and wind projects, is not a viable solution in most parts of the world, there are lots of other solutions mentioned in this book that truly are green and have proven successful—without destroying wildlife and the environment.

> *Maybe, just maybe, given the crisis of climate change, the world can understand that instead of spending $1.8 trillion a year collectively on weapons of destruction designed to kill each other — maybe we pool our resources and fight our common enemy, which is climate change.*
> **—US Senator Bernie Sanders**

Voters and Scientists Starting to Align

Individual climate scientists have defined the current state of the environment as an "emergency" before, and so have some governments in places like New York City and the United Kingdom. In November, 11,000 scientists took things up a notch by **banding together to declare a climate emergency** in the journal *BioScience*.

This is leaving Teirstein feeling confident that things will get better in the 2020 decade: "I feel comfortable making that wildly optimistic prediction because I actually do believe that nothing and no one can hold a candle to the burning conviction of millions of angry kids." The World Wildlife Fund adds two more successes to that list. In 2019, Scotland committed to a net-zero emissions target by 2045, and the wider UK committed to the same target by 2050.

Climate and the Law

Rachel Ramirez with *The Beacon* reports that residents of South Philadelphia are fuming after the Trump administration announced they want to keep the country's worst benzene polluter in business. Non-profit watchdog group the Environmental Integrity Project wants them to be held accountable, and points out that this crude oil refinery was producing some of the highest levels of benzene pollution of any refinery in the country. Their findings showed that ten refineries across the U.S. were releasing cancer-causing benzene into nearby communities at concentrations above the federal maximum in the year ending in September 2019.[250]

250 Rachel Ramirez, The Beacon , *This Philadephia refinery is the country's worst benzene polluter. Trump wants to keep it open,* Feb 7, 2020, https://grist.org/justice/this-philadephia-refinery-is-the-countrys-worst-benzene-polluter-trump-wants-to-keep-it-open/?utm_source=newsletter&utm_medium=email&utm_campaign=beacon

Ramirez adds, "Under 2015 EPA rules, facilities are required to investigate where their toxic emissions are coming from, then take immediate action to reduce impacts—both of which PES failed to do. The refinery had an annual average net benzene concentration that was more than five times the EPA standard, beating a long line of refineries in the oil-friendly state of Texas."

Philly Thrive, a grassroots environmental justice group that has been raising awareness about the public health costs of living near a fossil fuel facility since 2015, has been organizing community members from South Philadelphia to fight against the refinery and to ensure that they have a seat at the decision-making table. Despite the Trump administration's efforts to keep the refinery in operation, the fate of the project is still up in the air.

A decade of protests against fossil fuel companies is having a more positive outcome in Canada. In February 2020, Teck Resources withdrew its application for the $20B Frontier oilsands project—a major blow for the oilsands industry in Alberta, Global News confirmed.[251]

The company says it is "disappointed" to have made this "difficult decision." It was made in light of the broader conversation around climate change in Canada. And just like that, the largest tar sands mine ever proposed, and Canada's most polluting industry, is dead in the water. This mine would have created six megatonnes of carbon pollution and been twice the size of Vancouver, putting downstream ecosystems at risk.

The next major hurdle for Canada is to call off the Trans Mountain pipeline, which the Canadian federal government purchased from Texas-based Kinder Morgan in May 2018. But perhaps there is help on the way. The strong anti-pipeline movement formed by mostly Indigenous nations is now geared up to fight it. However, according to Living Oceans, it is possible that the global market will save us this effort.[252]

Living Oceans cites a CBC interview with Regan Boychuck, a researcher with the Alberta Liabilities Disclosure Project. Boychuck stated, "Every company operating outside of the oil sands in Alberta, at least their Alberta operations, is insolvent. They **have more financial and environmental debt than they can ever realistically expect to generate in profits**."

CBC also reported online that OPEC had failed to reach an agreement with Russia to make further cuts to supply; there's fear of an all-out price war as Saudi Arabia ramps up supply and discounts its prices. West Texas intermediate oil prices are down to $30 per barrel and falling. That puts the prices of Canadian heavy oils down to about one-third of the cost of production. And this at a time when Jason Kenney, in all his wisdom, has taken the constraints off Alberta producers, putting further downward pressure on prices.

Living Oceans notes, "Add in the shocking impact of reaction to the Covid-19 crisis on global oil demand, and the oil industry is facing a perfect storm. Growth projections for the sector were already modest and being revised downward month by month by the International Energy Agency.

---

251 Mercedes Stephenson , Maryam Shah , Karen Bartko and Allison Bench, Global News, *Teck Resources withdraws application for $20B Frontier oilsands mine,* February 23, 2020 https://globalnews.ca/news/6586908/teck-resources-withdraws-bid-frontier-mine/

252 Living Oceans, Energy and Climate Change: Will the market save us from ourselves?, March 10, 2020 , https://livingoceans.org/media/news/energy-and-climate-change-will-the-market-save-us-ourselves

While coronavirus impacts will hopefully be short-lived, escalating geopolitical tension and the climate crisis are definitely driving a global investment trend away from oil."

This shines light on the recent decision by Teck Resources to withdraw its massive tarsands project, as they would have lost their shirt on that deal. At this same time, Teck announced the purchase of a solar energy plant—perhaps an omen of what the future holds. While solar has problems of its own, this decision shows a willingness to invest in other climate solutions.

It seems that isn't the only problem fossil fuel is facing in 2020. The *Beacon*'s Emily Pontecorvo reports that when Exxon's law firm tried to recruit Harvard students, **they protested instead**. She explains, "More than 100 first-year Harvard law students gathered at a restaurant in Cambridge, Massachusetts, for a reception hosted by the corporate law firm Paul, Weiss, Rifkind, Wharton & Garrison LLP. The opulent affair ... was one of many regular functions held by elite law firms to draw elite aspiring attorneys into the fold. But about 30 students put their job prospects at risk when they interrupted the event with a demonstration."[253]

During a speech, a small group of students unfurled a banner that read #DropExxon and sang a protest song that alluded to the company's defence of ExxonMobil in several ongoing lawsuits over the oil giant's role in climate change. The students claim, "Exxon knew about climate change 35 years ago, yet continued to wreck the planet and fund climate denial that led us to this crisis. That's what this firm is enabling, and the tactics they are using are extreme and unethical."

A month later, Pontecorvo adds, "What started as a single protest is now growing into a movement. Paul, Weiss recently helped ExxonMobil win a case brought by the New York district attorney alleging that the company misled investors about the costs of climate change to its business. The firm is also representing Exxon in a similar case in Massachusetts, as well as other climate cases brought by the cities of San Francisco and Oakland, California, and Baltimore, Maryland. In those cases, the cities are seeking damages from multiple fossil fuel companies to pay for impacts of climate change they are already experiencing and to fund adaptation measures."[254]

Pontecorvo reports that now Harvard and Yale law students have launched a #DropExxon pledge that asks law students around the country to refuse to interview with or work for Paul, Weiss until it drops ExxonMobil as its client. This movement brings up an age-old ethical dilemma about the role of lawyers in society and who they choose to represent, since they do have a choice. Organizers of the protest emphasize that Paul, Weiss has given priority to a company that is **sabotaging humanity's chance to address climate change.**

The lawsuit will call on the federal government to protect young Canadians, do its fair share to stabilize the climate system and avert the catastrophic consequences of climate change.
—David Suzuki Foundation

Also planning on taking legal action are 15 children and teens who want to hold the Canadian government

253 Emily Pontecorvo, The Beacon, *Exxon's law firm tried to recruit Harvard students. Instead, they protested* Jan 16, 2020, https://grist.org/climate/exxons-law-firm-tried-to-recruit-harvard-students-instead-they protested/?utm_source=newsletter&utm_medium=email&utm_campaign=beacon

254 Emily Pontecorvo, Grist, *Calls for law firm to #DropExxon go national with law student boycott,* Feb 10, 2020 https://grist.org/climate/calls-for-law-firm-to-dropexxon-go-national-with-law-student-boycott/?utm_source=newsletter&utm_medium=email&utm_campaign=beacon

accountable for perpetuating the climate change crisis that violates their rights to life, liberty, and equality. Acording to CBC News, these youth were planning to sue the federal government in 2019, claiming they had "suffered specific, individualized injuries" due to climate change. The David Suzuki Foundation is backing the teens by supporting the case's legal framework and acting as a communications lead.[255]

The lawsuit is the latest in a series of legal actions filed by youth around the world in recent years. A group of young Americans sued the U.S. federal government in 2015, accusing federal officials and oil industry executives of violating their due process rights by knowing for decades that carbon pollution poisons the environment, but doing nothing about it.

"We have a global responsibility to address this problem; to fix the problem that has been created over the years. The federal government is uniquely positioned to do that—no province, or even group of provinces, could tackle this problem," said Tollefson, a law professor at the University of Victoria. The lawsuits, including the Canadian effort, are supported by Our Children's Trust, a non-profit organization that says it's dedicated to protecting natural systems for present and future generations.

Climate and Finance

When one country has a climate emergency, the whole world has a climate emergency.
—Cathy Orlando, Citizens' Climate Lobby International

Putting pressure on the finance world may prove to be one of the most effective ways to fight climate change, and a coalition of environmental groups called *Insure Our Future* is doing just that. Increasingly, organizers are deciding that after decades of failed government negotiations, they're also going to have to go after other power centres. Washington and Wall Street are deeply linked, but they're also distinct, and both need to shift dramatically.

> *Around the world political systems simply aren't responding to the greatest crisis they've ever faced—they're so corrupted by fossil fuel money, so overcome by inertia, so preoccupied with the next election or coup.*
> **—Bill McKibben,** *Time*

In 2019, *Time* reporter Bill McKibben stated, "Last week the UN climate talks in Madrid essentially collapsed, even as scientists were reporting that the 2010s had been by far the hottest decade since records began. Most of the blame fell on countries like the U.S., Brazil, and Saudi Arabia." But, amid the desolation, there were a couple of signs of hope. They came from the world of high finance.[256]

---

255 CBC News, *Canadian youths to sue Ottawa over government's role in climate change*, Oct 23, 2019, https://www.cbc.ca/news/canada/british-columbia/canadian-teens-lawsuit-federal-government-over-climate-change-1.5331773

256 Bill McKibben, Time.com, *Putting Pressure on the Finance World Could Be One of the Most Effective Ways to Fight Climate Change, Dec 18, 2019*, , https://time.com/5752188/financial-world-pressure-climate-change/

McKibben noted that while young people and Indigenous leaders were being kicked out of the UN talks for their protests, the Liberty Mutual insurance company became one of the first big U.S. insurers to announce it would stop aiding coal companies.

In December 2019, Goldman Sachs announced that it would restrict its lending to the coal industry, and further that it wasn't interested in funding the drilling of the Arctic. Again, a coalition of groups—from the Gwich'in tribe in Alaska's Arctic National Wildlife Refuge to the Sierra Club and the Rainforest Action Network—had pressed hard.

Another hopeful sign that the finance world will stop funding fossil fuel projects, former Bank of England Governor Mark Carney has been appointed United Nations special envoy for climate action and finance starting in 2020.

Carney said the UN climate change conference "provides a platform to bring the risks from climate change and the opportunities from the transition to a net-zero economy into the heart of financial decision-making. To do so, the disclosures of climate risk must become comprehensive, climate risk management must be transformed, and investing for a net-zero world must go mainstream." Yes, this does promise to be the decade of change.

In 2020, we continue to demand that finance be held accountable, and until the law of ecocide has been established to show criminal intent, we still have to rely on the outrage of activists and shareholder groups—and one very persistent U.S. Senator Elizabeth Warren. She is calling on big banks in America to disclose their plans to identify and offset climate-change risks.

"To protect themselves and the economy from climate-driven catastrophes, large financial institutions must act quickly to address risks," Warren said. She sent her queries to Bank of America, Bank of New York Mellon, Citigroup, Goldman Sachs, JPMorgan Chase, Morgan Stanley State Street, and Wells Fargo. Findings reveal U.S. banks dominated fossil fuel funding, with **JPMorgan Chase the worst** on climate change by an astonishingly wide margin—**providing $196 billion in funding of fossil-fuel projects between 2015 and 2018**, according to the Rainforest Action Network. JPMorgan Chase has responded that they plan to end or phase out loans to some fossil-fuel interests, namely Arctic drilling and coal mining, but the ongoing funding of major oil firms for the most extreme fossil fuel projects on Earth leaves activists saying this doesn't go far enough.[257]

In 2020, MarketWatch reporter Rachel Koning notes, "The bank said at its annual investor day that it will aim to facilitate **$200 billion in environmental and economic development deals**. It will put restrictions on financing new coal-fired power plants, phase out 'credit exposure' to the industry by 2024 and stop funding new oil and gas drilling projects as part of protecting the Arctic National Wildlife Refuge."[258]

Climate groups have been welcoming bank policy changes but warn that the goals from the finance sector aren't enough to limit climate change to 1.5 degrees Celsius, the target laid out in the

257 Rainforest Action Network, *Report Finds Global Banks Poured $1.9 Trillion into Fossil Fuel Financing Since the Paris Agreement was Adopted, with Financing on the Rise Each Year*, Mar 19 2019, https://www.ran.org/press-releases/bankingonclimatechange_2019/

258 Rachel Koning Beals, MarketWatch, *JP Morgan Chase — the oil industry's bank of choice — will withdraw support for some fossil fuels*, Feb. 25, 2020, https://www.marketwatch.com/story/jp-morgan-chase-the-oil-industrys-bank-of-choice-to-withdraw-support-for-some-fossil-fuels-2020-02-25

Paris Agreement. So, will some of the new climate initiatives be funded by a carbon tax? *Intellinews* reporter Ben Aris states, "Europe wants to be carbon neutral by 2050, and is about to introduce a carbon import duty that is already forcing Russia's biggest companies to speed up their efforts to go green."[259]

> *Even if we can get politicans to make real change, it will come slowly, and one national capitol at a time. But if these financial giants begin to move, the effects will be both quick and global—and those are the two things most required for effective progress on the climate.*
> **—Bill McKibben,** *Time*

As part of the EU's Green Deal launched in 2019, the European Commission (EC) published its European Climate Law that includes a "Carbon Border Adjustment Mechanism"—in effect a carbon import tax. Renewables still only account for 0.1 percent of Russia's installed power-generating capacity, but pressures to go green are mounting fast, Aris points out. The European Union is working to make Europe the first climate-neutral continent by 2050.

EC President Ursula von der Leyen said, "The Climate Law is the legal translation of our political commitment, and sets us irreversibly on the path to a more sustainable future. It is the heart of the European Green Deal. It offers predictability and transparency for European industry and investors. And it gives direction to our green growth strategy and guarantees that the transition will be gradual and fair." Aris adds, "The spillover effects of the new law will be significant and Russia's economy could be hit hard by the changes. Russian goods exported to Europe will have to pay an extra import duty if their factories are not clean and green. This is a serious change for Russia, as the EU accounted for 45% of its trade turnover in 2019, according to Rosstat—even more than Ukraine, which has 41.5% of its trade turnover with the EU, and also could be badly affected."

Vladimir Sklyar, a utilities analyst at VTB Capital, said in a note, "We believe this might have a fundamental effect on the competitiveness of Russian exporters and significant consequences for Russian utilities, leading in the long run to a premium for green energy producers." [260]

Russia has been slow to start greening its economy. Russian President Vladimir Putin said in November, "Humanity will end up in caves if it abandons hydrocarbons." Russia only ratified the Paris Agreement in 2019, but gave itself ease to reach CO_2 emission targets.

However, Russian companies are already reacting. For example, Russian major aluminum producer Rusal—the biggest single consumer of power in the country—is already switching from coal-burning power supplies to hydro in an effort to reduce its carbon footprint, reports VTBC, and is considering completely ditching its coal-fired power plants and coal mines.

259 Ben Aris, Intellinews, *Europe's plan to introduce a carbon import tax is forcing Russia to go green,* March 8, 2020, https://www.intellinews.com/europe-s-plan-to-introduce-a-carbon-import-tax-is-forcing-russia-to-go-green-178003/
260 ibid

Plant-Based Solutions

Reforesting the planet could be the best way to achieve Global Goal 13 to combat climate change and protect the planet. Trees can restore ecosystems, provide food, and preserve water sources, making them a key part of sustainable development.
—Global Citizen

When it comes to plant-based solutions, the most promising solution has been around for centuries, yet surprisingly gets almost no attention. This is a humble plant that could solve our climate crisis on numerous levels, reduce the amount of plastic added to our landfills and oceans, replace fossil fuels in cars, and serve as a sustainable building product for many purposes.

In fact, this most remarkable plant was used to build ancient aqueducts in Egypt, to produce the sails on Columbus's ship, and to make the paper that the Declaration of Independence was drafted on. It is widely held to be one of the first plants ever cultivated by humanity and one of the first sources of spun textiles. U.S. founding fathers George Washington, Thomas Jefferson, and John Adams all grew **hemp**. Henry Ford even built his first hemp-fueled car out of hemp steel. **Yes, it is industrial hemp**.

Most of us likely think of rope, paper, oil, and clothing when we think of modern products that come from hemp. However, this most versatile plant on the planet is also a perfect alternative for petrochemicals in the production of plastics. Although fossil fuel companies are depending on the expanding use of plastics to bail them out of bankruptcy as we transition towards alternative energy and electric cars, their plans may be futile. In fact, many of the products presently made from petrochemicals could be made much more sustainably from biodegradable industrial hemp. We know this because this technology already exists.

Not only that, but this plant could help transition humanity to a plant-based diet, possibly the best way to fight climate change next to reducing the human population. I will be discussing the vegan solution in detail in Chapter 8, Food Solutions. *CBD Origin* reporter Aaron Cadena emphasizes, "The realization that our current population is terribly unsustainable is the biggest problem of our time, ultimately leading to a not-so-happy ending."[261] Hemp food products provide an excellent protein alternative to feed our growing population.

At one time, plastic was seen as the next best thing to sliced bread. However, now that this avalanche of toxic material is destroying our oceans and earth, we must find an alternative. Cadena notes, "Since it was first mass-produced just six decades ago, we have accumulated over 9 billion metric tons of plastic, with more than 6.9 billion tons of it becoming waste which takes over 400 years to decompose. Even worse, experts estimate that of that waste, 6.3 billion tons of plastic has not been recycled."

While it might be impossible to get rid of the existing plastic on our planet, we can be proactive against this problem by using biodegradable alternatives to plastic and urging others to do the

261 Aaron Cadena, *CBD Origin*, "4 Remarkable Ways Hemp Can Save the Planet," November 21, 2018: https://medium.com/cbd-origin/4-remarkable-ways-hemp-can-save-the-planet-423380e3ad82

same. We do know that the fibre inside hemp can be used to produce a plastic alternative that offers the same capabilities of synthetic plastic and it's *fully biodegradable*.

Cadena adds, "Many eco-conscious brands and companies have already started using hemp plastics in their products, and with global brands like Coca Cola experimenting with plant-based plastic alternatives, hemp plastic might just be the plastic of the future." These same extraordinary qualities of hemp can be utilized to make hempcrete (concrete), a steel alternative, a wood replacement for most applications, and an eco-friendly car fuel. According to the National Hemp Association, hemp can be used to produce two types of fuels:

- Biodiesel: produced from the oil of the pressed hemp seed
- Ethanol/methanol: produced from the fermented hemp stalk

"With the increasing problems surrounding traditional fossil fuels and oils, the need for a natural alternative is stronger than ever," notes Cadena. Hemp is the only alternative fuel that can run in any unmodified diesel engine, and provides a solution that is more efficient, more affordable, and most importantly, more sustainable than traditional fuel.

*While this humble plant[hemp] has undoubtedly served as an invaluable resource throughout our early history, it may serve an **even greater** purpose in our immediate future.*
—Aaron Caena, *CBD Origin*

As the human population continues to grow, so does its demands for resources, most significantly wood from our shrinking forests. A 2015 study published in *Nature* estimated that we cut down an astronomical 15 billion trees every year, and since the rise of human civilization, the population of trees has decreased 46 percent.

Cadena emphasizes that because trees provide oxygen into the atmosphere and absorb carbon dioxide, the steady decrease of trees would radically change our atmosphere and intensify climate change. Eventually, Earth could be uninhabitable. With the ability to produce paper, fibre, wood, building material, and other wood products, hemp fibre is the ideal alternative to wood and the ideal solution to this problem.

Not only can hemp be used to produce the same wood products as a tree, it's also more cost effective and efficient. With a harvest time of just four months and the ability to produce four times more paper per acre than traditional trees, hemp is not only the more sustainable choice but also the smarter choice. We must break through the stigma towards industrial hemp and take advantage of the amazing solutions this plant has to offer.

Mangroves

Grist reporter Rache Ramirez has another solution in mind. She remarks, "Storing carbon and saving the economy? Mangroves can do both. From Typhoon Hagibis in Japan to Cyclone Idai in Mozambique to Hurricane Dorian in the Bahamas, climate change–fueled tropical storms have been increasingly wreaking havoc on coastal regions across the globe."[262]

262 Rachel Ramirez, Grist, *Storing carbon and saving the economy? Mangroves can do both*, Dec 24, 2019, https://grist.org/climate/storing-carbon-and-saving-the-economy-mangroves-can-do-both/

Ramirez believes that mangroves, trees that grow in coastal waters, could be an inexpensive nature-based solution. "When you take a look at these mangroves, they're these really dense ecosystems that could essentially act as a wall or barrier for wind and storm surge," explained Alejandro Del Valle, an assistant professor of risk management and insurance at Georgia State University.

Analyzing these data sources, researchers found that hurricanes can lead to significant economic losses but that wide mangrove belts are capable of mitigating the damage. The wider the swath of mangroves, the lower the impact from hurricanes, and less economic damage. Mangroves offer other benefits, too. Like all trees, mangroves remove and store carbon that humans have emitted into the atmosphere. Mangrove ecosystems also protect shorelines from erosion, shelter coral reefs, and filter polluted water that runs from land into the ocean. These dense trees with freakishly long roots are one of the most effective and critical nature-based solutions, along with forests, grasslands, and wetlands.

The irony is that more and more mangroves are being cleared by deforestation for economic gain, as seen in the Amazon, Central America, and elsewhere. In Indonesia, for example, mangroves are threatened by slash-and-burn agriculture for palm oil production. According to the Nature Conservancy, the rate of mangrove destruction is increasing in many parts of the world, threatening drastic consequences for the climate.

Trees—The True Warriors

Fortunately, places like Scotland are making up for it, as trees are the true warriors in the battle against climate change. *The Independent* reporter Phoebe Weston tells us that Scotland planted 22 million trees in 2018 to tackle the climate crisis, while England fell 7 million short of its target. She states, "A total of 11,200 hectares of Scottish countryside were covered—well exceeding the current annual target of 10,000 hectares, according to government statistics." [263]

In 2019, Britain made up for its shortfall with its largest-ever tree-planting campaign. Global Citizen reporter Fiona Harvey stated, "Volunteers are being urged to do their bit to stop the climate emergency by grabbing a spade and signing up for the biggest mass tree-planting campaign in the UK's history." Starting during National Tree Week 2019, the Woodland Trust aimed to mobilize a million people to plant native broadleaf varieties, such as oak, birch, and hawthorn, up and down the country on school, business, and government property.[264]

But in order for Britain to meet its legally binding **target of net-zero emissions by 2050**, the Committee on Climate Change has suggested that **1.5 billion new trees must be planted** across the UK in that timeframe. In July, a massive mobilization campaign in Ethiopia saw 353 million trees reportedly planted in just 24 hours. Approximately a quarter of Ethiopia's 105 million population volunteered as part of a £1.1 billion government project. In 2020, Madagascar planted 60 million

263 Phoebe Weston, The Independent, *Scotland plants 22 million trees to tackle climate crisis while England falls 7 million short of target,* June 14 2019, https://www.independent.co.uk/environment/trees-planted-scotland-england-woodland-climate-a8958601.html

264 Fiona Harvey, Global Citizen, *Call for 1m people to join UK's biggest mass tree-planting campaign*, Sept. 23, 2019, https://www.theguardian.com/environment/2019/sep/23/call-for-1m-people-join-uk-biggest-mass-tree-planting-campaign

trees in its ambitious drive inspired by its president. Eco Watch reporter Malavika Vyawahare notes, "The island nation celebrates 60 years of independence this year, and the start of the planting campaign on Jan. 19 marked one year since the inauguration of President Andry Rajoelina, who has promised to restore Madagascar's lost forests."[265]

"The government has the challenge of making Madagascar a green island again. I encourage the people to protect the environment and reforest for the benefit of the future generations," Rajoelina told the hundreds of people who attended the launch. In a span of a few hours, about one million trees were planted over 500 hectares (1,235 acres), according to the environment ministry—an area one-and-a-half times the size of New York City's Central Park.

Africa wasn't being left out as their plan is to plant a 4,815-mile wall of trees by 2030. *Time* reporter Aryn Baker remarks, "The seedlings are ready. One hundred and fifty thousand shoots of drought-resistant acacia, hardy baobab and Moringa spill out of their black plastic casings. The ground has been prepared with scores of kilometer-long furrows leading to a horizon studded with skeletal thorn trees. It's early August, and in less than a week, 399 volunteers from 27 countries will arrive in this remote corner of northern Senegal to participate in one of the world's most audacious efforts to combat the effects of climate change: an $8 billion plan to reforest 247 million acres of degraded land across the width of Africa, stretching from Dakar to Djibouti."[266]

The Big Climate Fightback is being organised by the Woodland Trust, who wants to plant a tree for every person in the UK by 2025 — that's 66 million in total.
—Global Citizen

Baker explains that the **Great Green Wall** project, spearheaded by the African Union and funded by the World Bank, the European Union, and the United Nations, was launched in 2007 to halt the expansion of the Sahara by planting a barrier of trees running 4,815 miles along its southern edge. Now, as concerns mount about the impact of climate change on the Sahel, the semi-arid band of grassland south of the Sahara that is already one of the most impoverished regions on earth, the Great Green Wall is filling a new role. The goal now, say its designers, is to transform the lives of millions living on the front line of climate change by restoring agricultural land ruined by decades of overuse; when done, it should provide food, stem conflict, and discourage migration.

When the project is completed in 2030, the restored land is expected to **absorb some 250 million metric tonnes of carbon dioxide from the atmosphere**, the equivalent of keeping all of California's cars parked for 3.5 years. Environmentalists celebrate the Great Green Wall for its epic territorial ambition, but its biggest impact will come from allowing people to meet their needs without destroying nature in the process

265 Malavika Vyawahare, Valisoa Rasolofomboahangy, Eco Watch, *Madacascar Is Planting 60 Million Trees in Ambitious Drive Inspired by Its President*, Jan. 21, 2020, https://www.ecowatch.com/madacascar-tree-planting-2644879937.html
266 By Aryn Baker, Time, Can a 4,815-Mile Wall of Trees Help Curb Climate Change in Africa?, September 12, 2019, https://time.com/5669033/great-green-wall-africa/

Regenerative Farming

National Geographic reporter Stacey Cook tells us that the solution to climate change is just below our feet: IndigoAg is unlocking agriculture's potential to help reverse climate change. That's the vision behind the Terraton Initiative, a global movement with the goal of using regenerative farming practices to take one trillion tonnes of carbon dioxide from the atmosphere.[267]

Cook noted that Adam Chappell was in the fight of his life when an invasion of pigweed was threatening to destroy his cotton farm. Chappell said, "We were spraying ourselves broke just to fight this weed. We were spending more money than we could ever hope to make. So for the farm to survive, we knew we had to change the entire way we were doing things."

Chappell turned to YouTube, where he found a guy growing organic pumpkins in a cereal rye cover crop, and was awestruck by the clean, wide rows. "He hadn't put any herbicides down; all the weed control in that field was the cover crop," he says. Instead of rye, a "cocktail" of cover crops could be used—including radishes, turnips, and sunflowers. That fall, the Chappell brothers planted cereal rye with their cotton and soybeans, and they kept the farm.

Cook points out, "Chappell's triumph over pigweed made him a proponent of regenerative farming practices. He stopped tilling most of the soil, which depletes it, and he's nearly eliminated pesticide and synthetic fertilizer. His soil has become healthy and dark, alive with earthworms, rich with carbon—and has reduced the amount of carbon dioxide in the atmosphere."

Agriculture has played a major role in the climate crisis, Cook adds—about a quarter of the world's greenhouse gas emissions come from land use and agriculture combined—but farmers are uniquely situated to be part of the solution. While the amount of carbon dioxide in the atmosphere has reached its highest level in human history, plants can draw down the carbon and restore the soil's organic carbon content—in the right conditions. If enough farmers adopted regenerative farming practices, they could begin to reverse the effects of climate change.

Author Paul Hawken states, "The relationship between regenerative agriculture and the climate is an intimate one that has been forgotten. Really, it's a path to walking back the carbon we have placed in the air. We placed it by industrial agriculture, deforestation, and combustion of fossil fuels. Those three together have nearly [destroyed] the planet. What we're talking about is bringing carbon back home."

> *The Terraton Initiative is grounded in the idea that if we could restore organic carbon content to all 3.6 billion acres of crop-producing farmland worldwide, we'd effectively be drawing about a trillion tons of carbon dioxide out of the atmosphere and depositing it into the soil, where it can do some good.*
> **—Indigo Ag**

So how do we bring one trillion tonnes of carbon home? Cook explains that plants pull carbon out of the atmosphere through photosynthesis and deposit it in the soil through their roots. Regenerative farming practices like no-till cultivation, cover crops, and crop rotation keep

267 Stacey Cook, National Geographic, *The solution to climate change is just below our feet,* October 11, 2019, https://www.nationalgeographic.com/science/2019/10/partner-content-solution-to-climate-change-below-our-feet/

the carbon in the soil, where it builds over time. In turn, carbon-rich organic matter feeds healthy plants. Conventional agriculture, using synthetic fertilizer, chemical pesticides, and monoculture planting have the opposite effect.

Regenerative farming practices have real potential to change the course of climate change, but it's not only up to farmers. Businesses and consumers can help by buying sustainably grown food, pushing for policy changes, and spreading the message.

Climate and Business

The vexed issue of rising population is inspiring morally complex Marvel storylines and plots for movies such as Ready Player One and Downsizing.
—Nick Hilton, *The Guardian*, 2018

Hilton stated, "There are 7,622,000,000 people in the world today, and not all of them are superheroes in the Marvel Cinematic Universe. But even though rising population figures are good for box-office receipts, it is a real-world trend that has sparked alarm and controversy for decades. And, while it is still a somewhat peripheral concern in contemporary politics—unlike, say, climate change—overpopulation has nevertheless become the crisis du jour in modern blockbuster filmmaking. As a movie-plot issue, population crisis exists between a plausible future and an imagined dystopia, offering Hollywood a force of moral nuance that exceeds the brute power of pure evil's wrecking balls."[268]

The makers of *Avengers: Infinity War* (2018) decided only anxiety over population growth could provide sufficient moral complexity for the franchise's big boss, Thanos. However, in the real world, people are grappling with the real threats facing a population of almost EIGHT BILLION every day. This is a world where eco-anxiety and nature deficit are paralyzing kids. Sierra Club **notes, "Young people spend an average of four to seven minutes a day in unstructured outdoor play, and nine *hours* in front of a screen."**[269]

According to a 2017 Global Burden of Disease report, air pollution kills a child every three minutes in India, the highest rate of child mortality in the world. In Delhi, authorities have been forced to shut down schools, halt construction, and restrict vehicles on the roads. BBC News reports, "Delhi residents desperate for a breath of fresh air in the midst of the city's toxic pollution crisis have found somewhere to turn: an '**oxygen bar**.' But with a price tag of 300 rupees ($4; £3.20) for 15 minutes of oxygen, it is an extravagance for most in India's smog-filled capital."[270]

268 Nick Hilton, The Guardian, *From Infinity War to Okja: why overpopulation is cinema's crisis du jour*, May 25, 2018, https://www.theguardian.com/film/2018/may/15/infinity-war-okja-overpopulation-cinema-ready-player-one-downsizing

269 Mary DeMocker, Sierra Club Magazine, *So Your Kids Are Stressed Out About the Climate Crisis,* Jan 25 2020, https://www.sierraclub.org/sierra/so-your-kids-are-stressed-out-about-climate-crisis?utm_source=insider&utm_medium=email&utm_campaign=newsletter

270 BC News, *Delhi pollution: The bar selling oxygen to choking city*, Nov 21, 2019, https://www.bbc.com/news/av/world-asia-india-50502972/delhi-pollution-the-bar-selling-oxygen-to-choking-city

In the U.S., it is a different kind of bar people are seeking out—one that sells vodka made from captured carbon dioxide. "Now that's the spirit," exclaims *The Beacon*'s reporter Maria Gallucci. Air Company started selling spirits to bars and restaurants in New York in 2019. The vodka is made without any potatoes or grains, and the key ingredients are hydrogen and captured carbon dioxide.[271]

Vodka is just the start, said Gregory Constantine, the company's CEO. The team hopes to expand into alternative fuels, plastic-making chemicals, or other markets. Gallucci notes that hundreds of companies and research groups worldwide are working to capture, reuse, and sequester heat-trapping gases as emissions soar. In 2018, global carbon emissions hit an all-time high of 40.9 billion tonnes, but much of it could be captured at the source and recycled—or possibly sucked directly from the air.

"Products that can turn carbon from a liability to an asset can play a really critical role in meeting our climate goals," said Giana Amador, managing director of Carbon180, a non-profit based in Oakland, California. They say the potential worldwide market for waste-CO_2 products could be $5.9 trillion a year. Cement, plastics, and transportation fuels make up most of that potential market, with consumer goods expected to represent less than 1 percent.

Gallucci adds that scientists are developing similar approaches at Oak Ridge National Laboratory and Stanford University, though they're focusing on making biofuels and battery-like systems. These all promise to be the big businesses of the future in our race to reduce carbon emissions.

Final Note

The many solutions to climate change outlined above could collectively provide the most effective solutions to our soaring emissions, and without all the negative impacts of alternative energies like mega solar and wind projects. Most have chosen the "green energy" route because it is the biggest money-maker, and also no one would have to "give up" anything—except our clean air, water, and disappearing wildlife. However, in the last ten years, we have increasingly seen science speak to the climate crisis, and have started to understand the impacts of adding even more emitters. So what is holding us back?

1. **Misinformation**: There is enough good science about climate change and all its related topics out there now, so we need to seek it out. Remember that the climate deniers are a small minority now, so don't let them deter you.
2. **Misguided policies**: Weak policies for climate and population have led to inaction and disaster. We need to recognize that our present population count is not sustainable, as population experts agree that approximately two to three billion would be a sustainable number, leaving room for old growth and other species. So, that is where every person on the planet should be setting their sights. We must demand that our governments implement climate and population policies that include a carrying-capacity study and a meaningful public consultation process.

271 Maria Gallucci, The Beacon, *Vodka made from captured carbon dioxide? That's the spirit,* Nov 12, 2019, https://grist.org/article/vodka-made-from-captured-carbon-dioxide-thats-the-spirit/

3. **Mismanagement and corruption**: We need to understand the role the world's leaders and decision-makers, corporations, and media have played in misleading the public and mismanaging our trust. We can trace the ecocide inflicted on our environment back to these entities, and they must be held accountable for their part in these horrendous crimes. Of course, we have all participated in this travesty, so we citizens must also change our ways in a colossal manner.

At this point, climate change should be at the forefront of our thinking, so why are so many of us still uncomfortable talking about this topic? Unfortunately, most of us are still worried about offending someone, but aren't the alternatives even more disturbing? This should be the centre of our conversations at school, at the dinner table, and at town hall meetings, but what is the proper etiquette when talking about climate change?

The Beacon's Eve Andrews asks, **"Is there a right way to talk about climate change?"** She suggests, "Don't focus too much on facts, because people don't respond to them. Don't be negative, because people lose hope. But don't misrepresent anything, either, because people won't trust you. Don't make it about the planet, because humans care more about humans than they do about trees. Make people aware of the fact that climate change is already worsening the lives of many millions of people in other parts of the world, but don't make it sound *too* far away, because then it's hard to relate to. If you can, try to tie climate change back to something your audience cares about, like, ah, wine or something."[272]

So, when talking to people who feel that they are in the midst of a disaster, Andrews stresses it is important to actively listen to them, be sensitive, show compassion, and gently suggest ways out of this mess. Since the majority of humanity—especially Millennials and Gen Z-ers—now agree that climate change is a real threat, and most people are dealing with some sort of climate-related emergency locally, that is a good place to start. Talk about the drought, heat wave, disappearing wildlife, pandemics, overpopulation, or rising cost of food taking place in their world today. Find a way to connect local events to the global picture, and suggest something positive that we can all do, like joining an activist group, writing a letter to the editor, moving towards a plant-based diet, or having one fewer child.

Therapy for Activists

Of course, being an activist can be stressful too, so the secret is getting some balance so that you don't burn out. Grist contributor Claire Elise Thompson interviewed one of the solutionaries to see how she copes. Margaret Klein Salamon, a clinical psychologist turned climate mobilizer who directs the advocacy group The Climate Mobilization, had this to say about what motivates her:

> *We have this idea that we can't scare people, that fear doesn't work as a motivator. Fear is one of the most basic motivators.*
> **—Psychologist Margaret Klein Salamon**

272 Eve Andrews, *The Beacon, Is there a right way to talk about climate change?*, Oct 10, 2019, https://grist.org/article/is-there-a-right-way-to-talk-about-climate-change/

"There's this energy [in the climate movement], there's this very clear sense of, like, being on a boat on a river and being taken somewhere. It's a new movement, and it is distinct from the gradualist movements that came before it. It's a movement that says, This is not a problem. This is an emergency. We do not need incrementalism or tweaks to the system. We need total system transformation right now. It's really an amazing thing to be part of that."[273]

273 Claire Elise Thompson, Grist, *Struggling with climate anxiety? You're not alone*, Oct 15, 2019, https://grist.org/fix/struggling-with-climate-anxiety-youre-not-alone/

CHAPTER 8
FOOD AND POVERTY

Feeding the world isn't an easy task when the population is growing, the land available for crops is shrinking, and greenhouse gases are wreaking havoc on our weather.
—Zachary Lippman, plant biologist

Most of the world's almost EIGHT BILLION citizens eat every day, but few, at least in the developed world, give thought to where that food comes from—before it hits the store aisle. How much do you know about your food?

Looking at your last meal, say it was a ham and pineapple pizza—what do you know about the pig that produced the ham? Maybe you know that pigs often lead short, inhumane lives and suffer during final transport and slaughter. We know they are surprisingly intelligent, easily the equal of dogs. Maybe you know that the United Nations strongly promotes plant-based human diets to counter climate change. Do you know that processed meats like ham, bacon, and salami are classified by the World Health Organization as group 1 carcinogens, placing them alongside cigarettes and asbestos as cancer-inducing? Pineapples, like many fruits, must travel long distances, thus contributing to climate change. The pizza's wheat is probably genetically altered, and its gluten may cause medical problems. When examining the true cost of food, many aspects need considering in the journey to our plate. Every person plays a role as producer, distributor, or consumer.

Most people in developed countries exist in industrial agricultural societies that have separated food production from consumption, something new in human history. Less agricultural societies tend to retain more connection to land and food supply. Although food is essential for survival it also reflects norms and has symbolic meaning including harvest rituals and celebrations, Buddhist sects that are strictly vegan, or Jewish and Islamic communities that eschew certain foods like pork or require ritual food handling. It would serve humanity in numerous ways to reconnect with the land that feeds our EIGHT BILLION diners.

Food and Culture

> *Rising human numbers is a significant cause of higher food prices and impending, potentially catastrophic global shortages. Investing in family planning and women's empowerment would slow population growth and increase future food security.*
> —Population Institute Canada

Humanity's oldest system of interaction with the earth for crop and livestock production that supplies our most basic needs is the practice of agriculture. It requires that we harness and alter our resources in various formulas of earth, water, nutrients, and energy to produce our desired output. This would involve the basic modes of hunter/gatherer, tending livestock, family gardening, and agriculture—where people primarily obtain their food by purchasing it. There are not many hunters/gathers left today, not even Inuit and Indigenous peoples in North America, who are heavily subsidized by our modern food system.

The UN tells us that agriculture is the world's largest industry; it occupies about **40 percent of the Earth's terrestrial surface** and uses 70 percent of fresh water. Our present population is already overwhelming the capacity of agriculture to fulfill food requirements without compromising other natural resources—by causing pollution of our water, extinction of one million species of wildlife, and climate change. Facing the twin challenges of food security and environmental degradation, increasing urban populations will only increase that threat and further disconnect humans from the land.

We also need to look at how much we eat in a day, as compared to people in other countries like Africa. Our habit of great excess is unhealthy, seriously contributing to an extreme obesity epidemic and food waste. Just imagine the amount of energy that went into all the excess food consumed to create this obesity crisis, and all the waste produced by this overconsumption—in the form of human waste, garbage, and CO_2 emissions from producing and transporting twice the amount of food actually required per person.

Then, hidden from sight, there is the tragic suffering and cruelty inflicted on animals in an attempt to produce enough food for the majority of people who are presently meat eaters.

However, we must remember that when it comes to feeding the planet, it is not simply a trade-off over hungry people versus nature. It is a matter of finding a win-win situation, which is problematic at this advanced stage of decline; our urge to make it seem OK—win-win—is now unrealistic, and some activities and industries now in existence will have to be modified or eliminated. We must take measures to provide for people's needs in the short term without destroying the resource base that is essential to providing for those needs. If the alleviation of human suffering always results in an increase in population, we are simply creating a larger problem—and more human suffering—in the future. This is exactly what we have done every time we have provided food aid in Africa with no effort to stem the rapid population growth. The number of hungry people in each crisis simply gets larger, and the environment becomes more degraded. The protection of nature and human well-being both depend on our acknowledging the population crisis. A perfect example today is

President Biden's raising immigration quotas to the U.S.; no one wants to look—few dare to—at the ecological devastation this creates because the social justice movement is too powerful!

No matter how unpleasant the solutions may seem, they are far more appealing than the alternatives. If we continue to destroy our planet and our life-support systems, we will suffer far greater sorrows than limiting our family size. The time for procrastination has run out—we must attack the urgent population crisis with all the foresight and wisdom and political might we can muster, if we aim to eliminate hunger. It is a travesty that millions of people dying from starvation and air pollution is not a powerful enough incentive. This is going to take people willing to bear the outright attacks and condemnation that someone like Martin Luther King did, battling for equal rights!

The **Universal Declaration of Human Rights** recognizes our "rights" to clean air, sanitation, water, food, and all the other things that make life possible and worth living, and the **UN Sustainability Development Goals** recognize our "responsibilities" toward our fellow beings and the planet and its vast diversity. However, these rights and responsibilities cannot be fulfilled if we do not include population in the conversation.

Industrial Agriculture

Modern agriculture is the use of land to convert petroleum into food.
—Professor Al Bartlett

It has been argued that affluence is not our greatest achievement but our biggest problem. This predicament of excess in developed countries and scarcity in developing countries was guaranteed to be solved by the Green Revolution, which began in the 1960s. It would supposedly allow us to feed the world's poor and better distribute the world's wealth. The thinking was that to avoid the mass starvation predicted by Thomas Malthus, technology could be used to artificially produce high-yielding crops with high-intensity growth from artificial fertilizers. Consequently, by sacrificing the environment and funded by the Rockefeller Foundation and the Ford Foundation, farmers were able to increase their yields by up to ten times, thus naming the phenomenon the "Green Revolution"—which would come to define the 20th century. Its most dramatic effects, almost doubling rice harvests, were felt in Asia where the population soared. This is the equivalent of an ecological trap used by wildlife researchers, where long-term survival is sacrificed, unknowingly, for short-term gain!

According to *The Atlantic* reporter Charles Mann, the worries were, "The results will be catastrophic: erosion, desertification, soil exhaustion, species extinction, and water contamination that will, sooner or later, lead to massive famines."[274] This reality that humans were overwhelming Earth's biological limits was understood and publicly recognized by writers and scientists like Rachel Carson (author of *Silent Spring*) and Paul Ehrlich (author of *The Population Bomb*): their

274 Charles C. Mann, The Atlantic, *Can Planet Earth Feed 10 Billion People?*, March 2018, https://www.theatlantic.com/magazine/archive/2018/03/charles-mann-can-planet-earth-feed-10-billion-people/550928/

writing fueled the environmental movement of the 1970s. They predicted that unless humankind drastically reduced consumption and limited population, it would ravage global ecosystems.

Now in 2020, we have run out of technological fixes; their fears, and those of thousands of worried citizens (like you), have come true. Now we see that without taking measures to stop population growth, the Green Revolution could never have been a permanent solution, merely a way to postpone the inevitable hunger crisis. We have driven most remaining species to the brink of extinction, drained aquifers, and contaminated most of our water supplies with fertilizer runoff and aquatic dead zones, and still our runaway human population continues to increase at an alarming rate of 80 million a year.

How can Planet Earth feed ten billion people? Or will we come to our senses and stop population growth in its tracks at EIGHT BILLION? We have barely ten years to decide: it's entirely up to us.

This frightening level of growth has caused severe degradation of land—the backbone of agriculture. The Union of Concerned Scientists, consisting of over 1,600 of the world's leading scientists, states, "Industrial agriculture is currently the dominant food production system in the United States. It's characterized by large-scale monoculture, heavy use of chemical fertilizers and pesticides, and meat production in CAFOs (confined animal feeding operations).... From its mid-20th-century beginnings, industrial agriculture has been sold to the public as a technological miracle."[275]

We are told it will allow food production to keep pace with a rapidly growing global population, and somehow still remain a profitable business. Something crucial was left out of this story: the price tag. In fact, our industrialized food and agriculture system comes with steep costs, many of which are picked up by taxpayers, rural communities, farmers themselves, other business sectors, and future generations. When we include these "externalities" in our reckoning, we can see that this system is not a cost-effective, healthful, or sustainable way to produce the food we need.

We are told the global population in 2050 will be **just** 25 percent higher than it is now, but typical projections claim that farmers will have to boost food output by 50 to 100 percent. The main reason: increased affluence tends to increase the demand for animal products such as dairy, fish, and especially meat—and growing feed for animals requires much more land, water, and energy than just growing plants to feed people.

A critical step is a shift towards plant-based meats, which makes the transition to vegetarian much easier for those meat-eating diehards. "Globally, per gram of edible protein, beef and lamb use around 20 times the land and generate around 20 times the greenhouse gas emissions of plant-based proteins. Fortunately, companies such as Impossible Foods and Beyond Meat are already making headlines by creating plant-based "beef" that looks, sizzles, tastes, and even bleeds like the real thing.[276]

---

275 Report, Union of Concerned Scientists, *The Hidden Costs of Industrial Agriculture* , Jul 11, 2008, https://www.ucsusa.org/resources/hidden-costs-industrial-agriculture

276 Tim Searchinger, Craig Hanson, Richard Waite and Janet Ranganathan, *10 Breakthrough Technologies Can Help Feed the World Without Destroying It*, July 17, 2019, https://www.wri.org/blog/2019/07/10-breakthrough-technologies-can-help-feed-world-without-destroying-it

The chart below shows how animal-based foods are more resource intensive than plant-based foods.

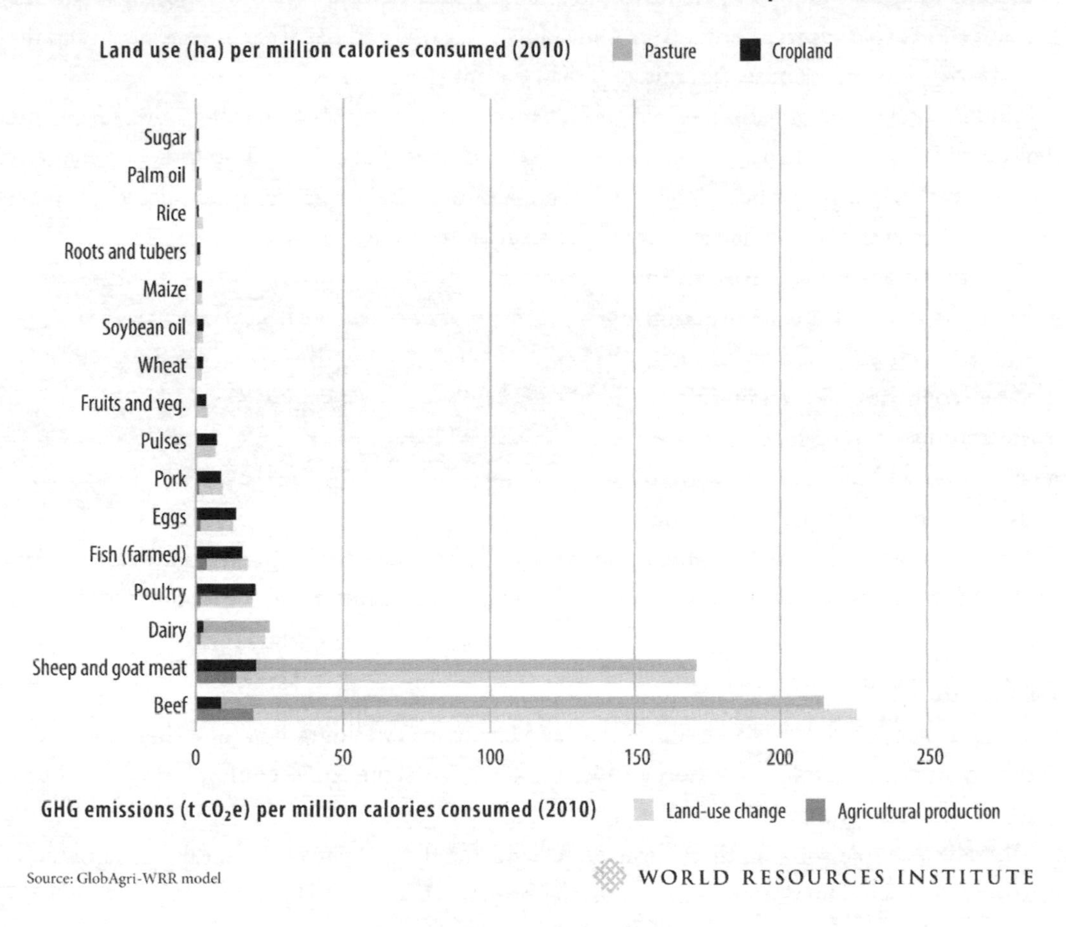

Figure 9. World Resources Institute

Recognizing the Need for New Options

We no longer have the options we had 50 years ago. We can't expand our agricultural land base much more, as most arable land is already in use. The forested land that remains is desperately needed as a carbon sink. Our available land is shrinking due to rising oceans and expanding deserts. We can't increase the amount of fertilizers we are using, as we are already using too much and polluting our water sources. We can't increase our irrigation much more, as many of our rivers are running dry before they reach the ocean and most of our aquifers are being pumped dry. The United Nations asserts, "Family planning could bring more benefits to more people at less cost than any other single technology available to the the human race."[277]

Nobel Prize–winning physicist Steven Chu surveyed the world's carbon-polluting industries and concluded that **animal agriculture and meat production contribute more greenhouse gas emissions than power generation**. Chu added, "If cattle and dairy cows were a country, they would have more greenhouse gas emissions than the entire EU 28."

Chu described the unnatural effects of industrial agriculture, what he called "oversexed corn" that devotes all its life energy to making giant kernels, pigs that gain 280 pounds in a matter of months, turkeys so breast-heavy they can't mate and must be artificially inseminated—a planet dominated by animals modified and raised and slaughtered to feed humans.[278]

> *Cutting beef consumption in favor of nonruminant meats or plant proteins like beans and pulses could liberate almost 11 million square miles of land—the size of the United States, Canada, and China combined.*
> **—IPCC, Climatic Change Journal**

According to the IPCC, **around 30 percent of global food production is wasted each year**, mostly in high-income countries. By ending food waste and distributing food surpluses more fairly, we can put an end to hunger while actually *reducing* global agricultural output. Scientists estimate that this could liberate several million square miles of land and cut global emissions by 8 to 10 percent, taking significant pressure off the climate.

We Can Do This!

In South Korea, households are required to pay a fee for every kilogram of food they toss. France and Italy have banned food waste by supermarkets. The same thing could be done for farms, going further upstream to the point of production."[279]

There is a simple solution that hinges on recognizing that 60 percent of industrial agricultural land globally is used for the single most wasteful product: **beef.** Hickel points out, "Far from being essential to human diets, beef accounts for only 2 percent of the calories that humans consume. Calorie for calorie, nutrient for nutrient, it's one of the most inefficient and ecologically destructive foods on the planet, and the pressure to find new land for pasture and feed crops is the single greatest driver of deforestation. In terms of total climate impact, **each kilogram of beef entails net emissions equivalent to a round-trip trans-Atlantic flight."**[280]

This simple shift in diet would allow us to return vast swaths of the planet to forest and wildlife habitat, creating new carbon sinks and cutting net emissions by up to eight gigatonnes of carbon

277 Population Institute Canada, Quotes, UNICEF 1992, https://populationinstitutecanada.ca/pic-resources/relevant-quotes/

278 Jeff McMahon, Forbes, *Meat And Agriculture Are Worse For The Climate Than Power Generation, Steven Chu Says* Apr 4, 2019, https://www.forbes.com/sites/jeffmcmahon/2019/04/04/meat-and-agriculture-are-worse-for-the-climate-than-dirty-energy-steven-chu-says/#706edd7d11f9

279 Jason Hickel, Foreign Policy, *The Nobel Prize for Climate Catastrophe,* Dec. 6, 2018 https://foreignpolicy.com/2018/12/06/the-nobel-prize-for-climate-catastrophe/

280 ibid

dioxide per year, according to the IPCC. That's around 20 percent of current annual emissions, Hickel explains.

Food Security

I have—over the last five years—quite rapidly become a Malthusian. I have been won over by the data, and I have been won over by the logic of the math. Population is the multiplier of everything we do wrong.... Continual increases in population and consumption cannot continue forever on a finite planet.
—Jeremy Grantham, investment strategist

I wish I had a loonie for every time I've read that Malthus was wrong. Thomas Malthus published his "Essay on the Principle of Population" in 1798. His premise was that there must be a "strong and constantly operating check on population." This would take two forms: "misery" in the form of war, famine, and disease, and "vice," by which he meant contraception and abortion (he was, after all, an Anglican minister).

His predicted food shortages and mass starvation did not materialize as quickly as he had anticipated, but there is good reason for this. The succession of revolutions in global agriculture, culminating in the misguided postwar "Green Revolution" and the current wave of genetically modified crops delayed, temporarily, the inevitable.

At that time, how could anyone have possibly predicted the ecologically brutal measures that would be taken to feed the planet? How could anyone have possibly foreseen the ruthless ruin of our environment and wildlife that would occur to avert hunger in the 20th century? Today it is evident how right Malthus truly was.

Achieving food sustainability by 2030 is the focus of Goal #2 of the 17 Sustainable Development Goals being promoted by the United Nations in 2020. "After decades of steady decline, world hunger has slowly been on the rise since 2015," states the UN. "An estimated 821 million people in the world suffered from hunger in 2018."[281] The immense challenge of achieving the zero hunger target by 2030 will not be achieved. At the same time, overweight and obesity continue to increase in all regions of the world.

Population Institute Canada adds, "There is rising anxiety over food security. Potential food supplies are being lost to waste, to bio-fuel production and to depletion of agricultural lands by development and land overuse. Fisheries are in decline. Climate change induced weather patterns and limits to water, fertilizer and oil reserves loom menacingly over food production prospects. None of these issues offers easy solutions. We must devote far more attention to helping people to have smaller families if food security into the future is to be ensured."[282]

If global demand for food continues to increase steadily along with **population growth**, food stocks are likely to become depleted and **food prices will continue to rise**. When climate

281 United Nations, *Food,* https://www.un.org/en/sections/issues-depth/food/index.html

282 Population Institute Canada, Overpopulation – a growing threat to global food security, https://populationinstitute-canada.ca/overpopulation-a-growing-threat-to-global-food-security/

instability and increasing crop failures are added to the mix, this will create the perfect storm for a food disaster. The coronavirus will pale in comparison to the food crisis that is looming.

But the UN has been sending a contradicting message to the world: Keep more people alive! And ... we need fewer people!!! "Hunger and malnutrition are the biggest risks to health worldwide—greater than AIDS, malaria and tuberculosis combined," the UN warns. Their 2000 Millenium Development Goals were successful in reducing malnutrition by half between 2000 and 2010. However, with **coping mechanisms now overwhelmed and hunger rising again**, it is more important than ever that **population be included in today's plans to fight poverty**.

"It is necessary to slow down and ultimately reverse population growth to guarantee food security for all in the long term.... Global food demand is expected to grow significantly in upcoming decades. This will be caused mostly by population growth, though rising incomes of people in developing countries will also influence demand," explains Population Matters. They add that the UK currently imports about 40 percent of all its consumed food, and this proportion is on the rise.[283]

Food-Borne Pandemics

It is not prudent to rely on science and technology alone to solve problems created by rapid population growth, wasteful resource consumption and harmful human practices.
—U.S. National Academy of Sciences and Royal Society of London

COVID-19 has exposed our short-sighted exploitation of anything that fuels our consumption! Expanding trade in wildlife and domesticated animals—for food, for pets, for medicine—on a planet of almost EIGHT BILLION people is ever-increasing the risks of pandemic diseases. The situation became so bad in 2020 that immediate steps to reduce population growth and move towards a plant-based diet were recommended by the UN.

COVID-19 brutally exposed the interconnectedness of humans and non-human species. "We share the same planet and breathe the same air, and we also exchange microbes including germs. Now, with our burgeoning human population and global economy, we face new threats from a wider distribution of diseases like this new strain of coronavirus," states Gene Baur, opinion contributor to *The Hill*.[284]

"Coronaviruses (CoV) are a large family of viruses that cause illness ranging from the common cold to more severe diseases such as Middle East Respiratory Syndrome (MERS-CoV) and Severe Acute Respiratory Syndrome (SARS-CoV).... Coronaviruses are zoonotic, meaning they are transmitted between animals and people. COVID-19 has been confirmed **in nearly 80 countries** and declared a public health emergency of international concern by the World Health Organization. It has characteristics similar to the bird flu, known as the Spanish Flu, which killed millions during World War One," notes the World Health Organization.

283 Population Matters, 2020, https://populationmatters.org/sites/default/files/population-and-food-security.pdf
284 Gene Baur, The Hill, *Coronavirus and the karmic interconnectedness of humans, animals,* 03/07/20 , https://thehill.com/opinion/energy-environment/486398-coronavirus-and-the-karmic-interconnectedness-of-humans-animals

Urgent action needs to be taken to curtail these deadly diseases, **including banning live animal markets and the wildlife trade.** The Centers for Disease Control and Prevention (CDC) warns, "... 3 out of every 4 new or emerging infectious diseases in people come from animals." These include viruses, bacteria, fungi, and parasites, and they infect millions of U.S. citizens every year.

For example, the pangolin (sometimes known as scaly anteater) has been identified as a likely coronavirus host. The endangered pangolin is considered the most trafficked animal on the planet, and more than one million have been snatched from Asian and African forests in the past decade, according to the International Union for Conservation of Nature (IUCN). They are destined for markets in China and Vietnam, where their scales are used in traditional medicine—despite having no known benefits—and their meat is bought on the black market. Also, to make some of the COVID-19 vaccines, the blood from horseshoe crabs is being used, causing untold suffering—even when other alternatives are available.

Baur adds, "In the U.S., **almost ten billion animals are exploited and slaughtered every year.** Most live short miserable lives in overcrowded factory farms, which are a breeding ground for disease, including emerging pathogens and virulent strains of antibiotic-resistant bacteria."

Animal agriculture and the trade in endangered wildlife continue to breed global pandemics. The agribusiness and governments refuse to acknowledge the connection. Many animal protection groups, like Farm Sanctuary and PETA in the U.S., have been working to increase public awareness, change government policies, and advocate for humane treatment of animals.

Individuals can play a major role here by moving towards a plant-based diet and urging our governments to support plant agriculture and stop subsidizing the meat industry. We could all benefit by eating more nutritious foods that will increase our ability to withstand these deadly viruses. We also need to reduce international travel, as a person can be in a forest in central Africa one day and in a city like central London the next, spreading a deadly virus.

It's important to note that these animals alone do not cause the transfer of the coronavirus to humans. COVID-19 (as well as SARS) was transferred by human interference with the animals. The capturing, slaughtering, selling/trading, and consumption of wildlife are to blame, not the animals themselves.[285] Illegal activity is often involved in the capture of these animals, many of which are endangered or threatened with extinction and are intended to be protected under international law.

COVID-19 is revealing that the main drivers of the extinction crisis—exploitation of wildlife and habitat loss—are fueling global pandemics. ***It is humans that are spreading disease.*** Our own survival depends on ending global trade in wildlife immediately. Although China has imposed a ban on the trade and consumption of wild animals, it is unclear if this is a temporary measure and what species will be protected. Will the ban include the use of exotic animal parts such as tiger bones, bear gall, and pangolin scales, as well as live foxes and parrots, or elephant and giraffe parts?

Population-driven issues like climate change, sprawling cities, human encroachment, and habitat destruction are creating a dangerous perfect storm for disease hosts, such as bats. With the loss of more and more wildlife habitat, bats are being driven into populated areas where

---

285 Nature, PBS, *The Novel Coronavirus and Its Connection to Wildlife,* March 3, 2020, https://www.pbs.org/wnet/nature/blog/the-novel-coronavirus-and-its-connection-to-wildlife/

cross-contamination is more likely, and with people capturing bats to be sold in wet markets, people are setting themselves up for transfer as well. All of these situations are market- and greed-driven, with little respect for the wildlife we are displacing.

Humans and their cattle already make up 94 percent of mammal biomass on the planet! It should be evident that we need to reduce these numbers to give other critters a bit of their land back. If we had a sustainable human population eating mainly plant-based diets, we would not be seeing this rate of increase in infectious diseases. Although these diseases show how interconnected humans and other animals are, they also show how little humans understand or care about the wildlife apocalypse occurring on the planet right now. It is bad enough that these bats are being slaughtered by the millions by mega wind farms and solar projects, but now their habitat is also being destroyed at an alarming rate, and they are being blamed for human-caused pandemics.

Bats are mammals like us, feel stress like we do, and feel pain and terror when they endure agonizing confinement and deaths in wet markets or the animal trade. Bats are the only mammal that can fly; they pick up pathogens from many different sites on their path, and distribute those along the way or once they have landed—or been captured and transported.

"Human activities are causing this. When a bat is stressed—by being hunted, or having its habitat damaged by deforestation—its immune system is challenged and finds it harder to cope with pathogens it otherwise took in its stride. We believe that the impact of stress on bats would be very much as it would be on people.... It would allow infections to increase and to be excreted—to be shed," explains Andrew Cunningham with the Zoological Society of London.[286]

Cunningham cautions that in the likely epicentre of the virus—the so-called wet markets of Wuhan, China—where wild animals are held captive together and sold as delicacies or pets, a terrifying mix of viruses and species can occur. All of the animals held in these markets are extremely stressed, so large numbers of viruses are likely being shed and possibly spread between animals and also between humans and other animals.

> *The underlying causes of zoonotic spillover from bats or from other wild species have almost always—always—been shown to be human behavior.*
> **—Andrew Cunningham, Zoological Society of London**

CNN Health reporters, Walsh and Cotovio, note that ultimately diseases like coronavirus could be here to stay, as humanity grows and spreads into places where it's previously had no business. The ultimate lesson is that damage to the planet can also damage people more quickly and severely than the generational, gradual shifts of climate change. Since destroying wildlife habitat is a cause of spread, restoring habitats is a solution.[287]

Changing human behaviour is likely to be an easier fix than developing a vastly expensive vaccine for each new virus. That will be the **most cost-effective way to protect humans**. The coronavirus

286 Helen Briggs, BBC News, *Coronavirus: The race to find the source in wildlife*, 25 February 2020, https://www.bbc.com/news/science-environment-51496830

287 Nick Walsh and Vasco Cotovio CNN Health, *Bats are not to blame for coronavirus. Humans are.* March 20 2020, https://www.cnn.com/2020/03/19/health/coronavirus-human-actions-intl/index.html

is perhaps humanity's first widespread, in your face, indisputable sign that environmental damage can kill humans fast. With **tens of thousands of viruses likely to be discovered**, it can happen again for the same reasons.

"It's not OK to transform a forest into agriculture without understanding the impact that has on climate, carbon storage, disease emergence and flood risk. You can't do those things in isolation without thinking about what that does to humans," said Kate Jones, chair of ecology and biodiversity at University College London.[288]

We can prevent more disease outbreaks in the future! Here's how:

1. Promote the reduction of human numbers to a sustainable level to allow us to cohabitate on the planet with other species while respecting each other's space.
2. Restore at least half of the planet's wildlife habitat as recommended in the Half Earth Project and by the UN.
3. Ban wet markets, animal exploitation, and the trade in wildlife or wildlife parts. Likewise, to improve human and animal health, it is imperative that we reduce our dependence on meat, and move towards a plant-based diet.
4. Consider all the impacts before pursuing major development projects involving wildlife, and include gender-equal input from the science and population communities as well as the general public. Decisions must consider the Greater Good.

Women and Food Production in the Face of COVID-19

In 2020, the United Nations emphasized, "This crisis has for the most part, unearthed a human spirit, a level of solidarity, a new unity of purpose that has ebbed in this 21st century—within countries and communities and across our diverse world. The post-COVID world needs to take that spirit and energy forward. It needs to become the zeitgeist of our times. Not because it is preferable to unilateralism or isolationism, but because it is the only way we can stay together in this world—safe, free, healthy, prosperous and in harmony with the planet."[289]

In March 2020, the United Nations announced findings and actions we must take: women and girls, often the primary food providers for families, must have a face in the response to the COVID-19 pandemic:

- Recognize that young women and girls are at higher risk of intimate partner violence and other forms of domestic violence during the pandemic.
- Ensure gender expertise in national, regional, and global response teams and task forces.
- Ensure the continued delivery of sexual reproductive health services, such as access to contraceptives without prescription during the crisis.

---

288 Helen Briggs, BBC News, *Coronavirus: The race to find the source in wildlife,* 25 February 2020, https://www.bbc.com/news/science-environment-51496830

289 United Nations, March 2020, *SHARED RESPONSIBILITY, GLOBAL SOLIDARITY:Responding to the socio-economic impacts of COVID-19,* https://www.un.org/sites/un2.un.org/files/sg_report_socio-economic_impact_of_covid19.pdf?utm_source=Unknown+List&utm_campaign=41b9ac5f35-

Women's organizations, operating with meagre resources, are often on the front line of community response—supporting those most affected economically by the crisis, ensuring shelters remain open for domestic violence victims, and channelling public health education messages to women. It is crucial that we continue and scale up funding and support for these initiatives during this pandemic.

UNITED NATIONS CALL TO ACTION: "The COVID-19 pandemic is a defining moment for modern society, and history will judge the efficacy of our response not by the actions of any single set of government actors taken in isolation, but by the degree to which the response is coordinated globally across all sectors to the benefit of our human family."

As Greta Thunberg emerged onto the world stage in 2019, leading a climate revolution, so has COVID-19 emerged on the world stage in 2019, leading a silent revolution against human overpopulation and animal cruelty. How many warnings will it take, each getting more formidable?

Science tells us that it is a sustainable society, not depending on a never-ending growth ethic, that would be best equipped to deal with diseases, disasters, or disturbances to our climate. It is an economic system based on the genuine progress indicator (GPI), not the GDP, that will be better able to withstand the economic turmoil caused by these disasters.

In our attempts to feed an unsustainably large human population, we are also causing the unnecessary suffering of billions of animals and putting them at greater risk of agonizing deaths due to infectious diseases. At the same time that COVID-19 was starting to take off in China in 2019, China was just recovering from the African swine fever (ASF) that took the lives of a quarter of the world's pigs living in China. It has also highlighted the struggle of a country of 1.4 billion meat eaters to feed itself.

"The pig population of China fell by 40 percent in 2018–9 after an epidemic of African swine fever.... A crisis that might have been manageable quickly became a small catastrophe because of how the Chinese state operates," states global health expert Yanzhong Huang.[290]

Shifting Views on Food and Poverty

Scientists radically changed their views on how to feed the world between 1970 and 2020. Where the focus of most scientific papers in 1970 was on **limiting demand through limiting human numbers**, recent papers focus on **accommodating demand through increased production**.[291]

Tamburino explains, "The problem is obvious: human societies continue to try to fit continuously growing food demands within a physically finite system. The **solution almost surely will have to involve limiting demand**: either through limiting human numbers, limiting per capita food demands, or both," Tamburino explains. [292]

290 Yanzhong Huang, New York Times, *Why Did One-Quarter of the World's Pigs Die in a Year?*, Jan 1, 2020, https://www.nytimes.com/2020/01/01/opinion/china-swine-fever.html

291 The Overpopulation Project, *From population to production*, https://overpopulation-project.com/from-population-to-production-scientists-changing-views-on-how-to-feed-the-world/

292 ibid

When it comes to generating food security in the 2020 decade, the International Food Policy Research Institute recognizes the role women play as food producers, gatekeepers, and shock absorbers:

> Meeting world food needs in the year 2020 will depend even more than it does now on the capabilities and resources of women. Women are responsible for generating food security for their families in many developing countries, particularly in Sub-Saharan Africa. Population growth, urbanization, and the limited potential for increasing production through the expansion of cultivated area imply that, for food needs to be met in the future, yields will have to increase. This brief examines the key roles that women play in maintaining the three pillars of food security—food production, food access, and food utilization—and it looks at how strengthening these pillars through policies that enhance women's abilities and resources provides a solution to meeting world food needs in the year 2020."[293]

> *Women not only process, purchase, and prepare food, but they also play a significant role in national agricultural production, producing both food and cash crops.*
> **—International Food Policy Research Institute**

While women are preparing to play a more prominent role in food production, human-caused forces are conspiring to derail progress in food security. Climate change is increasing the incidence of extreme weather events such as droughts, floods, fires, and heat waves. Deserts are expanding, and we are losing land to rising sea levels, inflicting permanent damage on some farm land. In a move towards biofuels, we are giving up large parcels of land from food production, and 30 percent of prime farmland has been seized for cattle and livestock operations. Add to this the devastating pandemics like COVID-19 that are expected to increase in the future, and our population which continues to grow at an alarming rate, and this makes the woman's role of providing food for children overwhelming.

Food and Climate

We can't frack our way back to economic prosperity; nor can we unplug a coal plant, plug in a solar panel, and go on expanding population and consumption.
—Richard Heinberg, Post Carbon Institute

According to the Australian Academy of Science, "In 2020, global agricultural production amounted to more than eight and a half billion tonnes of grains, vegetables, meat and other bio-products. Slightly more than eight billion tonnes (that's more than eight trillion kilograms!) of this was food. Despite this staggering number, more than 870 million people in the world are still

293 Lynn R. Brown, Hilary Sims Feldstein, Lawrence James Haddad, Christine Peña, Agnes R. Quisumbing International Food Policy Research Institute, *Generating food security in the year 2020*, https://www.ifpri.org/publication/generating-food-security-year-2020

hungry."[294] The Academy warns:

- Changing climatic conditions are going to affect the world's agricultural production.
- Increased average global temperatures may allow some production of crops in areas that were previously too cold, but this is likely to be outweighed by crop reductions in areas that will become too warm.
- Increased CO_2 can enhance plant growth in some cases, but this tends to increase only the leafy matter of the plant, not the roots, seeds or fruit which are usually more important food sources.
- Livestock may be affected by heat stress and also possible reduction in feed sources.
- Marine fisheries may be affected by warmer ocean temperatures and increasing ocean acidification.
- An increase in extreme weather events may also compound the other stresses upon agricultural production.

In 2019, the Global Hunger Index focused on the impacts of climate change on hunger, and called attention to those areas of the world where hunger levels are highest and where the need for additional efforts to eliminate hunger is greatest. It reports that **43** countries out of **117** countries have levels of hunger that remain *serious,* **4** countries—Chad, Madagascar, Yemen, and Zambia—suffer from hunger levels that are *alarming,* and **1** country—Central African Republic—from a level that is *extremely alarming.*[295]

The report warns that human actions have created a world in which it is becoming ever more difficult to adequately and sustainably feed and nourish the human population. Ever-rising emissions have pushed average global temperatures to 1 degree Celsius above pre-industrial levels. Climate change is affecting the global food system in ways that increase the threats to those who currently already suffer from hunger and have the least capacity to adapt.

Nature reporter Jeff Tollefson tells us about a landmark United Nations–backed IPBES report that finds agriculture is one of the biggest threats to Earth's ecosystem. He emphasizes, "Up to one million plant and animal species face extinction, many within decades, because of human activities."[296]

He notes that according to the report, agricultural activities have had the largest impact on ecosystems that people depend on for food, clean water, and a stable climate. **The loss of species and habitats poses as much a danger to life on Earth as climate change does**, says a summary of the work, released in 2019. The analysis distills findings from nearly 15,000 studies and government reports, integrating information from the natural and social sciences, Indigenous peoples, and

> *We are eroding the very foundations of our economies, livelihoods, food security, health and quality of life worldwide.*
> **—Robert Watson, IPBES chair**

294 Australian Academy of Science, *Feeding a hot, hungry world*, 2020, https://www.science.org.au/curious/earth-environment/feeding-hot-hungry-world

295 Global Hunger Index 2019, https://www.globalhungerindex.org/results.html

296 Jeff Tollefson,Nature, *Humans are driving one million species to extinction,* May 6 2019, https://www.nature.com/articles/d41586-019-01448-4

traditional agricultural communities. It is the first major international appraisal of biodiversity since 2005. Representatives of 132 governments met in 2019 in Paris to finalize and approve the analysis.

Without "transformative changes" to the world's economic, social, and political systems to address this crisis, the IPBES panel projects that major biodiversity losses will continue to 2050 and beyond. Globally, urban areas have more than doubled since 1992, half of agricultural expansion occurred at the expense of forests, and there are now more than two billion people who rely on wood fuel to meet their primary energy needs.

Agricultural activities are also some of the largest contributors to human emissions of greenhouse gases. They **account for roughly 25 percent of total emissions** due to the use of fertilizers and the conversion of areas such as tropical forests to grow crops or raise livestock such as cattle.

The IPBES report is solid on the science. Gains from societal and policy responses, while important, have not stopped massive losses. This lack of initiative has produced a truly global and generational threat to human well-being and a devastating loss of wildlife. The UN IPBES report warns, "Past and ongoing rapid declines in biodiversity, ecosystem functions and many of nature's contributions to people mean that most international societal and environmental goals, such as those embodied in the Aichi Biodiversity Targets and the 2030 Agenda for Sustainable Development will not be achieved based on current trajectories."[297]

However, the UN claims that **the world can reverse this biodiversity crisis**, but doing so will require proactive environmental policies, the sustainable production of food and other resources, and a concerted effort to reduce greenhouse-gas emissions. If we fail to act, the ensuing lack of food could have a violent outcome.

In 2008, we saw what rocketing food prices could do to the planet as riots broke out in numerous countries around the world. As *Time* reporter Vivienne Walt stated, "Millions are reeling from sticker shock and governments are scrambling to staunch a fast-moving crisis before it spins out of control. From Mexico to Pakistan, protests have turned violent. Rioters tore through three cities in the West African nation of Burkina Faso last month [2008], burning government buildings and looting stores. Days later in Cameroon, a taxi drivers' strike over fuel prices mutated into a massive protest about food prices, leaving around 20 people dead. Similar protests exploded in Senegal and Mauritania late last year. And Indian protesters burned hundreds of food-ration stores in West Bengal last October, accusing the owners of selling government-subsidized food on the lucrative black market."[298]

The International Food Policy Research Institute noted that it was a serious security issue, and food analysts said that it could take a decade to bring down food prices. One reason was that billions of people were buying ever-greater quantities of food—especially in booming China and India, where many had stopped growing their own food and had the cash to buy a lot more of

---

297 United Nations, *IPBES Sustainable Development Goals*, 2019, https://www.un.org/sustainabledevelopment/blog/2019/05/nature-decline-unprecedented-report/

298 Vivienne Walt, Time, *The World's Growing Food-Price Crisis*, Feb. 27, 2008, http://content.time.com/time/world/article/0,8599,1717572,00.html

it. Increasing meat consumption, for example, had helped drive up demand for grain, and with it the price.

The push to produce biofuels as an alternative to hydrocarbons was further straining food supplies, especially in the U.S., where generous subsidies for ethanol had lured thousands of farmers away from growing crops for food. These issues continue to be an obstacle in 2020. The area used for biofuels and livestock operations is increasing each year, along with our human family's ever-increasing demand.

The Independent reporter Johann Hari asks this question, "Is our planet over-stuffed with human beings? Are we breeding to excess? These questions are increasingly poking into public debate, and from odd directions. Philip Mountbatten—husband of the British monarch Elizabeth Windsor—said in a documentary screened this week [2008]: 'The food prices are going up, and everyone thinks it's to do with not enough food, but it's really [due to] too many people."[299]

Hari adds, "They say with a frown that this global swarming is driving global warming. How can you be prepared to cut back on your car emissions and your plane emissions but not on your baby emissions? Can you really celebrate the pitter-patter of tiny carbon-footprints? ... They argue that although the swelling billions are not now emitting large amounts of greenhouse gases, they will see that we are doing it and will (totally understandably) want to join in the carbon bonfire. We should be helping them by building a global anti-Vatican, distributing the pill and the words of Mary Wollstonecraft [women's advocate]. To achieve this green goal, it's necessary to mix some estrogen into the environmentalist palette."

Food Production in Times of Climate Change

It's not just that climate change is ravaging the world's agriculture.
Agriculture is also ravaging the climate.
— Jason Hickel, ForeignPolicy.com

It is safer politically to point to extreme weather events or bats or monkeys rather than human-caused disasters, for then we might have to change our behaviour or give up something. The media is not showing images of throngs of people on overcrowded streets and warning that we need to reduce our population; nor are they showing large swaths of forests being logged to make room for more cattle operations and saying we need to eat less meat. Most people do not want to be told that they will have to eat half as much meat, have fewer children, drive half as much, or reduce the amount of waste they produce.

Foreign Policy reporter Jason Hickel points out, "But while extreme weather poses a real threat to human societies (consider what Hurricane Maria did to Puerto Rico), some of the most worrying aspects of climate change are much less obvious and almost even invisible. A new 1,400-page report from the Intergovernmental Panel on Climate Change (IPCC) is a case in point. It explores the

299 Johann Hari, The Independent/UK, *Are There Just Too Many People in the World*, 15 May 2008, https://www.commondreams.org/views/2008/05/15/are-there-just-too-many-people-world

impacts of climate breakdown on the most fundamental, even intimate feature of human civilization—our food system. Consider the mighty Himalayan glaciers. When we think about melting glaciers, we mourn the loss of a natural wonder and worry about sea level rise. We don't think much about what glaciers have to do with food. But that's where the real crunch is coming."

The water that flows from Himalayan glaciers provides half of Asian's population with drinking water and other household needs. But more importantly, their food production absolutely depends on it. With these glaciers now melting much faster than they are being replaced, this present trajectory could cripple the region's food system, leaving 800 million people in crisis.

Hickel warns, "And that's just Asia. In Iraq, Syria, and much of the rest of the Middle East, droughts and desertification will render whole regions inhospitable to agriculture. Southern Europe will wither into an extension of the Sahara. Major food-growing regions in China and the United States will also take a hit. According to NASA, intensive droughts could turn the American plains and the Southwest into a giant dust bowl. Today all of these regions are reliable sources of food. Without radical emission reductions, that will change. Under normal circumstances, regional food shortages can be covered by surpluses from elsewhere on the planet. But models suggest there's a real danger that climate breakdown could trigger shortages on multiple continents at once."

Livestock production accounts ***for 70 percent of all agricultural land and 30 percent of the land surface of the planet*** *...70 percent of previously forested land in the Amazon is occupied by pastures, and feedcrops cover a large part of the remainder.*
—United Nations

According to the IPCC, **agriculture contributes nearly a quarter of all anthropogenic greenhouse gas emissions and is rapidly degrading the planet's soils.**

Famine

The truth is this: the Earth cannot provide enough food and fresh water for 10 billion people, never mind homes, never mind roads, hospitals and schools.
—Richard Branson, entrepreneur and philanthropist

Changing weather patterns and the ineffectiveness of medieval governments in dealing with crises, wars, and epidemic diseases like the Black Death helped to cause hundreds of famines in Europe during the Middle Ages, including 95 in Britain and 75 in France. During the 20th century, an estimated 70 million people died from famines across the world, of whom an estimated 30 million died during the famine of 1958–61 in China.[300]

In the 21st century, Canadians have been cushioned from the recent famines. For the first time in recent memory, there were food riots in 2007 in a host of countries ranging from Zimbabwe and Morocco to Mexico, Hungary, and Austria. Russia and Pakistan introduced food rationing for the first time in decades (and Pakistani troops were sent out to guard imported wheat). To conserve

300 Wikipedia, *Famine,* https://en.wikipedia.org/wiki/Famine

dwindling stocks, India banned the export of rice, and other big rice-eating countries, notably the Philippines, are talking of a "rice crisis" and promoting drastic measures to guarantee supply.

The world has seen large-scale food shortages and spikes in commodity prices before, but this time could be different. **We have made obtaining food much more income-dependent than ever before.** Even in the Depression in the 1930s, most people tended to know someone on a farm so they could barter for food. Doctors were often paid in eggs or potatoes. In today's world, if you don't have cash, you could starve.

The World Resource Institute (WRI) points out, "As the global population grows from 7 billion in 2010 to a projected 9.8 billion in 2050, and incomes grow across the developing world, overall **food demand is on course to increase by more than 50 percent,** and demand for animal-based foods by nearly 70 percent."[301]

The American Farmland Trust calculates that the United States loses more than one million acres of productive farmland to urban sprawl each year—an area larger than the state of Delaware. And as humanity continues to expand its footprint, wetlands and other environmentally sensitive areas are further threatened by conversion to cropland.

New technology in agriculture has also brought new health risks, warns Population Connection. Pesticide use has increased to more than 800 million pounds a year in the United States alone, and pesticides and fertilizers used in agriculture are among the largest sources of U.S. water pollution. In addition to polluting water supplies, modern agriculture's intensive water usage is depleting groundwater aquifers, resulting in water shortages and fostering reliance on energy-intensive irrigation. Unfortunately, sustainable agriculture appears to be a long ways off, as many developing countries are following North America's misguided example.

The WRI notes that technology may continue to push back the limits, but 50 percent of plants and animals are already harvested for our use, creating a huge impact on our partner species and the world's ecosystems.

Many of us remember the food shortages and ensuing food riots that occurred in 30 countries around the world in 2008, but researchers predict that these are just a coming attraction for what the future holds. This is partly because we have failed to become self-sufficient, and our average meal travels 1,000 miles before it arrives on our plate. Our breakfast smoothie may contain bananas from South America, soy milk from Japan, and orange juice from the U.S. Food from halfway around the world can encounter numerous disruptions, from droughts and wars to frosts and foods.

Although China, with its 1.4 billion people, has dazzled the world with food-technology advances, it may be in grave danger of massive food shortages in the coming years. With their cities growing and deserts expanding, their population and pollution could soon overwhelm the country, creating millions of environmental refugees. A Chinese proverb says unless we change direction, we will wind up where we are going. Unfortunately, in 2020 we still have not changed direction, and population continues to grow at an alarming rate of over 200,000 additional people per day to feed.

---

301 World Resource Institute, *Executive Summary*, 2020, https://wrr-food.wri.org/executive-summary-synthesis

Pesticides and Pollination

The Independent reporter Phoebe Weston warns us a new study has found that DDT continues to damage ecosystems 50 years after it was last used. She states, "DDT was previously used widely in industry and agriculture and was banned in the UK in 1986. Scientists studied sediments at the bottom of five remote lakes in north-central New Brunswick in Canada where planes sprayed more than 6,000 tonnes of DDT between 1952 and 1968 to manage insect outbreaks. The chemical was applied to the forests but washed into remote lakes where it has caused permanent damage, according to the study published in *Environmental Science & Technology*."[302]

Studies also show that pesticides are having a devastating impact on insects, including our most vital pollinators. Even the new "safe pesticides" designed to replace banned chemicals coud be causing harm when combined with other chemicals being applied to crops. The alarming disappearance of bees has led to anti-pesticide protests around the world in recent years. Ironically, those who discovered the wonder penicillin drug and the miracle DDT pesticide in the 20th century were both awarded the Nobel Prize. In reality, they turned loose the unprecedented human growth rate now threatening the survival of much of humankind.

As much as humans like to think that they are the centre of the universe, they will now have to step aside to the wisdom of science. Instead, **bees have been declared the most important living thing on Earth** by the Earthwatch Institute. UNILAD reporter Lucy Connolly states, "As per *The Science Times*, 70% of the world's agriculture depends exclusively on bees, while pollination—perhaps the most important function of bees—enables the plants to reproduce. Without them, the fauna would soon begin to disappear."[303]

Bees are the only living being that does not carry any type of pathogen—regardless of whether it is a fungus, a virus, or a bacterium. Despite this, bees have now been declared an endangered species, with recent studies showing a dramatic decline in the insects' numbers; **almost 90 percent of the bee population has disappeared** in the past few years due to uncontrolled use of pesticides and destruction of habitat, Connolly notes.

Basically, Connolly suggests, "Ecological farming is needed to preserve wild habitats and protect the bees. This works by avoiding large mono-crops and preserving ecosystem diversity, therefore resisting damage to insects. In turn, bee populations will be restored and so pollination will be improved, which will then improve crop yields to feed humans."

302 Phoebe Weston Science Correspondent, INDEPENDENT, *Banned pesticide DDT still damaging ecosystems 50 years after it was sprayed*, June 2019, https://www.independent.co.uk/environment/ddt-ban-pesticide-damage-ecosystems-new-brunswick-study-a8955171.html

303 Lucy Connolly, UNILAD, *Bees Have Been Declared The Most Important Living Thing On Earth*, Oct 14 2019, https://www.unilad.co.uk/science/bees-have-been-declared-the-most-important-living-thing-on-earth/

Poverty and Inequality

Poverty is the worst form of violence.
—Mahatma Gandhi

According to the UN, children in developing countries are more than twice as likely as adults to live in extreme poverty. This in itself makes the case for funding family-planning programs around the world. Extreme poverty impacts hundreds of millions of people all over the world. Global Sustainable Development Goal #1 aims to eradicate global poverty by 2030 so that every person can achieve their full potential.[304]

As discussed in earlier chapters, the **2030 Agenda for Sustainable Development**, adopted by all United Nations member states in 2015, provides a shared blueprint for peace and prosperity for people and the planet, now and into the future. At its heart are the 17 Sustainable Development Goals (SDGs), which are an urgent call for action by all countries—developed and developing—in a global partnership.291

Goal #2 is to eradicate hunger. Although fewer people are currently living in extreme poverty today than in 2000, in sub-Saharan Africa, the number of undernourished people increased from 195 million in 2014 to 237 million in 2017.

We must rethink how we grow, share, and consume our food to accommodate almost EIGHT BILLION people. If done right, agriculture, forestry, and fisheries can provide nutritious food for all and generate decent incomes, while supporting people-centred rural development and protecting the environment. The key aim is to end both hunger and population growth, recognizing the synergy between the two. The UN also recommends moving towards a plant-based diet to stop further destruction of the environment.

A profound change of the global food and agriculture system is needed if we are to nourish the 821 million people who are hungry today and the additional 2 billion people expected to be undernourished by 2050. The more sensible option would be to stop population growth at EIGHT BILLION, and reduce the increasing demand for food. Investments in family planning are crucial to helping to eradicate hunger and restore a portion of wildlife habitat to stop the alarming loss of species and forests. The UN points out:

Agriculture is the single largest employer in the world, providing livelihoods for 40 percent of today's global population. It is the largest source of income and jobs for poor rural households.

- Five hundred million small farms worldwide, most still rainfed, provide up to 80 percent of food consumed in a large part of the developing world. Since the 1900s, some 75 percent of crop diversity has been lost from farmers' fields. Better use of agricultural biodiversity can contribute to more nutritious diets, enhanced livelihoods for farming communities and more resilient and sustainable farming systems.

304 291UN Sustainable Development Goals: Hunger, https://www.un.org/sustainabledevelopment/hunger/

- **If women farmers had the same access to resources as men, the number of hungry in the world could be reduced by up to 150 million.**

Some 840 million people have no access to electricity worldwide—most of whom live in rural areas of the developing world. Energy poverty in many regions is a fundamental barrier to reducing hunger.

Population Matters notes that an individual Briton produces 70 times more CO_2 than a person from Niger, and uses nearly three times the renewable natural resources its land can provide. [305]

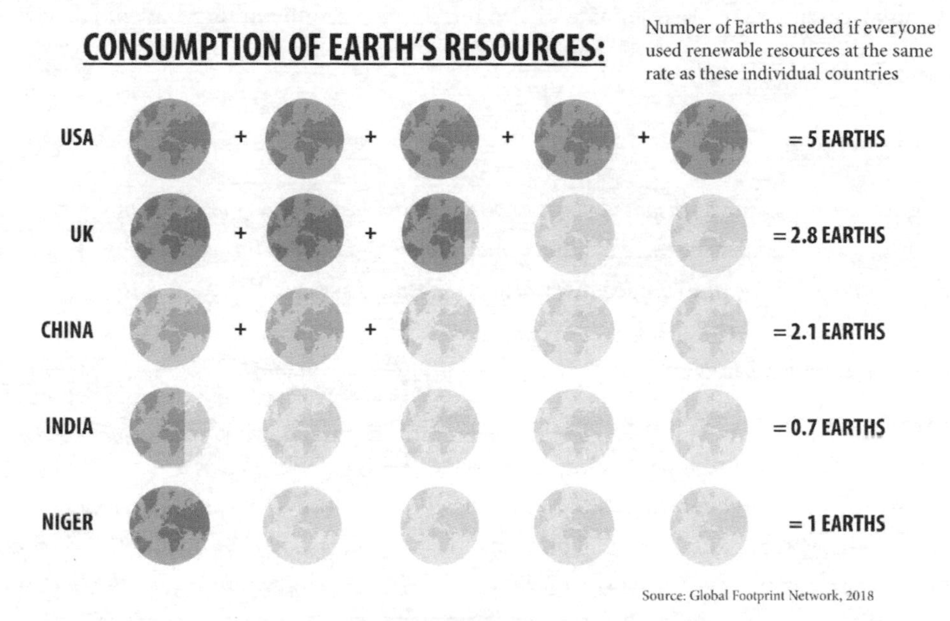

Figure 10. Global Footprint Network 2018

We also need to take into consideration the different climates. A person living in a cold climate requires far more resources just to survive. A person in the UK or Canada needs hats, coats, boots, gloves, and quilts just to survive winter, plus an insulated house and a furnace to provide heat. A person in Niger does not require any of these necessities because of their warm climate. Add to this the extremely short growing season for food crops and the enormous cost to import food in the winter, and this paints a different picture. None of these additional necessities have been taken into account in this chart. Yes, absolutely, developed countries need to reduce consumption, but the growing middle class in countries like China and India are striving towards a similar standard of living as developed countries, so more importantly we need to consider what number of humans this planet can sustain at that humane standard of living. It is certainly not ten billion, and not even seven billion, if we are going to allow any space for wild critters and old-growth forests.

More than two-thirds of the land needed to produce the UK's food and animal feed is abroad: 64 percent of the related greenhouse gases are emitted on foreign soil. The UK is one of the most "nature depleted" countries in the world, with more than one in seven species facing extinction and

305 Population Matters, *Overpopulation in the UK,* https://populationmatters.org/the-facts/uk

more than half in decline. Population density in Europe is just 34 people per square kilometre. At 426 people per square kilometre, England is the most overcrowded large nation in the EU.[306]

According to the Global Footprint Network, in 2020, Earth Overshoot Day was on July 29—so for the rest of the year we are operating in ecological deficit. They warn that in seven months, humanity has burned through the resources it takes Earth a full year to replenish.

This is overshoot, and yes it is a very big deal, because it means we are stealing from future generations and other species. Although the rate of population growth is slowing, population itself is still increasing at an alarming rate of two people per second and driving climate change to new heights.

Food and Garbage

Uncollected waste and poorly disposed waste have significant health and environmental impacts. The cost of addressing these impacts is many times higher than the cost of developing and operating simple, adequate waste management systems. Solutions exist and we can help countries get there.
—Silpa Kaza, World Bank

What are we going to do with the mountains of solid waste that come from food packaging and other garbage that is not being recycled? A 2018 World Bank report warns that global waste is predicted to grow by 70 percent by 2050 unless urgent action is taken. Their report states, "Driven by rapid **urbanization and growing populations**, global annual waste generation is expected to jump to 3.4 billion tonnes over the next 30 years, up from 2.01 billion tonnes in 2016."[307]

Mismanagement of waste is harming human health and local environments while adding to the climate challenge. It doesn't have to be this way. Our resources need to be used and then reused continuously so that they don't end up in landfills.
—Laura Tuck, World Bank

The report notes that plastics are especially problematic. If not collected and managed properly, they will contaminate and affect waterways and ecosystems for hundreds, if not thousands, of years.

Solutions and Successes

Nothing will benefit human health and increase chances for survival for life on earth as much as the evolution to a vegetarian diet.
—Albert Einstein, 1921

We all want to be responsible consumers, don't we? Jane Goodall offers her advice: "Consumers,

306 Henri de Ruiter et al, Royal Society, *Global cropland and greenhouse gas impacts of UK food supply are increasingly located overseas*, Jan 1,2016, https://royalsocietypublishing.org/doi/10.1098/rsif.2015.1001

307 World Bank PRESS RELEASE, *Global Waste to Grow by 70 Percent by 2050 Unless Urgent Action is Taken: World Bank Report*, September 20, 2018, https://www.worldbank.org/en/news/press-release/2018/09/20/global-waste-to-grow-by-70-percent-by-2050-unless-urgent-action-is-taken-world-bank-report

at least if they're not living in poverty, have an enormous role to play, too. If you don't like the way the business does its business, don't buy their products. This is beginning to create change. People should think about the consequences of the little choices they make each day. What do you buy? Where did it come from? Where was it made? Did it harm the environment? Did it lead to cruelty to animals? Was it cheap because of child slave labor? And it may cost you a little bit more to buy organic food, but if you pay a little bit more, you waste less. We waste so much. **And eat less meat.** Or no meat. Because the impact on the environment of heavy meat eating is horrible, not to mention the cruelty."

The United Nations is leading the way towards food security with their most ambitious goal for the 21st century. Food is Goal #2 of the Sustainable Development Goals: "End hunger, achieve food security and improved nutrition, and promote sustainable agriculture." Achieving this goal by the target date of 2030 will require a profound change of the global food and agriculture system.

We know that we must close the gap between demand for food and supply of food in a way that will also produce a decrease in climate change. The best ways to reduce the demand for food are:

1. Ending and reversing population growth
2. Shifting towards a plant-based diet and taxing meat and dairy globally
3. Reducing food loss and waste
4. Reducing the incidence of obesity in overdeveloped countries
5. Increasing the production of industrial hemp for a high-protein food source
6. Making fruit and nut trees a priority in our tree-planting initiatives
7. Eliminating subsidies to industrial and animal agriculture and projects that use harmful fertilizers and pesticides, and instead subsidize regenerative, organic plant-based solutions

We also know that efforts to increase supply have almost always led to more environmental damage, so these above actions should all be priorities in order to feed humanity in the future while also reducing climate change, animal suffering, and human health problems.

Supporting countries to make critical **solid waste management** financing, policy, and planning decisions must also be a key component. Since 2000, the World Bank has committed over $4.7 billion to more than 340 solid waste management programs in countries across the globe. Canada holds a record in this department, and not one to be particularly proud of. According to *USA Today*, **Canada produces the most waste in the world,** and the U.S. ranks third.[308]

While many countries now have laws pertaining to solid waste management, the amount actually salvaged from the waste stream is seriously inadequate and remains one of the most urgent and critical issues that needs to be addressed globally. More than half of waste is currently being openly dumped in **South Asia and sub-Saharan Africa**, yet they are **set to generate the biggest increase in garbage in the future**.

UN solutions include:

308 Hristina Byrnes and Thomas C. Frohlich, USA Today, *Canada produces the most waste in the world,* July 12, 2019, https://www.usatoday.com/story/money/2019/07/12/canada-united-states-worlds-biggest-producers-of-waste/39534923/

- Providing financing to countries most in need, especially the fastest growing countries, to develop state-of-the-art waste management systems
- Supporting major waste-producing countries to reduce consumption of plastics and marine litter through comprehensive waste reduction and recycling programs
- Reducing food waste through consumer education, organics management, and coordinated food waste management programs

Rethinking Economic Growth and Biodiversity

The 2019 UN Global Assessment Report on Biodiversity and Ecosystem Services warns, "Society should not fixate on economic growth and countries should base their economies on an understanding that nature is the foundation for development. Countries must begin focusing on restoring habitats, growing food on less land, stopping illegal logging and fishing, protecting marine areas, and stopping the flow of heavy metals and wastewater into the environment. Countries need to reduce their subsidies to industries that are harmful to nature, and increase subsidies and funding to environmentally beneficial programs. Restoring the sovereignty of indigenous populations around the world is also suggested, as their lands have seen lower rates of biodiversity loss. Additionally, it highlighted needed shifts in individual behaviours, such as reducing meat consumption."[309]

Protecting the diversity of bee species is also crucial to food production—as we rely on them to pollinate crops. Although their habitat is under threat from agriculture, deforestation, and climate change, there are many projects in the works to help.

Food security is also behind projects like Singapore's successful urban garden initiative. With farms atop malls, this city is getting serious about sustainable food production. The *Guardian* reporter Rina Chandran notes, "Comcrop's 600-square-metre (6,450-square-foot) farm on the roof of one of the malls uses vertical racks and hydroponics to grow leafy greens and herbs such as basil and peppermint that it sells to nearby bars, restaurants, and stores. The farm's small size belies its big ambition: to help improve the city's food security."[310]

Urban agriculture can build social cohesion among residents, improve economic prospects for growers, and have nutritional benefits. In addition, greening cities can help to transition away from traditional concrete jungles.
—Matei Georgescu, urban planner, Arizona State University

Countries across the world are battling the worsening impacts of climate change, shipping disruptions, water scarcity, and population growth to find better ways to feed their people.

With more than two-thirds of the world's population forecast to live in cities by 2050, urban agriculture is critical, and presently provides up to

309 United Nations, *Global Assessment Report on Biodiversity and Ecosystem Services,* 2019,https://en.wikipedia.org/wiki/Global_Assessment_Report_on_Biodiversity_and_Ecosystem_Services

310 Rina Chandran, Thomson Reuters Foundation, *With Farms Atop Malls, Singapore Gets Serious About Food Security,* Jan. 8, 2019, https://www.globalcitizen.org/en/content/rooftop-farm-food-security-singapore/?utm_source=twitter&utm_medium=social&utm_content=global&utm_campaign=general-content&linkId=80836792

10 percent of global output of pulses and vegetables. Additional benefits, such as reduction of the urban heat-island effect, avoid storm water runoff, nitrogen fixation, and energy savings could be **worth $160 billion annually**, reports say.

Mobilizing the World's Wealth

Although developing countries suffer unduly from poverty, every world city has a sector of society that is often homeless and food deprived. But it doesn't have to be this way. There is certainly enough wealth on this planet to eliminate this problem, as well as the overpopulation problem that usually accompanies hunger. According to their website, Move Humanity is a new global initiative aiming to mobilize each year at least 1 percent of the wealth of the world's super-rich for the Sustainable Development Goals, either through philanthropy or taxes.

It highlights the power and potential that the world's wealthiest individuals can have by donating just 1 percent of their wealth each year to addressing this century's most pressing challenges. The initiative aims to help close the SDG financing gap in the lowest-income countries (LICs) by mobilizing greater private funding for basic health and education, critical infrastructure, and environmental conservation priorities.

Move Humanity is a partnership of Human Act, Copenhagen and the UN Sustainable Development Solutions Network. With the billionaires' wealth estimated to be around $10 trillion, Move Humanity calls for mobilizing **at least $100 billion per year in philanthropy and wealth taxation of the world's estimated 2,208 billionaires** on behalf of the SDGs.[311]

The 2019 UN IPBES Global Assessment Report presents an illustrative list of other possible actions and pathways for achieving them. They assert that these actions and societal initiatives are helping to raise awareness about the impact of consumption on nature, protecting local environments, promoting sustainable local economies, and restoring degraded areas.

This report offers the best available expert evidence to help inform these decisions, policies, and actions—and provides the scientific basis for the biodiversity framework and new decadal targets for biodiversity, to be decided in 2021 in China, under the auspices of the UN Convention on Biological Diversity sectors.[312]

> *From the young global shapers behind the #VoiceforthePlanet movement, to school strikes for climate, there is a groundswell of understanding that urgent action is needed if we are to secure anything approaching a sustainable future.*
> **—Sir Robert Watson, IPBES chair**

Quantifying the **role that population plays** in food security is "an incredibly powerful piece of information," said Kathleen Mogelgaard, consultant

311 Move Humanity, *This Inequality Cannot Go On. Ask the World's Richest People to Help End Extreme Poverty,* 2020, https://movehumanity.org/

312 UN- IPBES report 2019, *Nature Decline Unprecedented - Report,* webitehttps://www.un.org/sustainabledevelopment/blog/2019/05/nature-decline-unprecedented-report/

for the Environmental Change and Security Program.[313] Mogelgaard argues that we must **incorporate population and reproductive health considerations into strategies to address future food security and climate adaptation** "in ways that will build the resilience of women and build the resilience of communities around the globe."

Going Plant-Based

In the Māori worldview, the rivers, lakes, and forests are our ancestors. They are part of us and we are part of them. Right now, our rivers and forests are sick, and intensive animal agriculture, and especially dairy in New Zealand, have played a huge role in that.

—Lauren O'Connell Rapira, a vegan Māori woman

Meat consumption is considered one of the primary contributors of the sixth mass extinction. A 2017 study by the World Wildlife Fund found that **60 percent of global biodiversity loss is attributable to meat-based diets**, in particular from the vast scale of feed crop cultivation needed to rear **tens of billions of farm animals for human consumption**, which puts an enormous strain on natural resources resulting in a wide-scale loss of lands and species. [314]

Currently, livestock make up 60 percent of the biomass of all mammals on earth, followed by humans (36 percent) and wild mammals (4 percent). In November 2017, 15,364 world scientists signed a Warning to Humanity calling for, among other things, drastically diminishing our per capita consumption of meat and "dietary shifts towards mostly plant-based foods." The 2019 IPBES report also recommends reductions in meat consumption in order to mitigate biodiversity loss.[315]

A July 2018 study in *Science* says that meat consumption is set to rise as the human population increases along with affluence, which will increase greenhouse gas emissions and further reduce biodiversity. The good news is that the UK National Trust has plans to remedy that by converting beef farms into forests of 20 million trees. Live Kindly senior editor Charlotte Pointing states, "In a bid to help the environment and become carbon neutral, The National Trust plans to plant trees on 44,000 acres of its land."[316]

Acknowledging that people need nature now more than ever, The National Trust is encouraging people to eat climate-friendly plant-based food as another way of reducing environmental impact. Richard Deverell, director of Britain's Royal Botanic Gardens, advises, "The campaign should start with consumers being made aware of what they're eating and the consequences of that. Do people know where their food is coming from? I was in San Francisco last autumn and I was surprised

313 Carolyn Lamere, New Security Beat, *Kathleen Mogelgaard on How Malawi Shows the Importance of Considering Population, Food, and Climate Together,* October 24, 2012, https://www.newsecuritybeat.org/2012/10/kathleen-mogelgaard-malawi-shows-importance-population-food-security-climate-change/

314 Anna Starostinetskaya, VegNews, *Meat-Based Diets Destroying 60 Percent of Biodiversity*, Oct 9, 2017, https://vegnews.com/2017/10/meat-based-diets-destroying-60-percent-of-biodiversit

315 Wikipedia, *Meat Consumption and Biodiversity Loss*, https://en.wikipedia.org/wiki/Meat

316 Charlotte Pointing, Live Kindly, The National Trust Is Converting Beef Farms Into Forests Of 20 Million Trees, January 17, 2020, https://www.livekindly.co/nation-trust-beef-farms-forests

to see the number of posters up urging a vegan diet—in 20 years time, will eating meat in San Francisco be as socially unacceptable as drinking and driving is now?"[317] Deverell suggests ways to eat more sustainably and avoid a global food crisis:

1. **Eat smaller portions.**
2. **Eat less meat, perhaps fewer times per week.**
3. **And try to reduce the miles that your food has had to travel.**
4. Grow your own food on a balcony, a roof, or in a garden. Not only is it really enjoyable, but it also **reconnects individuals with nature**.

Some people would like to take that thinking even further. The EU wants to introduce a sustainability tax on meat to reduce contributions to the climate crisis and to correct the artificially low prices of meat. Around the world, **meat and dairy have been subsidized** for decades **by our tax dollars.** This change could reduce the EU's carbon dioxide emissions by 120 million tonnes. The proposal would be added to the European Commission's *Farm to Fork Strategy,* which aims to create a more sustainable food system by implementing further restrictions on the use of pesticides, fertilizes, and antibiotics. As age-old methods of providing food fail, **almost one in four food products launched in UK in 2019 were labelled vegan.**[318]

As a dietary caution, both the World Health Organization and the Cancer Council warn that meat can increase our risk of cancer. Current research shows that there are certain chemicals in red and processed meats that cause these foods to be carcinogenic.[319]

Switzerland is also taking action with plans to ban all factory farms, so animal agriculture in Switzerland could look very different in the future. "The fact that we were able to submit this initiative so quickly shows how much people in Switzerland care about animals. The majority of them are unaware of the significant amount of individuals that are still being raised in unbearable conditions. Factory farming is unacceptable—and we should use the power of direct democracy to also make it illegal," states Live Kindly.[320]

The time has come for us to act decisively with policy on the ***environmental consequences of animal protein,*** *the price of which has been kept artificially low for far too long.*
—Philip Mansbridge, Executive Director of ProVeg

Supporters of the ban argue that factory farming is linked to poor animal welfare. "**Fifty percent of all piglets raised in Switzerland are slaughtered without ever seeing the sky.**" More research is uncovering animal agriculture's impact on the planet. The United Nations Environment

317 Joseph Hincks, Time.com, *The World Is Headed for a Food Security Crisis. Here's How We Can Avert It,* March 28, 2018, https://time.com/5216532/global-food-security-richard-deverell/

318 The Vegan Kind News, 17 January 2020, *Almost one in four food products launched in UK in 2019 labelled vegan,* https://www.thevegankind.com/news/2020-01-17-almost-one-in-four-uk-food-products-launched-in-2019-labelled-vegan

319 Cancer Council, How red and processed meat increase cancer risk, 2020, https://www.cancercouncil.com.au/1in3cancers/lifestyle-choices-and-cancer/red-meat-processed-meat-and-cancer/

320 Kat Smith, Live Kindly, *Switzerland May Soon Ban All Fatory Farms,* Feb 3 2020, https://www.livekindly.co/switzerland-ban-factory-farms/?utm_source=ActiveCampaign&utm_medium=email&utm_content=Victory+for+Animals&utm_campaign=LIVEKINDLY+Sun+Newsletter+February+9+2020

Programme (UNEP) named meat "the world's most urgent problem" in September 2018. "Our use of animals as a food-production technology has brought us to the verge of catastrophe," UNEP said in a statement.

Scotland also sees the elegance of this idea as they too move towards a vegan lifestyle. "Don't think the vegan trend has hit the mainstream yet? It's so popular even prisons are ditching animal products. This Christmas, ten of Scotland's prisons will serve up vegan options to prisoners for Christmas Dinner and on New Year's Day. Scotland also recently just got another vegan supermarket and scores of vegan restaurants have opened in Edinburgh and Glasgow.... Some 60 percent of Scottish dairy farmers said they see the industry as being in 'full decline.'"[321]

That survey followed up on an Oxford Farming Conference poll earlier this year. "**It's time for the farming community to get their head around veganism**," journalist and farmer Darragh McCullough wrote in the *Irish Independent* after the conference.

"Farmers need to stop claiming that if we had to rely on vegans to feed the world, we'd all go hungry. It's a lame attempt to ignore the real point," he wrote. "Equally, claims that you cannot get enough nutrients from a vegan diet ignores the fact that a big chunk of the world's population in places like India have effectively been living off a vegetarian diet for centuries."

Scotland's prisons aren't the only correctional facilities to embrace vegan food, either. UK prisons served vegan Christmas dinner last year to thousands of inmates. New York City just passed a resolution requiring its prisons to offer inmates vegan options. And California has been working to increase its state vegan options in prisons as well. Actress Pamela Anderson recently encouraged Justin Trudeau to serve Canada's prisoners vegan meals.

"Canada could be a true leader in saving our planet for future generations," she wrote. "Beans, rice, lentils, pasta, and potatoes and other vegetables, as well as fruits, supply all the nutrients that anyone needs."

> *Fortunately, foods that are health-promoting tend also to be those that are climate-friendly.*
> **—Dr. Alex Macmillan, University of Otago**

Around the world, there is a rising chorus from animal rights groups, youth, science, and enlightened government leaders proclaiming that we must end the cruelty of animal agriculture. According to Live! Be Vegan, a new short film called *The Nightmare Before Christmas* exposes animal cruelty on UK turkey farms.[322]

There are other reasons that we should be turning vegan, and New Zealand has found that they could drop GHGs by 42 percent with a shift to a vegan diet, according to a study of animal agriculture's impact on climate change by University of Otago. Live Kindly senior editor Kat Smith explains, "This includes farming and processing, packaging, transportation, warehouse and

321 Jill Ettinger, Live Kindly, *10 Scottish Prisons to Serve Up Vegan Christmas Dinner*, December 23, 2019, https://www.livekindly.co/scottish-prisons-vegan-christmas-dinner/

322 Live Be Vegan, Dec 20, 2019, *The 'Nightmare Before Christmas' Exposes Holiday Dinner Animal Cruelty*, https://livebevegan.com/vegan-news/the-nightmare-before-christmas-exposes-holiday-dinner-animal-cruelty/

distribution, refrigeration needs, and supermarket overhead. The team then assessed the climate, health, and health system cost impacts of different diets."[323]

Population Density Stress

Dr. Gregg Miklashek with Millennium Alliance for Humanity and the Biosphere believes that population density stress is killing us **now!** He notes, "You may not want to hear it, and when I started medical practice 46 years ago, I certainly did not plan on finding it, but human overpopulation and our modern life-style choices are causing physiological changes responsible for our top ten killing 'diseases of civilization.' My training was in medicine and psychiatry, and the majority of my patients suffered from 'anxiety' and 'depression,' but I became increasingly aware of the direct association of their psychiatric problems with other general medical conditions."[324]

According to NutritionFacts.org, "A single meal containing animal products may double the amount of cortisol—a steroid hormone—in the blood, as opposed to cortisol levels actually **going down after a vegan meal.**"[325]

Miklashek explains that eventually he came to realize that nearly all psychiatric conditions, and most general medical problems as well, could be explained as resulting from our overactive stress response. Our chronically overactive stress response was generating abnormally elevated blood levels of the adrenal stress hormone cortisol, and crowded animal research dating back over 100 years indicated a direct connection between these elevated cortisol levels and the comparable diseases of civilization in crowded, overpopulated research animals. Animals have been proven to also experience high levels of anxiety and stress from overcrowding, along with elevated cortisol levels. Ironically, almost none of our myriad and rapidly increasing "diseases of civilization" are to be found in traditional living hunter-gatherer migratory, sparsely populated, physically active clans/bands living on their marginal lands worldwide.

"Our chronically over-active stress response, due to **population density stress**, is quite likely responsible for our epidemics of anxiety and depression, which often spill over into the other epidemic health problems including suicide, alcoholism, and addiction, newly named *the diseases of despair*!" Miklashek adds.

Miklashek notes, "The solution to the problem of our rapidly increasing diseases of civilization, our increasing cortisol levels or their collapse, and the human overpopulation driving them is simple. We need to voluntarily restrict our future worldwide reproduction to one-child per couple on average, which will bring our worldwide population down to the 1950 level of two and a half billion by 2,100."

---

323 Kat Smith, LiveKindly, *New Zealand Could Drop GHGs by 42% With Shift to Vegan Diet*, February 3, 2020, https://www.livekindly.co/new-zealand-could-drop-ghgs-shift-vegan-

324 Gregg Miklashek, Millennium Alliance for Humanity and the Biosphere, *Population Density Stress is Killing Us Now!*, November 14, 2019, https://mahb.stanford.edu/blog/population-density-stress-killing-us-now/

325 Nutrition Facts.org, *Cortisol*, https://nutritionfacts.org/topics/cortisol/

Vegan Foods: a Booming Business

The U.S. is also noticing benefits as their vegan meat sales are set to hit $1 billion for the first time ever. *Plant Based News* reporter Maria Chiorando stated, "US vegan meat sales skyrocketed by 18 percent in 2019—taking the value of the sector to close to $1 billion ($939 million), according to data released by the Good Food Institute (GFI) and the Plant Based Foods Association. Plant meat is the second-largest category in the plant-based sector, behind plant-based milk. Category sales grew more than six times faster than conventional meat in the same time period." [326]

The Good Food Institute Associate Director of Corporate Engagement Caroline Bushnell said, "This is really only the beginning: plant-based foods will continue to expand rapidly across the store in response to demand as consumers increasingly switch to foods that match their changing values and desire for more sustainable options. We expect further acceleration in 2020."

This tipping point in sales is primarily due to health concerns, availability, and the curiosity of flexitarians, as a major shift claims the appetites of citizens around the world. According to *Numerator,* 80 percent of U.S. shoppers intend to substitute some or all meat with plant-based foods.

There is also a growing awareness of where our food comes from that is causing this shift. The use of the Internet to distribute information regarding the treatment of animals in the food industry has created alarm over inhumane practices. It is not always easy to obtain this footage. *Food Safety News* reporter Dan Flynn notes, "Animal activists are challenging state laws because they inhibit their undercover investigations from identifying and documenting animal abuse. With evidence, they do 'damage an enterprise,' mostly by associating farms with consumer brands. Brands often drop farms involved in animal abuse controversies."[327]

Finally in 2020, the oldest "ag-gag" law in the U.S. was ruled unconstitutional. Flynn adds, "Federal judges did not much like the newer laws to protect animal agriculture from prying eyes. So-called 'ag-gag'" laws adopted during the last decade in Utah, Idaho, and Iowa were struck mainly down as unconstitutional when challenged by animal rights activists. Now a federal judge has ruled that Kansas cannot bar the public from taking pictures or recording videos of animal agriculture even if the intent is to damage an enterprise."

To do so, according to U.S. District Judge Kathryn Vratil, unconstitutionally criminalizes free speech. Stephen Wells—the executive director for Animal Legal Defense Fund—said in a statement, "for 30 years, Kansas lawmakers have suppressed whistleblowers from investigating cruel conditions on factory farms with this unconstitutional law." He added that the decision is "a victory for the millions of animals" on factory farms.

Preventing Exploitation of Our Wildlife

But it isn't just domesticated animals that have people concerned, as the sale of wildlife for food and trophies is just as disturbing. Live Kindly reporter Charlotte Pointing asserts, "Buying giraffe

326 Maria Chiorando, Plant Based News, US Vegan Meat Sales Grow Six Times Faster Than Conventional Meat Sales, Mar 5, 2020, https://www.plantbasednews.org/news/-us-plant-based-meat-sales-skyrocket-1-billion
327 Dan Flynn, Food Safety News, Kansas becomes 4th state with unconstitutional "ag-gag" law, January 29, 2020, https://www.foodsafetynews.com/2020/01/kansas-becomes-4th-state-with-unconstitutional-ag-gag-law/

body parts is a thing, and the state of New York just banned it." The Humane Society of the United States (HSUS) notes that New York is the first in the world to do so.[328]

Giraffes are a vulnerable species, with only 68,000 mature individuals currently alive in the wild. Habitat destruction, bushmeat hunting, and trophy hunting have contributed to the decline in the species. The U.S. has contributed significantly to the latter threat. According to the Humane Society International, every day at least one giraffe trophy is imported into the U.S. In the years between 2006 and 2015, around 40,000 giraffe articles were brought into the country.

"We thank Gov. Cuomo and New York for standing tall for giraffes today," wrote Kitty Block, the president and CEO of HSUS, in a press release earlier this month. "No giraffe should have to die for a knife handle or a jacket, and we cannot act soon enough to end our nation's role in the trafficking of giraffe body parts."

Let's End Factory Farming

Pointing has also reported that vegan presidential candidate Cory Booker has a plan to end factory farming forever in America. In 2019, the New Jersey senator put forward a new bill titled the Farm System Reform Act (FSRA), aimed at transitioning agriculture away from the factory farming system.[329]

She points out that in the U.S., 99 percent of all animal products currently come from factory farms. These conditions are not only harmful to the animals, but they're a threat to the planet too. **One farm of 5,000 pigs can produce as much waste as a town of 20,000 people**, according to animal protection organization Make It Possible. This waste can pollute soil and make its way into water systems, including oceans, rivers, and streams.

If passed, FSRA will place an immediate moratorium on new large-scale facilities known as CAFOs (confined animal feeding operations). It will also place limits on existing CAFOs. It will devote $100 billion over a period of ten years to helping CAFO owners transition to more sustainable forms of agriculture. By 2040, the bill hopes to phase out the largest CAFOs. Booker, who follows a vegan lifestyle, hopes the new bill will help small farmers and hold larger corporations to account for their environmental impact.

Joining in this movement, Bernie Sanders also slammed factory farms as a threat to America, writing in a tweet in May, "Factory farms are responsible for 1.4 trillion pounds of animal waste in America. They are a threat to the water we drink and the air we breathe," he continued. "It is unbelievable to me that Republicans in Congress have been working overtime to exempt factory farms from environmental laws."

The greenhouse gas footprint of animal agriculture rivals that of every car, truck, bus, ship, airplane, and rocket ship combined.
—United Nations

It is the elephant in the room here that every person added to the planet will require food, and is **another**

328 Charlotte Pointing, Live Kindly, *Buying Giraffe Body Parts Is a Thing and New York Just Banned It*, December 20, 2019, https://www.livekindly.co/giraffe-body-parts-new-york-banned/

329 Charlotte Pointing, Live Kindly, *Cory Booker has a plan to end factory farming forever*, December 20, 2019, https://www.livekindly.co/cory-booker-end-factory-farming/

person who has to be convinced not to eat meat. Reducing population growth is by far the easiest and most effective way to reduce the number of animals being slaughtered for food. Converting to plant-based food is the second-most efficient way, and we are starting to do that in Canada. During COVID-19, the demand for beans skyrocketed, according to a Canadian supplier.

"Grocery store demand for beans during the COVID-19 pandemic is up between '500 and 1,000 per cent,' according to Canada's biggest supermarket supplier, Agrocrop Exports," reported CTV News reporter Cillian O'Brien. Canada is "by far the world's largest grower of beans and lentils," and Agrocrop is the "largest private label processor and exporter of consumer packaged beans and lentils to Europe and the USA."[330]

Hard-working vegan groups have brilliantly introduced the world to the benefits of going vegan in the last decade, and now we are celebrating a surge in vegan products being consumed during this pandemic. This fresh dose of realism has attracted celebrity vegan activists like Joaquin Phoenix and Rooney Mara to support the cause. Live Kindly reporter Audrey Enjoli tells us the duo intends to expose the horrors of factory farming, since the animal agriculture industry breeds virus outbreaks.[331]

> *When we overcrowd animals by the thousands, in cramped football-field-size sheds, to lie beak to beak or snout to snout, and there's stress crippling their immune systems, and there's ammonia from the decomposing waste burning their lungs, and there's a lack of fresh air and sunlight—put all these factors together and you have a perfect storm environment for the emergence and spread of disease.*
>
> **—Dr. Michael Greger, plant-based physician and author**

Enjoli adds that during an episode of HBO's *Real Time With Bill Maher*, Maher said, "America's factory farming is just as despicable as a wet market and just as problematic for our health. Factory farming has a lot more lobbyists, but ecological timebombs tick the same," he continued. "Americans should not get too high and mighty about wet markets while we are [factory farming. Animals on factory farms also live miserable lives in crowded, unsanitary conditions."

A surprising new turn is also taking place in Australia, known as one of the biggest meat-eating countries in the world. *Guardian* reporter Calla Wahlquist relays Craig Buchanan's tale:

> There is a scene in the documentary *Blue Planet II* where a harlequin tuskfish, a 30cm long inhabitant of the Great Barrier Reef, uses a bump on a coral to break open a clam.The coral is littered with broken clam shells. The fish clearly use this spot a lot. It strikes the coral repeatedly until the clam gives way. The scene lasts three minutes and ends with a quiet word from Sir David Attenborough. "So here is a fish that uses tools," he says. "Some fish are much cleverer than you might suppose."

330 Cilian O'Brien, CTV News, *Demand for beans skyrocketing during COVID-19 pandemic: Canadian supplier*, March 25, 2020, https://www.ctvnews.ca/health/coronavirus/demand-for-beans-skyrocketing-during-covid-19-pandemic-canadian-supplier-1.4868039

331 Audrey Enjoli, Live Kindly, *Joaquin Phoenix and Rooney Mara Expose Horrors of Factory Farming*, April 28, 2020, https://www.livekindly.co/joaquin-phoenix-rooney-mara-horrors-factory-farming/

> Sitting on his couch in Rockingham, Western Australia, Craig Buchanan was astonished. "It blew me away that fish could do something as complex as that," he says. This is the kind of awareness it will take to convince billions of people that **fish are intelligent, have feelings, and are as worthy of living as we are.** Despite the growing threats from the fishing industry producing over a third of the plastic pollution in the ocean, and humanity dumping an avalanche of toxic chemicals in our seas, it is not too late to save the remaining fish species if humanity becomes plant based eaters."[332]

Perhaps "Gone Fishing" will become just a reminder of a tragic era now left behind. All those signs and fish trophies mounted on walls will end up in fish museums. I hope I live to see that day!! If you are looking for a little vegan fun to celebrate your cruelty-free lifestyle, Australia offers eight vegan festivals, with the most notable one held in October on World Vegan Day in Melbourne—and it's even free! Cruelty Free reporter Michael Olei states, "This festival has a lot going on including live music, presentations, fashion shows, food stands, art exhibition and a kids area."[333]

Animas Australia notes that an additional 400,000 Australians have chosen kinder, more sustainable vegetarian meals since 2016 alone! This great news reflects the exciting shift towards plant-based eating across Australia—**with one in three people** already eating vegetarian or actively reducing meat. This works out to **one person deciding to eat less meat or go meat-free every five minutes!**[334]

This comes with some amazing benefits: Every year, up to **11 million *more* farmed animals** will be spared from slaughter, and many millions more marine animals. Every day, an additional **760 million litres of water** will be saved—in other words, more than 300 Olympic swimming pools worth of fresh water each day.

> *Animals raised for human consumption not only contribute more to global warming than all of the world's transport combined, they consume more food than they produce, use more water and create more waste than plant-based foods.*
> **—2019 Oxford study**

This study found that if the whole world stopped eating animals, the impacts on public health and global warming would be astronomical. The research showed that by eating plant-based foods, **by 2050 we could save US$1 trillion in health care and cut greenhouse emissions by 70 percent**. Add to this the countless animals' lives that would be spared, and the world looks like a much cleaner, healthier, kinder place.[335]

More great news: in the U.S., more than 3,200 dairy farms were shut down in 2019 because people are drinking less dairy milk. Live Kindly's Audrey Enjoli states, "The US Department of

---

332 Calla Wahlquist, The Guardian, *It just didn't make any sense why Australians are turning away from meat*, Jul 22 2019, https://www.theguardian.com/lifeandstyle/2019/jul/20/it-just-didnt-make-any-sense-why-australians-are-turning-away-from-meat

333 Michael Olei, *Cruelty Free, 8 Vegan Festivals to Book in Your Calendar Each Year*, October 4, 2018, https://www.crueltyfreesuper.com.au/blog/8-vegan-festivals-to-book-in-your-calendar-each-year

334 Animals Australia, *Surge in Aussies eating vegetarian continues*, February 4 2020, https://www.animalsaustralia.org/features/study-shows-surge-in-Aussies-eating-veg.php

335 Animal Australia, *Oxford study shows what a vegan world could look like*, 16 October 2019, https://www.animalsaustralia.org/features/this-could-save-8-million-lives.php

Agriculture (USDA) reports the number dropped nearly nine percent since the year before. The consumption of milk has plummeted 40 percent since 1975."[336]

Contributor to *The Island Now*, Rabbi Tara Feldman, is also considering humanity's threat to nature and how Jews can turn the tide. He states, "As Tu B'Shevat grows closer, Jews around the world pause to **consider our connection to the land, trees in particular**. But this 'birthday' is a somber one. It's been a difficult year for trees—and the Earth—which reminds us of the connection between what we eat and the environment. Thousands of fires destroyed more than 2 million acres of unique, biodiverse Amazon rainforest. Many of those **fires were intentionally set to clear land for cattle grazing** and feed-crop production."[337]

Consider these points that Feldman makes:

1. Meat and dairy production is the leading cause of deforestation, taking up an astounding 30 percent of the Earth's surface.
2. In the United States alone, 80 percent of agricultural land is used for raising animals and growing feed crops.
3. Meat and dairy production is responsible for at least 15 percent of global greenhouse gas emissions, 80 percent of antibiotic use, and 37 percent of pesticide use in the United States.
4. Factory farms and slaughterhouses are a major cause of air and water pollution, as well as dangerous, unjust working conditions and animal cruelty.
5. Producing meat uses about ten times more water than soybeans, as 2,400 gallons are necessary to produce one pound of meat, while only 200 gallons are needed to grow one pound of soybeans.
6. By giving up one hamburger a week, you can save enough water to take a five-minute shower every day for nearly five years.

Feldman urges, "What we eat is about more than personal choice—it's about family, community, and culture. When we choose an Earth-friendly diet, we can inspire those around us to do the same and **create new traditions that change markets and agricultural systems**."

Solutions to Food Waste

It is also about what is on our plates that gets thrown into the garbage, and all the other ways we are producing food waste. Not only are we wasting a lot of money, but all of this food waste is a nightmare to dispose of properly. There are many opportunities to reduce this waste through the use of food apps, from finding cheaper restaurant meals, swapping and sharing food, or getting reduced prices at your supermarket. This puts the power to change in your hands. Or you can focus on changing your household regimen to prevent excessive buying and waste—also compost and reduce overeating.

336 Audrey Enjoli, LiveKindly, *More than 3,200 dairy farms shut down in 2019, March 4 2020,* https://www.livekindly.co/dairy-farms-shut-down-2019/

337 Rabbi Tara Feldman, The Island Now, *Readers Write: Considering our impact on TuB'Shevat,* February 18, 2020, https://theislandnow.com/opinions-100/readers-write-considering-our-impact-on-tu-bshevat/?

As we saw earlier in this chapter, **Canada produces the most waste in the world per person**, and food is a big part of that waste. From the producers in the field, to the transporters, to the manufacturers, to the retailers, to the consumers, we all have a role to play in reducing food waste. In times of anxiety, such as during the COVID-19 pandemic, it might be worth taking a moment to think about preventing food waste and resisting panic buying. When food gets thrown out and ends up in a landfill, it releases greenhouse gases as it decays. A 2019 report by Second Harvest states that more than half of all food produced in Canada is lost or wasted, and about a third of that wasted food could be "rescued" and sent to communities in need across the country.[338]

The report, entitled "The Avoidable Crisis of Food Waste," was released in 2019 by Second Harvest, a Toronto agency that collects surplus food throughout the supply chain and distributes it to agencies ranging from shelters and drop-in centres to breakfast programs and summer camps.

According to the research, some 4.82 million tonnes of food, or nearly $21 billion worth, is lost or wasted during the processing and manufacturing process. Some 2.38 million tonnes of food, or more than $10 billion worth, is lost at the consumer level each year. In total, the value of all food that is lost or wasted in **Canada is a staggering $49 billion annually**, the report says. The amount of food is enough to feed every Canadian for five months, Second Harvest CEO Lori Nikkel told reporters. The annual cost of avoidable food loss and waste in Canada is $1,766 per household.

"The outcomes of this report make it very clear that we need to radically change how we as Canadians value food," Nikkel said during a news conference. "The abundance of food we produce has led us to dismiss its **intrinsic** value."

This waste comes with environmental impacts. Each year, food waste in Canada creates some 56.6 million tonnes of carbon dioxide–equivalent emissions, according to the report. Food in landfills also creates methane gas, which is "25 times more damaging to the environment than carbon dioxide," the report says.

> *From 2010 to 2016, global food loss and waste contributed eight to 10 per cent of human-caused GHG emissions.*
> **—UN Intergovernmental Panel on Climate Change**

Second Harvest dubs the research as a "world first" because it measures weight using "a standardized system across the whole food value chain," and includes all food types from both land and water. The report identifies a number of "root causes" of avoidable food loss and waste:

- Consumers buying food at the grocery store, particularly when there's a sale, and throwing the surplus away.
- Consumers and retailers throwing out food near or past its best-before date, despite the fact product dating practices "have no correlation to food safety" and the food can often still be eaten or donated.
- Produce being left to rot in the field due to labour shortages, or low prices creating an environment in which it is no longer worth it for farmers to harvest.
- Thousands of acres of produce being "plowed under" due to cancelled orders.

338 Second Harvest, *More than half of all food produced in Canada is lost or wasted, report says,* January 17, 2019, https://secondharvest.ca/news/more-than-half-of-all-food-produced-in-canada-is-lost-or-wasted-report-says/

- Fish being caught and tossed back into the water to die if they don't match a quota.

Second Harvest also outlined dozens of recommendations for industry and government aimed at preventing food loss and waste at the source, improving the redistribution and donation of edible food, and managing waste when it does occur. Much of this food could be rescued and eaten by the consumer—you. There's an app for that, in fact more than one.

Flashfood is a start-up that partnered with grocery giant Loblaw in 2020 to sell surplus food at a discount. Second Harvest also suggests an app called Feedback. It offers discounts of 15 to 50 percent at restaurants at off-peak times. In the U.S., there are similar apps, such as the Food For All restaurant app, OLIO which helps neighbors swap food and is now available in 49 countries, and Download and Dig which is particularly popular in Europe. YourLocal can get you up to 70 percent off!!

Globe and Mail reporter Ann Hui points out that the average Canadian family will pay about $480 more for groceries in 2020 when compared to 2019. If we were to reduce our wasteful habits, produce more fruits and vegetables locally, and move towards a plant-based diet, we could avert this price hike and benefit the environment at the same time."[339]

This rise in food prices helps to dispel the myth of green growth. According to *Foreign Policy* reporter Jason Hickel, growth can't be "green." **New data proves you can support capitalism or the environment—but it's hard to do both.**

Hickel emphasizes:

> Warnings about ecological breakdown have become ubiquitous. Over the past few years, major newspapers, including the *Guardian* and the *New York Times*, have carried alarming stories on soil depletion, deforestation, and the collapse of fish stocks and insect populations. These crises are being driven by global economic growth, and its accompanying consumption, which is destroying the Earth's biosphere and blowing past **key planetary boundaries that scientists say must be respected t**o avoid triggering collapse.[340]

New data argues that with the right economic and government policies, we can end infinite growth and reverse overpopulation and environmental destruction. Instead of using the flawed GDP to define policy and prosperity, we must adopt the genuine progress indicator (GPI), which would measure success by human and planetary health. Michael Moore points out in his compelling 2020 movie, *Planet of the Humans*, that "green growth" simply isn't working. Almost all of the world's critical issues are getting worse instead of better.

In this chapter, numerous UN and scientific studies have revealed the shocking reality of what will happen to global resources if economic growth continues on its current trajectory. We have breached our limits, and it will take a monumental effort to reverse this trend. The good news is that we know what the solutions are, and COVID-19 has shown us that humanity does have the ability to make radical changes when in a crisis.

---

339 Ann Hui, Globe and Mail, *Average Canadian family will pay about $480 more for groceries in 020, major study predicts*, December 4, 2019, https://www.theglobeandmail.com/business/article-average-canadian-family-will-pay-about-480-more-for-groceries-in-202/

340 Jason Hickel, Foreign Policy, September 12, 2018, *Why Growth Can't Be Green* https://foreignpolicy.com/2018/09/12/why-growth-cant-be-green/

CHAPTER 9
WATER

We live in a water-challenged world, one that is becoming more so each year as 80 million additional people stake their claims to the Earth's water resources.
—Lester Brown, Worldwatch Institute

Fresh Water

How would you feel if you woke up one day and there were almost a quarter of a million more people around you than the day before? What if this happened the next day too, and the next? Would you start to panic just a little? Well, this is exactly what is happening. Every day, the world's population increases by over 220,000 people. In 2020, there are **2.2 billion people who still do not have access to clean drinking water**, according to the United Nations.[341]

Because water has traditionally been an abundant resource, and essentially free, we have been taking it for granted. However, it is now increasingly clear that with a population of almost EIGHT BILLION, Earth's fresh water supplies are becoming a scarce resource. The Worldwatch Institute recommends that the time has come when we need to begin thinking about trying to stabilize or reduce population everywhere, limiting families to two children.

Lester Brown at the Worldwatch Institute warns that **population growth is sentencing millions of people to water poverty**. He states, "Unfortunately, nearly all the projected 3 billion people to be added over the next half century will be born in countries that are already experiencing water shortages. Even now many in these countries lack enough water to drink, to satisfy hygienic needs, and to produce food."[342]

In 2020, world leaders have yet to remedy global water shortages. This is partly because the demand for water has tripled since 1950, causing increasing conflict and diminished quality of life. We are also seeing water tables fall on every continent as **aquifers are being exhausted**. Some experts point out that when the demand exceeds the supply, the problem can't be solved.

About 70% of the water pumped out of underground aquifers worldwide is used for agriculture while much of the remainder quenches the thirst of cities.
—Eric Roston, Bloomberg

341 UN Water, *WHO and UNICEF launch updated estimates for water, sanitation and hygiene*, June 20, 2019, https://www.unwater.org/who-and-unicef-launch-updated-estimates-for-water-sanitation-and-hygiene/

342 Lester R. Brown, Earth Policy Institute, *Population Growth Sentencing Millions to Hydrological Poverty*, June 21, 2000, http://www.earth-policy.org/plan_b_updates/2000/alert4

Bloomberg reporter Eric Roston warns that in 2019 **river flows all across the globe are dropping**. He states, "As industrial development spreads at a speedy clip, the rate at which those critical reservoirs are emptied is far outpacing the rate at which they are naturally replenished.... It's causing surface waters to fall at an alarming rate."[343]

Aquifer depletion is a relatively new problem that has emerged in the last half-century. The Punjab aquifer in India is falling by half a metre per year, and the aquifer under the North China plain is falling by 1.5 metres per year. Closer to home, it is estimated that within 20 years the Ogallala aquifer, extending under eight U.S. states, will be pumped dry to the point where it can no longer supply any serious quantity of water.

Today, many of the world's major rivers fail to make it to the sea as a result of high population density, intensive agriculture, and climate change. The Colorado River in the United States rarely reaches the Gulf of Mexico. The Nile has little water left when it reaches the Mediterranean, and the Ganges in India is almost dry when it reaches the Bay of Bengal. China's Yellow River runs dry for part of each year, and in 1997 it ran dry for more than half the year. Over half of China's 50,000 rivers have completely dried up in the last 20 years, and some areas have among the most serious water scarcity problems in the world.[344]

Many countries lack enough water to drink, satisfy hygienic needs, and produce food. Fresh water only makes up 2.5 percent of the total volume of the world's water, but considering 70 percent of that freshwater is in the form of ice and permanent snow cover, we know that demand for water already exceeds supply. So, in many cases, water shortages are becoming food shortages. We must recognize the synergistic relationship between food and water, especially when we consider the impacts of floods, droughts, and climate disruption on these vital resources. With a lack of sufficient water to produce crops, many countries, such as China and Egypt, are importing record amounts of grain.

Water and Climate Disruption

> *By 2025, two-thirds of people worldwide are expected to face water shortages as businesses, agriculture and growing populations compete for the ever more precious commodity.*
>
> —Lester Brown, Earth Policy Institute

Scientists are telling us that climate change is already affecting our water supply. As the average global temperature rises, the rate of evaporation from rivers is also increasing. Climate change also causes rainfall patterns to become more unpredictable and prolongs droughts.

We have long known that the fresh water boom is over and we are now borrowing water from future generations. But can we turn this around? In 2006, *Guardian* contributor George Monbiot warned, "We can avert global thirst—but it means cutting carbon emissions by 60%. Sounds

343 Eric Roston, Bloomberg, *River Flows All Across the Globe Are Dropping*, October 2, 2019, https://www.bloomberg.com/news/articles/2019-10-02/river-flows-are-falling-worldwide-as-groundwater-is-depleted

344 Physics World, *How vulnerable is China's water*?,24 Sep 2019, https://physicsworld.com/a/how-vulnerable-is-chinas-water/

ridiculous? Consider the alternative. Many parts of the world, for reasons that have little to do with climate change, are already beginning to lose their water."[345]

In 2020, these losses became even greater due to climate change, and millions of people are facing a waterless future.

Every thirst-driven person added to our planet of almost EIGHT BILLION has claimed a stake to the world's fresh water supply, and their food has claimed an even bigger stake. The groundwater boom is turning to bust, and for most, the Green Revolution is over. Monbiot explains that the great famines predicted for the 1970s were averted by new varieties of rice, wheat, and maize, through the Green Revolution. They produce tremendous yields, but require plenty of water. This has been provided by irrigation, much of which uses underground reserves, which are being exploited much faster than they are being replenished.

In China, 100 million people live on crops grown with underground water that is not being refilled. Many more rely on the Yellow River, which already appears to be drying up, partly due to climate change. Around 90 percent of the crops in Pakistan are watered by irrigation from the Indus. Almost all the river's water is already diverted into the fields—it often fails now to reach the sea. The Ogallala aquifer that lies under the western and southwestern United States, and which has fed much of the world, has fallen by 30 metres in many places. It now produces half as much water as it did in the 1970s.

Monbiot adds, "Even this account—of rising demand and falling supply—does not tell the whole grim story. Roughly half the world's population lives within 60 kilometers of the coast. Eight of the 10 largest cities on earth have been built beside the sea. Many of them rely on underground lenses of fresh water, effectively floating, within the porous rocks, on salt water which has soaked into the land from the sea. As the fresh water is sucked out, the salt water rises and can start to contaminate the aquifer. This is already happening in hundreds of places. The worst case is the Gaza Strip, which relies entirely on underground water that is now almost undrinkable. As the sea level rises as a result of climate change, salt pollution in coastal regions is likely to accelerate. As these two effects of climate change—global drying and rising salt pollution—run up against the growing demand for water, and as irrigation systems run dry or become contaminated, the **possibility arises of a permanent global food deficit**. Nothing I could write would begin to describe what a world in deficit—carrying 9 billion people—would look like."

Drying Soils, Disappearing Wetlands

Australia's University of New South Wales (UNSW) has announced, "A global study has found a paradox: our water supplies are shrinking at the same time as climate change is generating more intense rain. And the culprit is the drying of soils, say researchers, pointing to a world where drought-like conditions will become the new normal, especially in regions that are already dry.[346]

345 George Monbiot, The Guardian, 10 Oct 10 2006, *The freshwater boom is over. Our rivers are starting to run dry,* https://www.theguardian.com/commentisfree/2006/oct/10/comment.water

346 University of New South Wales, Science Daily, *The long dry: Why the world's water supply is shrinking,* December 13, 2018, https://www.sciencedaily.com/releases/2018/12/181213090004.htm

The study—the most exhaustive global analysis of rainfall and rivers—relied on actual data from 43,000 rainfall stations and 5,300 river monitoring sites in 160 countries, instead of basing its findings on model simulations of a future climate, which can be uncertain and at times questionable.

"This is something that has been missed," said Sharma. "We expected rainfall to increase, since warmer air stores more moisture—and that is what climate models predicted too. **What we did not expect is that, despite all the extra rain everywhere in the world, the large rivers are drying out.** We believe the cause is the drying of soils in our catchments. Where once these were moist before a storm event—allowing excess rainfall to run-off into rivers—they are now drier and soak up more of the rain, so less water makes it as flow. It is extremely concerning."

The report emphasizes that less water into our rivers means less water for cities and farms. And drier soils means farmers need more water to grow the same crops, and this pattern is repeating all over the world. For every 100 raindrops that fall on land, only 36 drops are "blue water"—the rainfall that enters lakes, rivers, and aquifers—and therefore, all the water extracted for human needs. The remaining two-thirds of rainfall is mostly retained as soil moisture—known as "green water"—and used by the landscape and the ecosystem. As warming temperatures cause more water to evaporate from soils, those dry soils are absorbing more of the rainfall when it does occur, leaving less "blue water" for human use.

"It's a double whammy," said Sharma. "Less water is ending up where we can store it for later use. At the same time, more rain is overwhelming drainage infrastructure in towns and cities, leading to more urban flooding."

Engineers have begun a search for solutions, which must go much further than just more dams, and could include water catchments and re-engineering water systems on a massive scale with an enormous price tag. Or perhaps they will see the wisdom in reducing demand instead of a costly scheme. If that funding were to be invested in family planning programs instead, it would be far more effective in the long run. Since this dilemma was caused by humanity breaching our planet's boundaries, it can only be sensibly solved by reversing our population trend.

A 2019 United Nations IPBES report only adds to this dilemma, stating that 85 percent of wetlands present in 1700 had been lost by 2000—loss of wetlands is currently three times faster, in percentage terms, than forest loss.[347]

Combine this with expanding deserts and rising coastal waters, and this is only intensifying water poverty for the growing masses and the next generations. It is now believed that future wars will more likely be fought over water than oil, since nearly 47 percent of the land area of the world has international water basins that are shared by two or more countries.

Brian Merchant with *Tree Hugger News* tells us that 120 nations met in 2009 in an attempt to avert a global water crisis. He states, "Half of the world's population will be living in areas of 'acute water shortage' by 2030. That's the grim finding of a report issued by the UN last week, and it's one of the main reasons that 120 countries have convened in Istanbul for a World Water Forum in order to address the burgeoning crisis. There's more at stake than trying to ensure countries

347 United Nations, *IPBES Sustainable Development Goals*, 2019, https://www.un.org/sustainabledevelopment/blog/2019/05/nature-decline-unprecedented-report/

have an adequate water supply—they've also got to prevent full on wars from breaking out over water resources. Another major catalyst of the summit was the growing tension between dozens of countries, who are fighting over rivers, lakes and glaciers—an issue that must be addressed not only for the well-being of the water-deprived populace, but to maintain peace in general."[348]

Former UN Secretary General Ban Ki-moon, who attended the meeting, said that water scarcity is a "potent fuel for wars and conflict." A particularly striking example is the Sudan, where water shortages have been an underlying cause of the gruesome conflict in Darfur. The dry countries Tajikistan and Kyrgyzstan are currently embroiled in a dispute over water shortages, and have asked the World Water Forum to moderate. Countries in the Middle East and Central Asia are having trouble watering crops—crops like cotton and grain that support many of the regions' economies—and it's leading to uneasy international relations.

Lester Brown with the Earth Policy Institute states, "As the world's demand for water has tripled over the last half-century and as the demand for hydroelectric power has grown even faster, dams and diversions of river water have drained many rivers dry. As water tables have fallen, the springs that feed rivers have gone dry, leading to the disappearance of some rivers."[349]

Water and Government Policy

As governments get up to speed on ways to deal with our water shortages, it is apparent that the process of identifying the most crucial locations and implementing the necessary policies and actions could be a long ways off. In the meantime, the clock is ticking, and we are adding over 80 million thirsty citizens to the planet each year.

In 2020, *EOS* contributor Margaret Shanafield warned, "New research and science-based policies are needed to ensure the sustainability of these long-overlooked waterways. Australia's third-longest river, the Darling, normally experiences periods of medium to low flow, punctuated by flood events. But vast stretches of the river in New South Wales have been bone dry for the past two summers, and in 2019 the river was dry by early spring."[350]

> *Nonperennial rivers are a major—and growing—part of the global river network.*
> **—Margaret Shanafield, Entrepreneurial Operating System**

Shanafield explains that the lack of flows has left communities along its banks in dire straits, with many trucking in water to serve even basic domestic water requirements. Millions of dollars have been spent building pipelines to distant reservoirs, while groundwater resources have also been put under increased stress to fill gaps. River ecosystems have also felt the impacts acutely, with mass fish deaths being just one example. High consumption levels, drought, and damming are largely responsible for the water scarcity.

---

348 Brian Merchant, Tree Hugger, 120 Nations Met in Attempt to Avert Global Water Crisis, March 16, 2009, https://www.treehugger.com/corporate-responsibility/120-nations-meet-in-attempt-to-avert-global-water-crisis.html
349 Lester R. Brown, Earth Policy Institute, *Plan B: Rescuing a Planet Under Stress and a Civilization in Trouble* 2014, http://www.earth-policy.org/books/pb/pbch2_ss3
350 Margaret Shanafield, et al., Entrepreneurial Operating System (Eos), *Science gets up to speed on dry rivers*, Feb 13 2020, *Eos*, https://eos.org/opinions/science-gets-up-to-speed-on-dry-rivers,.

"Despite scientists' clear calls for enhanced legislative protection of the flows and ecosystems of nonperennial rivers, **legal protections of these waterways have actually decreased in recent years** in many places. The challenges posed by the increasing frequency and duration of no-flow periods in rivers are not unique to arid regions: **Over half of the world's streams and rivers are dry for some part of the year**, and the geographic extent of nonperennial waterways is forecast to increase because of climate change and increasing water use," says Shanafield.

The damage to many of these most crucial world rivers is irreversible and is causing great hardship for both wildlife and people. At present, the rate of damage to our river systems is far outpacing any positive action being taken. Meanwhile, researchers with the Dry Rivers Research Coordination Network and other multidisciplinary research groups are piecing together the science of nonperennial rivers.

In the form of solutions, Shanafield urges, "Policy makers and the public must start placing the sustainability of river ecosystems first, so that the rivers can continue to sustain us in the long term.... We call for the formation of a global, collaborative network led by the scientific community and intended to foster emerging interdisciplinary links among scientists and to accelerate such promotional efforts."

Owlcation reporter Kelley Marks states, "Dramatic environmental change has hit Africa in recent decades, and the shrinking of Lake Chad is a primary aspect of this pending catastrophe. Once the size of the Caspian Sea, Lake Chad, located in west-central Africa, has lost about 95 per cent of its water since the 1960s."[351]

Dams Impacting Water Reserves

Marks adds that Lake Urmia in Iran, formerly the largest saltwater lake in the Middle East, has shrunk to only 10 percent of its original size. The reasons are many: the 13 rivers entering the lake have been dammed, increased groundwater pumping has reduced flows into the lake, water diversions, climate change, overpopulation, and drought. Unfortunately, if Lake Urmia vanishes, the lake's marshes will dry up too, which support 226 species of birds and many other animals. Lake Puzhal in India, Owen's Lake in California, and Poyang Lake in China—reduced due to the notorious Three Gorges Dam—all have similar sad stories. Unfortunately, imminent water distress or drought is already a fact of life on every continent and continues to worsen as populations grow.

Lester Brown with the Earth Policy Institute has been warning about this water crisis for decades. One of his concerns is the devastating dams built to generate energy and their reservoirs, which greatly increase evaporation in arid areas, with water loss typically near 10 percent of a reservoir's water. There are **57,000 large dams** in the world as of 2020.

Brown notes that the Tigris and Euphrates rivers, which originate in Turkey and flow into Syria and Iraq en route to the Persian Gulf, are prime examples. He states, "This river system, the site of

351 Kelley Marks, Owlcation, 15 *Worst Drying Lakes in the World,* Oct 6 2019, https://owlcation.com/stem/10-Worst-Drying-Lakes-in-the-World

Sumer and other early civilizations, is being used at near capacity. Large dams erected in Turkey and Iraq have reduced water flow to the once 'fertile crescent,' helping to destroy more than 90 percent of the formerly vast wetlands that enriched the delta region."[352]

In Southeast Asia, the flow of the Mekong is being reduced by the dams being built on its upper reaches by the Chinese. The downstream countries, including Cambodia, Laos, Thailand, and Viet Nam—countries with 160 million people—complain about the reduced flow of the Mekong, but this has done little to curb China's efforts to exploit the power and the water in the river.

Brown adds that the Helmand River, which originates in the mountains of eastern Afghanistan, flows westward across the country and into Iran, where it empties into Lake Hamoun. When the Taliban built a new dam on the Helmand during the late 1990s, in violation of a water-sharing agreement between the two countries, they effectively removed all the remaining water. As a result, Lake Hamoun, which once covered 4,000 square kilometres, is now a dry lakebed. The abandoned fishing villages on its shores are being covered by sand dunes coming from the lakebed itself. Most other major rivers in the world face similar fates as population soars.

As the average global temperature rises, the rate of evaporation from rivers is also increasing. *USA Today* reporter Ian James tells us that about 40 million people get water from the Colorado River in the U.S., and studies show it's drying up and rarely makes it to the sea. He states, "Scientists have documented how climate change is sapping the Colorado River, and new research shows the river is so sensitive to warming that it could lose about one-fourth of its flow by 2050 as temperatures continue to climb."[353]

With the states of Colorado, Utah, Arizona, Nevada, and, most importantly, California depending heavily on the Colorado River, this is destroying the river's ecosystem, especially its fisheries. Scientists with the U.S. Geological Survey found that the loss of snowpack due to higher temperatures plays a major role in driving the trend of the river's dwindling flow. They estimated temperatures were behind about half of the 16 percent decline in the river's flow during the stretch of drought years from 2000 to 2017, a drop that has forced Western states to adopt plans to boost the Colorado's water-starved reservoirs.

James points out, "The research has major implications for how water is managed along the Colorado River, which provides water for more than 5 million acres of farmland from Wyoming to Southern California. Researchers estimated that more than half of this lost flow was attributable to higher temperatures. That equates to a loss of roughly 1.5 million acre-feet of water per year, which is more than half of the annual water allotment for the entire state of Arizona."

352 Lester R. Brown, Earth Policy Institute, Plan B 2.0: Rescuing a Planet Under Stress and a Civilization in Trouble, 2006, http://www.earth-policy.org/books/pb2

353 Ian James, USA Today, About 40 million people get water from the Colorado River: Studies show it's drying up. Feb 22, 2020, https://www.usatoday.com/story/news/nation/2020/02/22/climate-change-drying-up-colorado-river-studies-say/4842148002/

Water Quality

The Reality Check Team with BBC News states, "In 1990, more than 90% of the population in Kenya's urban areas had access to clean water, according to the United Nations. Now, it is estimated that just 50% of Nairobi's four and a half million residents have direct access to piped water."

This severe problem with the supply and quality of its water supply is driving people to buy it from unsafe and potentially contaminated sources, and since 2017, a system of rationing has been in place. Now projects are underway to expand local dam capacity and construct a new water transfer tunnel. Officials say these projects were due to be finished more than a decade ago, but faced problems in raising the financing.

Poor water quality is caused when water from drains or sewage seep into the supply due to the deterioration of pipes and much of it is undrinkable, according to the World Health Organization (WHO). The Reality Check Team notes that a study of water provision in 15 cities in lower-income countries around the world was carried out by the U.S.-based World Resources Institute, which found that Nairobi is certainly not the worst for water provision. The study also says that the cost of water is high in many developing cities, with a high proportion of household income going to purchase piped or bottled water.

Contamination of drinking water is also a threat from gas pipelines, as seen in Montana. Grist reporter Naveena Sadasivam tells us, "The Keystone XL pipeline's proposed path cuts underneath the Missouri River at the edge of the Fort Peck Reservation's southwestern border. The Fort Peck Reservation is home to several bands of the Assiniboine and Sioux tribes, and a spill from the pipeline could contaminate their main source of drinking water."[354]

> *The demand (for water in Nairobi) is higher than supply. We are 20 years behind ... so we need to develop new sources.*
> **—Nahashon Muguna, managing director, Nairobi City Water**

Although U.S. federal agencies are required by law to work with Native American tribes that could be affected by oil and gas projects, there hasn't been a single public hearing on the Fort Peck Reservation as of 2020. This project has been a source of controversy for about a decade, as the U.S. government races to push through the project.

Sadasivam adds, "There are similar stories in Virginia, New Mexico, and elsewhere across the country, according to Indian law experts tracking oil and gas projects. Tribal officials say they try to contact federal agencies and don't hear back, or that agencies make key decisions before contacting them. Sometimes an agency sends letters asking for tribes' input to the wrong address or never contacts them at all."

---

354 Naveena Sadasivam, The Grist, *Federal agencies are required to consult with tribes about pipelines. They often don't,* Jan 2, 2020, https://grist.org/energy/federal-agencies-are-required-to-consult-with-tribes-about-pipelines-they-often-dont/

Oceans

We are entering a phase of extinction of marine species unprecedented in human history.
—International Program on the State of the Ocean, UK

The 2019 United Nations (IPBES) report tells us that nature is declining globally at rates unprecedented in human history—and the rate of species extinctions is accelerating, with grave impacts on people around the world now likely.[355]

"The overwhelming evidence of the IPBES Global Assessment, from a wide range of different fields of knowledge, presents an ominous picture," said IPBES Chair Sir Robert Watson. "The health of ecosystems on which we and all other species depend is deteriorating more rapidly than ever. We are eroding the very foundations of our economies, livelihoods, food security, health and quality of life worldwide."

"The member States of IPBES Plenary have now acknowledged that, by its very nature, transformative change can expect opposition from those with interests vested in the status quo, but also that such opposition can be overcome for the broader public good," Watson said.

The report notes two ways in which oceans are being impacted:

- In 2015, 33 percent of marine fish stocks were being harvested at unsustainable levels; 60 percent were maximally sustainably fished, with just 7 percent harvested at levels lower than what can be sustainably fished.
- Plastic pollution has increased tenfold since 1980, 300 to 400 million tonnes of heavy metals, solvents, toxic sludge, and other wastes from industrial facilities are dumped annually into the world's waters, and fertilizers entering coastal ecosystems have produced more than 400 ocean "dead zones," totaling more than 245,000 square kilometres (94,595 square miles)—a combined area greater than that of the United Kingdom.

Greenland's Ice Is Melting

But we may have an even bigger problem on our hands. Greenland's ice is melting at an unprecedented rate, causing sea levels to rise, the ocean's circulation to change, and Earth's gravitational field to change as well. Scientists are warning that CO_2 emissions must be sharply reduced to avoid further melting and inevitable catastrophe.

Yale Environment 360 reporter Jon Gertner states, "Greenland's ice sheet covers about 80 percent of the island, and measures about 660,000 square miles; in its center, it runs to a depth of about two miles. According to the most recent NASA studies, the ice sheet holds enough water to raise sea levels by about 24 feet, should it ever disappear completely."[356]

In the modern era, we take for granted satellite data that can tell us instantly—and almost precisely—what percentage of Greenland's surface is melting, Gertner points out. NASA images

355 UN IPBES Report, Nature Decline Unprecedented, May 2019, https://www.un.org/sustainabledevelopment/blog/2019/05/nature-decline-unprecedented-report/

356 Jon Gertner, Yale Environment 360, *In Greenland's Melting Ice, A Warning on Hard Climate Choices,* June 27, 2019, https://e360.yale.edu/features/in-greenlands-melting-ice-a-warning-on-hard-climate-choices

show that the ice sheet is melting more, and melting earlier in summer, and melting in ways that computer models suggest will ultimately threaten its long-term existence—not only that but the **melting is accelerating**. Why?

Konrad Steffen, who has built up a record of meteorological readings around Greenland over the course of the past 30 years, has calculated that between 1990 and 2018, average **temperatures on the ice sheet have increased by about 2.8 degrees Celsius**, or 5 degrees Fahrenheit. Scientists believe that there are many reasons for this exceptional melting, and they are human-caused. Besides warming air temperatures, there are warming ocean temperatures and rushing streams cutting through the ice.

It isn't only the loss of ice, but also its effect on Earth's ocean currents and gravitational field that is a concern. CBC reporter Sima Sahar Zerehi warns, "Greenland is losing about 8,300 tonnes of ice per second each day—ice that is melting on land and running into the water, as well as icebergs that are being discharged into Baffin Bay said William Colgan of Toronto's York University."[357]

"It was well known that Greenland's ice was melting, it was well known that melting was accelerating, and it was well known that extra melting was changing the salinity of the North Atlantic Ocean—what was not known is what effect if any that would have on ocean circulation," said Tim Dixon, who recently co-authored a study published in *Nature Communications*.[358]

As ocean currents are disrupted, this can wreak havoc on the ocean's very structure, causing dramatic global temperature changes and making some parts of Earth inhabitable. Dixon warned, "These changes are some of the first alarming signals of the possible effects of climate change. This is the first hint that these effects are starting a bit faster than people had imagined and implying that we need to get our act together and do something about this. Which means we have to stop putting so much CO_2 into the atmosphere."

York University's William Colgan stated, "That's a rapid, rapid mass loss that's occurring in Greenland right now and it's actually changing the Earth's gravity field so quickly that we can detect it with satellite (imaging)." Colgan explained that just like the moon pulls tides around Earth with its gravity, by being relatively massive, Greenland pulls water towards it. As Greenland gets smaller, the ocean water flows farther away towards the equator, in what is called a gravitational far-field. So sea level is decreasing around Greenland and increasing around the equator, which could have drastic impacts on our weather patterns.

Jonathan Overpeck, professor of geosciences at the University of Arizona, told the Associated Press that the discovery that humans are making the North Pole move through climate disruption "highlights how real and profoundly large an impact humans are having on the planet. Scientists have previously found that humans aren't just making the Earth spin on a slightly different axis—they're also making it spin more slowly."[359]

---

357 Sima Sahar Zerehi, CBC News, *Melting Greenlasn ice changing ocean circulation, Earth's gravitational field*, Feb 08, 2016, https://www.cbc.ca/news/canada/north/melting-greenland-ice-changing-ocean-circulation-earth-s-gravitational-field-1.3437904

358 ibid

359 CBC News, *North Pole is headed east and humans are the cause*, March 14, 2017, https://www.cbc.ca/news/technology/north-pole-drift-1.3530656

National Observer reporter Fatima Syed states that studies show marine life is migrating by 52 kilometres per decade toward the poles in search of cooler waters and suitable areas to survive—a migration set to disrupt Earth's ecosystems as we know them.[360]

Impacts of Our Disappearing Cryosphere

William Cheung was one of the Canadian authors of the latest Intergovernmental Panel on Climate Change report, in which scientists released staggering findings about the impacts of human activity on our oceans and the **cryosphere**, the ecological realm of sea ice, snow, and glaciers.

"Under high-emissions scenarios, we're looking at unprecedented changes in ocean conditions," said Cheung. "This will implicate people who rely on our fisheries systems, especially Indigenous people who rely on marine species for food, livelihood and culture."[361]

Syed points out that the sweeping report finds a failure to dramatically reshape "all aspects of society" to limit global warming to less than 2 degrees C above pre-industrial levels will mean devastation for coastal cities, Arctic communities, small island-states, anyone who relies on glaciers for water, and virtually **all marine life.**

The report warns that the world's oceans have already absorbed "more than 90 per cent of the excess heat in the climate system," and that sea levels are rising more than twice as fast and accelerating over time. If global heating increases by 2 degrees Celsius or more, sea levels could rise by between 61 and 110 centimetres by 2100. Many glaciers are "projected to disappear regardless of future emissions," the scientists wrote. And 20 to 90 percent of the world's current coastal wetlands are projected to disappear by 2100.

Cheung explains that off Canada's East Coast, Atlantic cod are swimming toward the Arctic. In the west, the Pacific Ocean's high temperatures are threatening Atlantic salmon. Coral reefs and kelp forests are already disappearing. Cheung urges Canada and countries around the world to implement a low-emissions pathway and move away from overfishing toward more sustainable water-based activities to mitigate the serious biodiversity loss in Earth's waters.

"Although the oceans' future is under serious threat, we know how to respond," he said. "We need protective pathways to save marine life." So now both on land and in the sea, displaced wildlife has to try to wind their way through our burgeoning human presence to survive. But don't despair—lots of solutions will be addressed later in this chapter.

---

360 Fatima Syed, National Observer, *Unprecedented damage to oceans has Canada's marine life on the run,* Sept 25th 2019, https://www.nationalobserver.com/2019/09/25/news/unprecedented-damage-oceans-has-canadas-marine-life-run
361 ibid

The Plastic Predicament

The earth, the air, the land, and the water are not an inheritance from our forefathers but on loan from our children. So we have to handover to them at least as it was handed over to us.

—Mahatma Gandhi

By now we've all figured out that plastic is becoming an additional menace to our oceans, landfills, and food—some saying it's as critical an issue as climate change. As this crisis escalates, we are looking back at an industry portrayed as an answer to all of our food and household woes just 50 years ago—now seen as the world's most menacing trash tragedy. The United Nations points out that ingestion of plastic kills an estimated one million marine birds and 100,000 marine animals each year. In addition, more than 90 percent of all birds, fish, and many other sea animals are believed to have tiny pieces of plastic particles in their stomach—and humans too.

Business Insider reporter Rebecca Harrington tells us that by 2017, the ocean was filled with about 165 million tonnes of plastic. That's 25 times heavier than the Great Pyramid of Giza. She adds, "By 2050, **plastic in the oceans will outweigh fish**, predicts a report from the Ellen MacArthur Foundation, in partnership with the World Economic Forum."[362]

The report projects the oceans will contain at least 937 million tonnes of plastic and 895 million tonnes of fish by 2050 if we don't radically change our ways. We also don't reuse or recycle as much plastic as we should, causing them to go into landfills, end up in the ocean, or be strewn on beaches. Recycling plastic packaging could help businesses, too. Today, they're losing at least $80 billion dollars a year because they have to make so many new plastics from scratch, says CNN.

The report helps quantify just how much plastic this is: It's equivalent to dumping the contents of one garbage truck into the ocean every minute.
—Ellen MacArthur Foundation

CNN's Marian Liu warns, "According to a three-year study published in Scientific Reports Friday, the mass known as the Great Pacific Garbage Patch is about 1.6 million square kilometers in size—up to 16 times bigger than previous estimates. That makes it more than double the size of Texas. Ghost nets, or **discarded fishing nets, make up almost half** the 80,000 metric tons of garbage floating at sea, and researchers believe that around 20% of the total volume of trash is debris from the 2011 Japanese tsunami."[363]

Yet Grist reporter Rebecca Leber tells us that fossil fuel companies are counting on plastics to save them. She states, "Fossil fuel companies are staring down a time when their signature product will no longer be so critical in our lives. As the world transitions slowly but surely away from

362 Rebecca Harrington, Business Insider, By 2050, the oceans could have more plastic than fish, Jan 26, 2017, https://www.businessinsider.com/plastic-in-ocean-outweighs-fish-evidence-report-2017-1

363 Marian Liu, CNN, Great Pacific Garbage Patch now three times the size of France, March 23, 2018, https://www.cnn.com/2018/03/23/world/plastic-great-pacific-garbage-patch-intl/index.html

fuel-guzzling cars, gas-powered buildings, and coal-fired power plants, industry execs must count on growth that comes from somewhere else—and they see their savior as plastics."[364]

With society's addiction to plastic bags, food containers, headphones, fertilizers, polyester clothing, and bottled water, the petrochemical industry is in fact betting on increased demand for plastic in coming years. Since plastics are made from gas, oil, and coal—plus a load of other toxic additives including dangerous endocrine inhibitors—they are taking a huge toll on all life on Earth. Our alarming population growth fuels this addiction, occurring in both overdeveloped and developing countries. And with the human addiction to fish, rather than a plant-based diet, this habit alone is causing **half of the plastic at sea.**

Leber adds, "In the last decade, petrochemicals have moved from a sideshow for the oil and gas industry to a major profit machine, and the trend is expected to accelerate: The energy research group International Energy Agency predicts that plastics' consumption of oil will outpace that of cars by 2050. In a recent report about its 20-year growth, ExxonMobil executives assured shareholders that the company could offset losses from the transition to electric cars with growth in petrochemicals."

> *Plastics keep us on the fossil fuel treadmill.*
> **—Judith Enck, EPA**

Although traditionally most plastics came from foreign countries, the U.S. is now producing its own by **using ethane, a biproduct of gas from fracking**. In 2020, as our population races towards EIGHT BILLION, the U.S. market is saturated with plastic, and it is now exporting the manufacturing materials around the world. There are also climate impacts at every point of the lifecycle of plastics, and they release greenhouse gases, toxic chemicals, and micro plastics when they break down. Although activists are protesting against further fracking projects in the U.S., the Trump administration had publicly celebrated the ascent of plastics.

Leber warns, "Emerging research has shown how polyethylene releases greenhouse gases when it breaks down and might interfere with the tiny algae plants that play an essential role in helping the oceans absorb excess carbon. Even when recyclable plastics make it to blue bins, much of it ends up in landfills and about 12 percent is burned at an incinerator to generate energy—which vents toxic fumes into nearby communities and more carbon pollution into the atmosphere."

A 2019 report by the Center for International Environmental Law (CIEL) found that global **emissions from the plastics sector rose 15 percent from 2012 to 2018**. Last year alone, they estimated that plastic production contributed the equivalent of 189 large coal plants. If plastics production continues apace, the sector is on track to reach the equivalent annual pollution of 295 large coal plants in the next ten years, and double that by 2050.

Judith Enck, a former U.S. Environmental Protection Agency (EPA) administrator and founder of the environmental coalition Beyond Plastics, believes that focusing on how plastic makes cars lighter is a distraction. The real problem, she argues, is that **the glut of gas has made plastics incredibly cheap**, intensifying the world's growing hunger for more single-use plastics. Ethane

364 Rebecca Leber, The Grist, Fossil fuel companies are counting on plastics to save them, Mar 8, 2020, https://grist.org/climate/fossil-fuel-companies-are-counting-on-plastics-to-save-them/?utm_source=newsletter&utm_medium=email&utm_campaign=beacon

crackers are not an offramp from oil, she says—instead, they're another way of embedding fossil fuels in our daily lives.

On the other end of the plastics dispute is how to dispose of it safely, if that is even possible. With most overdeveloped countries exporting much of their recycled plastics to developing countries for disposal, countries like China and Poland are starting to say NO, leaving the EU and North America fumbling to unload their glut of plastic trash.

DW News reporter Jo Harpe states, "Racing to meet an EU recycling target, the waste market in Poland is booming. But Warsaw is increasingly fed up with illegally imported foreign waste, much of it from the UK. Poland has become a dumping ground for UK plastic waste and a so-called trash mafia has allegedly grown up to manage its illegal incineration. Poland's Deputy Minister of the Environment, Slawomir Mazurek, promised recently that the government would come down hard on illegal incineration of imported plastic waste."[365]

Poland's relative self-regulatory backwardness was highlighted vividly in 2018 when over 60 fires started in landfills nationwide in May, Harpe adds. The biggest lasted two days and covered the town of Zgierz, in central Poland, in smoke, where fragments of burnt waste originally from the UK were found. In January 2018, China introduced an import ban on 24 types of solid waste and as a result other countries were targeted by shippers of illegal waste. Malaysia, Vietnam, Thailand, Indonesia, Taiwan, South Korea, Turkey, India, and Poland reportedly took up the slack. Greenpeace has accused Western countries of exploiting poorer nations with inadequate regulatory frameworks that cannot deal with it properly. In reality, only 9 percent of the world's plastic ends up being recycled, *National Geographic* reported in 2017. Instead, much of it gets incinerated, which is the number one driver of greenhouse gas emissions from plastic waste management.

Harpe adds, "Three British waste disposal companies are under investigation by the UK's Environment Agency (EA) for sending 1,000 tons of falsely labelled recyclable waste to Poland. EA chief Sir James Bevan warned two years ago that **waste crime was becoming 'the new narcotics,' costing Britain 1 billion pounds ($1.2 billion) a year.**"

Where Is All This Plastic Coming From?

The *Guardian* reporters Sandra Laville and Matthew Taylor tell us that more than 480 billion plastic drinking bottles were sold in 2016 across the world, up from about 300 billion a decade ago. If placed end to end, they would extend more than halfway to the sun. By 2021, this will increase to 583.3 billion, over half a trillion bottles a year, according to the most up-to-date estimates from Euromonitor International's global packaging trends report.

On the other hand, a sustainable human population would reduce water contamination, our water would be pure enough to drink, and we wouldn't require bottled water at all. Also, if there were a meaningful refundable deposit scheme globally like we have in Canada on bottled water,

365 Jo Harpe, DW News, *Poland won't take UK garbage any more,* July 25 2019, https://www.dw.com/en/poland-wont-take-uk-garbage-any-more/a-49725035

more people would recycle them or use a reusable one. The solutions are really not that hard to figure out. They will be discussed in more detail later in this chapter.

Laville and Taylor state, "A million plastic bottles are bought around the world every minute and the number will jump another 20% by 2021, creating an environmental crisis some campaigners predict **will be as serious as climate change**. The demand, equivalent to about 20,000 bottles being bought every second, is driven by an apparently insatiable desire for bottled water and the spread of a western, urbanized 'on the go' culture to China and the Asia Pacific region."[366]

Scientists have been telling us for decades that plastic takes hundreds of years to decompose and produces toxins that are harmful for our health. Now, Laville and Taylor warn, "Scientists at Ghent University in Belgium recently calculated people who eat seafood ingest up to 11,000 tiny pieces of plastic every year. Last August, the results of a study by Plymouth University reported **plastic was found in a third of UK-caught fish**, including cod, haddock, mackerel and shellfish. Last year [2018], the European Food Safety Authority called for urgent research, citing increasing concern for human health and food safety 'given the potential for microplastic pollution in edible tissues of commercial fish.'"

A 2018 World Health Organization report warns that every day, humanity is exposed to the toxicity of plastic, even before we are born. In fact, **97 percent of today's human population tests positive for plastic additive BPA**. With petrochemical companies refusing to release ingredient lists for their plastics, there is no way to trace exposure to its origin. Microplastics, acting like poison pills on our endocrine systems, can cause birth defects, cancer, and other illnesses, and are considered a global threat. Lead, mercury, and now plastic—being passed from humans to fish—and now back to humans. Perhaps the karmic interconnectedness of the animal kingdom is about to be felt.

Besides this microplastic being found on remote islands and beaches in the Arctic, it has now been reported in the deepest place on the planet—**the Mariana Trench**—on the deepest dive ever made by a humans inside a submarine, nearly 11 kilometres below sea level, according to Charli Shield with DW News. He states, "A new report shows it's also accelerating global warming. **Stopping climate change also means curbing plastic**. Here is something you might not know: the plastic industry is the second largest and fastest-growing source of industrial greenhouse gas emissions, and **99% of what goes into plastic is derived from fossil fuels**."[367]

As Carroll Muffett, lead author of a Center for International Environmental Law (CIEL) report, points out, producing plastic is how the fossil fuel industry is able to "take what would otherwise be a waste stream and turn it into a profit stream." Over the next ten years, the report projected that emissions from the plastics lifecycle could reach 1.34 gigatonnes per year—equivalent to the emissions from more than 295 five-hundred-megawatt coal-fired power plants operating at full capacity.

366 Sandra Laville and Matthew Taylor, The Guardian, *A million bottles a minute: world's plastic binge 'as dangerous as climate change'*, 10 Apr 2019, https://www.theguardian.com/environment/2017/jun/28/a-million-a-minute-worlds-plastic-bottle-binge-as-dangerous-as-climate-change

367 Charli Shield, DW News, The plastic crisis isn't just ugly – it's fueling global warming, May 15, 2019, https://www.dw.com/en/the-plastic-crisis-isnt-just-ugly-its-fueling-global-warming/a-48730321

Shield explains, "Each stage in the 'plastic life-cycle' produces greenhouse gas emissions, the researchers found. This includes: the extraction and transportation of fossil fuels used to formulate plastic, its refinement and manufacture, the management of plastic waste and its disposal."

If the expansion of petrochemical and plastics production continues as currently planned, the report outlines that by 2050, plastic will be responsible for 10 to 13 percent of the total "carbon budget," **which is the amount of CO_2 we can emit globally and still remain below a 1.5 degree Celsius temperature rise.** And early research shows it could have an impact on the oceanic carbon sink. There's increasing evidence, the CIEL researchers found, of microplastic being consumed by plankton. These creatures not only "form the base of the oceanic carbon chains" but also provide the "single most important mechanism for absorbing atmospheric carbon and transporting it to deep-ocean carbon sinks," the report said.

Shield states, "The world's oceans provide the largest natural carbon sink for greenhouse gases, having absorbed 30-50 percent of atmospheric CO_2 produced since the start of the industrial era. If microplastics were to disrupt the ability for underwater ecosystems to absorb carbon, it could seriously compromise efforts to stop global warming."

How Is All This Plastic Getting Into the Oceans?

Global Citizen reporter Seneo Mwamba states, "According to World Economic Forum researchers, just 10 rivers across **Asia and Africa** carry 90% of the plastic that ends up in the oceans. The study states that eight of these rivers are in Asia: the Yangtze, Indus, Yellow, Hai He, Ganges, Pearl, Amur, and Mekong. Two of the rivers can be found in Africa: the Nile and the Niger. The WEF added that the two things all the rivers named have in common is a high population living in the area, as well as a poor waste management system."[368] Of course, North Americans ship huge quantities of plastic to Asia and Africa for disposal as well, only adding to this tragedy.

Mwamba notes, "Research from Plymouth University has found that close to 700 species of marine life are facing extinction due to the increase of plastic pollution. A study conducted by the University of Queensland in Australia, based on data collected since the late 1980s, found that Green sea turtles now ingest twice the plastic they did 25 years ago."

According to *National Geographic*, 73 percent of all beach litter is plastic.The litter includes filters from cigarette butts, bottles, bottle caps, food wrappers, grocery bags, and polystyrene containers (from pop and water bottles). In 1974, global plastic consumption per year was 2 kilograms (4.4 pounds) per capita. Today, this has increased to 43 kilograms (about 95 pounds)—and this number is still set to increase.

For example, Mwamba tells us that globally about two million plastic bags are used every minute, the average time a bag is used for is 12 minutes, and they take up to a thousand years to decompose. This is the legacy of EIGHT BILLION shoppers living in excess. Is this toxic avalanche

368 Seneo Mwamba, Global Citizen, 10 Plastic Pollution Facts That Show Why We Need To Do More, June 14, 2018, https://www.globalcitizen.org/en/content/plastic-pollution-facts/

of billions of tonnes of plastic in this time of crisis going to be allowed to continue and cause our demise? What of the turtles, porpoises, and fish who are threatened with extinction? We also know that biodegradable hemp plastic could provide most of our future needs with little environmental consequence and at a fraction of the cost.

Toxic Waters

Water is the lifeblood of our bodies, our economy, our nation and our well-being.
—Stephen Johnson

A 2019 United Nations report warns that the world's oceans now have 400 "dead zones"—places where pollution threatens fish, other marine life, and the people who depend on them—compared to 200 dead zones in 2006. The United Nations marine experts said the number and size of oxygen-deprived zones has grown each decade since the 1970s. Not all the dead zones persist year-round, as some return seasonally, depending on winds that bring nutrient-rich water to the surface. Officials warn that these toxic areas are causing low levels of oxygen, which are a major threat to sea grass beds, coral, fish stocks, and other marine life.

National Geographic reporter Jenny Howard explains, "Often encompassing large swaths of ocean (and even lakes and ponds), dead zones become oceanic deserts, devoid of the usual aquatic biodiversity. Though hypoxic zones can occur naturally, many more are caused by agricultural practices across the world—a big problem for wildlife and for people."[369]

We know that runoff from manure and fertilizers triggers dead zones in oceans or lakes, often killing any wildlife that once existed there, and they have been increasing since the Green Revolution emerged in the 1960s.

The UN report notes that the largest dead zone in the world lies in the Arabian Sea, covering almost the entire 63,700-square-mile Gulf of Oman. The second largest sits in the Gulf of Mexico in the United States, averaging almost 6,000 square miles in size. Dead zones appear annually, May through September, in the Gulf of Mexico, after tonnes of nutrients from fertilizer use and sewage in the Mississippi watershed wash downstream into the Gulf. Excess nutrients spark an algal explosion, giving rise to a dead zone.

> *The top five meat and dairy companies, including Tyson and Cargill, emit more greenhouse gases combined than ExxonMobil, Shell or BP.*
> **—Reynard Loki,** *Truthout*

Although it is possible to shrink a dead zone, this would require fertilizer runoff reductions of at least half to make any significant improvement and reach target goals. The Mississippi River/Gulf of Mexico Hypoxia Task Force stated, "The best way to accomplish this is to encourage retaining nutrients at their original source—on land. This is done through better farm management practices, like using less fertilizer and using crop covers to help anchor soil in place."

369 Jenny Howard, National Geographic, *Dead Zones, Explained,* July 31, 2019 https://www.nationalgeographic.com/environment/oceans/dead-zones/

Truthout reporter Reynard Loki notes that corporate food brands drive the massive dead zone in the Gulf of Mexico, which is more than 8,000 square miles, according to a report released in 2017 by the Institute for Agriculture and Trade Policy.[370]

Loki points out that CO_2 emissions aren't the only problem caused by the animal industry, as corporate meat giants are driving the massive dead zones in oceans around the world. She states, "Whole Foods bills itself as 'America's healthiest grocery store,' but what it's doing to the environment is anything but healthy. According to a new report, the chain is helping to drive one of the nation's worst human-made environmental disasters: the dead zone in the Gulf of Mexico. By not requiring environmental safeguards from its meat suppliers, the world's largest natural and organic foods supermarket—and most of its big-brand counterparts in the retail food industry, like **McDonald's, Subway and Target**—are sourcing and selling meat from some of the worst polluters in agribusiness, including **Tyson Foods and Cargill.** The animal waste and fertilizer runoff from their industrial farms end up in the Gulf of Mexico, where each summer, a growing marine wasteland spreads for thousands of miles, leaving countless dead wildlife in its oxygen-depleted wake."

A report released by Mighty Earth, an environmental action group based in Washington, D.C., states, "The major meat producers like Tyson and Cargill that have consolidated control over the market have the leverage to dramatically improve the supply chain. Yet **to date they have done little, ignoring public concerns** and allowing the environmentally damaging practices for feeding and raising meat to expand largely unchecked."

According to NASA, "The number and size of ocean dead zones is closely connected to human population density." It's basic math: **More people means more meat-eaters**, and more meat production means more and bigger dead zones. Also, antibiotics and hormones injected into livestock are pollutants that will run off into streams and could cause illness in people.

Mighty Earth concludes that eating meat is risky business. The main source of water contamination in the U.S. is the manure and fertilizer coming from industrial farms that grow feed to raise animals to be killed for human consumption. By converting rainforests and prairies into industrial farms, large-scale meat producers are responsible for the widespread destruction of many of the planet's native ecosystems, which threatens wildlife by destroying native habitats and releases stored carbon dioxide into the atmosphere, further exacerbating climate change. **Where is government regulation for toxic fertilizers?**

Yet, even with over 400 dead zones already destroying our oceans, African farmers are feeling left out and are building their own fertilizer plants to grow enough food for their teeming millions. Grist reporter Rupa Shenoy states, "For decades, fertilizer was too expensive for African farmers. It had to be imported, and transportation into the continent was expensive. Now, though, Africa is turning a corner toward producing more of it locally. A Moroccan company has signed a nearly $4 billion deal to build a fertilizer plant in Ethiopia. A Danish company is helping the Democratic

370 Reynard Loki, Truthout, *Corporate Food Brands Drive the Massive Dead Zone in the Gulf of Mexico*, August 28, 2018, https://truthout.org/articles/corporate-food-brands-driving-the-massive-dead-zone-in-gulf-of-mexico/

Republic of Congo build a $2.5 billion fertilizer plant. The African Development Bank Group helped fund a new fertilizer plant in Nigeria that's already boosting farmers' yields."[371]

And this August in Ghana, the vice president stood proudly before the largest fertilizer plant ever built in the country. Mahamudu Bawumia said the factory was the solution Ghana's farmers had been demanding. "This factory can meet all of Ghana's demand for fertilizer," he told a crowd that clapped enthusiastically. Is there really any hope for our oceans with this mindset?

Solutions and Successes

Population growth is the primary source of environmental damage.
—Jacques Cousteau

Governments Urging For Transformative Change

Nature is declining globally at rates unprecedented in human history—and the rate of species extinctions is accelerating, with grave impacts on people around the world now likely, warns a landmark new UN report.[372]

The report also tells us, "It is not too late to make a difference, but only if we start now at every level from local to global. Through '**transformative change**,' nature can still be conserved, restored and used sustainably—this is also key to meeting most other global goals. By transformative change, we mean a fundamental, system-wide reorganization across technological, economic and social factors, including paradigms, goals and values."

The UN notes that this kind of meaningful change can expect opposition from those with **interests vested in the status quo**, but also that such **opposition can be overcome for the broader public good**.

The report also presents a wide range of illustrative actions for sustainability and pathways for achieving them across and between sectors such as agriculture, forestry, marine systems, freshwater systems, urban areas, energy, finance, and many others. It highlights the importance of, among others, adopting integrated management and cross-sectoral approaches that take into account the trade-offs of food and energy production, infrastructure, freshwater and coastal management, and biodiversity conservation. A key element of more sustainable future policies is the evolution towards a **global sustainable economy**, steering away from the current limited paradigm of economic growth and the GDP. We know that the **genuine progress indicator** would help make this transition possible.

"IPBES presents the authoritative science, knowledge and the policy options to decision-makers for their consideration," said IPBES Executive Secretary Dr. Anne Larigauderie. "We thank the hundreds of experts, from around the world, who have volunteered their time and knowledge to

371 Rupa Shenoy, Grist.org, *Small farmers feel left out amid a boom in African fertilizer production,* Jan 22 2020, https://grist.org/food/small-farmers-feel-left-out-amid-a-boom-in-african-fertilizer-production/?

372 United Nations, UN Report: Nature's Dangerous Decline 'Unprecedented'; Species Extinction Rates 'Accelerating', May 5 2019, https://www.un.org/sustainabledevelopment/blog/2019/05/nature-decline-unprecedented-report/

help address the loss of species, ecosystems and genetic diversity—a truly global and generational threat to human well-being."

What we do know is that the United Nations and participating countries do have the data, science, and future predictions—they do understand the ominous threat posed by our global crisis—and some are finally making a concerted effort to confront this problem. However, it is up to all citizens to make critical world issues—like population, food, water, and climate disruption—a political issue when it comes to voting. If countries like the U.S. and China continue to opt out of being part of the solution, UN goals can never be achieved. Also, we know that Goal #5 for family planning and gender equality is pivotal to all of the other 16 UN Sustainable Development Goals to succeed, including *water management*, as stated on the UN website.

This UN report offers a number of solutions in the form of **policy tools, options, and exemplary practices. For marine systems**, they recommend ecosystem-based approaches to fisheries management, spatial planning, effective quotas, protecting key marine biodiversity areas, reducing run-off pollution into oceans, and working closely with producers and consumers.

For managing freshwater systems, their policy options and actions include, among others, more inclusive water governance for collaborative water management and greater equity; better integration of water resource management and landscape planning across scales; promoting practices to reduce soil erosion, sedimentation, and pollution run-off; increasing water storage; promoting investment in water projects with clear sustainability criteria; and addressing the fragmentation of many freshwater policies.

Movements for Marine Protection

All the water that will ever be is, right now.
—National Geographic

But it isn't only involved governments that are determined to improve the health of our oceans. SeaLegacy is a non-profit society in Canada that joined forces with the Blue Sphere Foundation in 2019 to form an innovative ocean agency. Their expanded team will bring together the world's best photographers, conservationists, scientists, storytellers, and strategists to lead a bold new movement to engage one billion people in ocean conservation.[373]

They say, "The core of each SeaLegacy mission is built around an expedition. We traverse the oceans, telling powerful visual stories that move people from apathy to action. With guidance from our diverse expert council, *The Compass*, we select locations and issues where our work creates a tipping point in favour of conservation."

> *There are whales alive in the Arctic today that were born before Moby-Dick was published in 1851.*
> **—Gillian Anderson, Greenpeace**

SeaLegacy has learned the **power of telling a story**, and this is the best way to motivate people. They assert, "The objective of our work is to create movements and revolutions. It is to put pressure on governments,

373 Sea Legacy website, *Turning the Tide for our Oceans*, 2020, https://www.sealegacy.org/about-us

politicians, corporations, and to empower individuals to be the change they want to see. We know that **science is important, but it has failed to "save the planet"** despite the vast resources that have been invested in scientific research over the past 50 years. We connect art with that science and then move into conservation."

Another project for a marine protected area is already in the works. According to Greenpeace Antarctic Ambassador Gillian Anderson, the group is working to create an **Antarctic Ocean Sanctuary**. In 2018, award-winning actor Gillian Anderson delivered a petition signed by two million people globally (350,000 in the UK) to the head of the UK's Diplomatic Service, Sir Simon McDonald. The petition calls for the creation of the largest protected area on Earth.[374] The petition was also delivered to all other governments involved.

This 1.8 million square kilometre area would be five times the size of Germany and offer protection to penguins, killer whales, leopard seals, and blue whales. Plans are for the sanctuary to ring-fence the waters and restrict commercial fishing and other human activity. This would include any future attempt to mine the seabed or drill for oil. Anderson points out, "Humans kill about 300,000 whales, dolphins and porpoises every year—most die indiscriminately when they get caught up in fishing gear. Our oceans and the life they sustain are under mounting pressure from multiple threats, including overfishing, climate breakdown, deep sea mining, oil drilling and plastic pollution. **Quite simply, they are in crisis.**"

In 2019, the Commission for the Conservation of Antarctic Marine Living Resources (CCAMLR) met in Hobart, Australia, to discuss the proposal. While 22 governments in the CCAMLR supported the plan for a new Antarctic Ocean sanctuary, Russia, China, and Norway derailed the process. All 26 members must consent for it to be approved, so work is continuing to get this area protected against exploitation, overfishing, and overhunting. The CCAMLR has previously established other major ocean Antarctic sanctuaries—including the world's largest spanning 1.55 million square kilometres in the Ross Sea—but governments and environmental groups say more action is needed to protect the last pristine continent.

> *The government will create 41 new Marine Conservation Zones (MCZs) across the UK—in addition to the existing 50.*
> **—Michael Gove, UK Environment Secretary**

While this new ocean treaty hangs in the balance, we must convince our leaders to act boldly and grasp the opportunity to protect these wild spaces. Ocean sanctuaries would provide protection for wildlife populations and ecosystems, allowing them to recover and thrive.

In 2019, Anderson added, "But the fight did not end there. Today, governments will begin negotiating the first draft text of a global ocean treaty at the UN in New York that would cover waters that lie beyond national borders. This vast expanse of sea covers almost 50% of the Earth's surface. If they get it wrong the treaty could entrench many of the worst practices that are impacting our

374 Press release, Greenpeace, *Gillian Anderson delivers 350,000-strong petition to Government to Protect the Antarctic,* October 9 2018, https://www.greenpeace.org.uk/news/10131/

oceans. But if they get it right the treaty could pave the way for the creation of a network of ocean sanctuaries, making 30% of our marine world off-limits to human activity."[375]

Meanwhile, the UK plans to almost double its number of marine conservation zones. Global Citizen reporter Imogen Calderwook explains that the zones are intended to protect rare sea life and threatened marine habitats around the British coast, and it's all part of the UK's "Blue Belt" of conservation areas.

"The UK is surrounded by some of the richest and most diverse sea life in the world," said UK Environment Secretary Michael Gove in a statement. "We must protect these precious habitats for future generations."[376] As of 2020, this project is still awaiting approval as well. This would be of great benefit to the Half Earth project, which aims to protect half of Earth for future generations of wildlife and humans. Government officials say this conservation project would protect fish, worms, prawns, reefs, lobsters, coral, sponges, as well as rare and threatened species such as short-snouted seahorse, the stalked jellyfish, and peacock's tail seaweed.

In 2018, Chile also announced three massive new marine sanctuaries encompassing more than 450,000 square miles, showing that economic prosperity and conservation can go hand-in-hand, announced Chilean President Michelle Bachelet. Global Citizen reporter Joe McCarthy stated, "The first reserve, Rapa Nui Marine Protected Area, spans 278,000 square miles and covers the economic zone of Easter Island, which is a UNESCO world heritage site because of its astounding biodiversity. Industrial fishing and extraction will no longer be permitted in the area, but traditional forms of fishing will be allowed, a concession to indigenous people who played a role in the reserve's formation, Smithsonian reports."[377]

The announcement comes at a time when the world's oceans face rapid decline from a range of threats, and establishes Chile as a leader in the growing field of marine conservation, according to Pew Charitable Trusts. For decades, scientists have been warning that the world's precious coral reefs are under attack, dying off due to ocean warming, acidification, climate change, and coral mining (for bricks, roadfill, or cement).

Fortunately, in 2018, the UK joined the fight to protect the world's coral reefs. Global Citizen reporter Imogen Calderwood states, "The world's coral reefs are bearing the brunt of global warming. They have been dying out in vast swathes, and scientists have recently warned nearly all coral reefs could have vanished by 2050. Rising sea temperatures and climate change are leading to what's known as 'coral bleaching'—when the organisms that make their home on the coral are literally cooked alive. But the UK has now joined the global effort to step up and save the coral reefs."

The UK's waters are home to coral reefs that are around 8,000 years old. The only known coral reef in English waters—the Canyons—covers an area of more than 650 square kilometres, and has

375 Gillian Anderson, Open Pods, *This is crunch point for our oceans: let's do the right thing*, August 19, 2019, https://www.openpods.com/news/2019/8/19/this-is-crunch-point-for-our-oceans-lets-do-the-right-thing

376 Imogen Calderwook, Global Citizen, *The UK Is Set to Protect a Marine Area 8 Times the Size of Greater London*, June 8, 2018, https://www.globalcitizen.org/en/content/uk-marine-conservation-zones-michael-gove-world-oc/

377 Joe McCarthy, Global Citizen, Chile announces 3 Massive New Marine Sanctuaries, FEB. 28, 2018, https://www.globalcitizen.org/en/content/chile-new-marine-reserves-protection/

been protected as a marine conservation zone (MCZ) since 2013. Calderwood notes, "As well as helping protect these reefs, the UK's involvement will also support research into the threats."

Plastic Cleanup

Water is the most critical resource issue of our lifetime and our children's lifetime. The health of our waters is the principal measure of how we live on the land.
—Luna Leopold

One scene that tells a powerful story is the mounds of plastic strewn across once pristine beaches—washed up with the ocean tides or intentionally dumped by burgeoning populations in developing countries. These images have motivated one special girl in the U.S. *Mother Nature* reporter Lindsey Reynolds tells us that 23-year-old Mimi Ausland got her start saving animals at the age of 11—now she's set her sights on saving the ocean from plastic. She comments, "Mimi Ausland has big plans for the planet.... Once honored as the 2008 ASPCA Kid of the Year, Ausland isn't a kid anymore, but she's still an activist."[378]

Ausland comments, "Looking back on it, I believe the experience showed me that people want to make a difference and that small actions can create big impact. Freekibble not only grew my interest in giving back but also in business—specifically in finding ways to combine impact with business. It gave me the belief that if you care about something enough, you can make a difference."

Free the Ocean is Ausland's new website and is similar to Freekibble in that it's a quiz website that "rewards" you (whether you're right or wrong) when you answer the daily trivia question. Your reward, this time, funds the removal of plastic from the ocean. An estimated 19 billion pounds of plastic enters the ocean each year. Free the Ocean has removed over 1,570,000 pieces of plastic through their online public participation project as of 2019. The good news is that she isn't the ony one who has noticed the plastic crisis killing so many of our threatened animals in the ocean.

National Geographic reporter Laura Parker states, "By 2016, the year sales of bottled water in the United States officially surpassed soft drinks, the world had awakened to the burgeoning crisis of plastic waste. The backlash against the glut of discarded bottles clogging waterways, polluting the oceans and littering the interior has been swift. Suddenly, carrying plastic bottles of water around is uncool."[379]

The more trendy fashion in 2020 is wearing refillable stainless steel bottles as an accessory. They are taken everywhere we go, from the gym to the beach—packed in our backpack or clipped to our hip. They let everyone know that we care about plastic destroying our oceans and wildlife.

Parker adds, "What sets bottles apart from other plastic products born in the post-World War II rise of consumerism is the sheer speed with which the beverage bottle, now ubiquitous around

378 Lindsey Reynolds , Mother Nature Network, *She got her start saving animals. Now she's set her sights on saving the ocean,* September 20, 2019, https://www.mnn.com/earth-matters/climate-weather/stories/freekibble-creator-sets-her-sights-saving-ocean

379 Laura Parker, National Geographic, *How the plastic bottle went from miracle container to hated garbage,* August 23, 2019, https://www.nationalgeographic.com/environment/2019/08/plastic-bottles/

the world, has shifted from convenience to curse. The transition played out in a single generation. Plastic bottles and bottle caps rank as the third and fourth most collected plastic trash items in the Ocean Conservancy's annual September beach cleanups in more than 100 countries. Activists are zeroing in on the bottle as next in line for banning, after plastic shopping bags. The tiny towns of Concord, Massachusetts and Bundanoon, Australia already have banned bottles, as have numerous public parks, museums, universities, and zoos in Europe and the United States."

The United Nations points out that in the developing world, there are 2.2 billion people who still don't have access to clean drinking water, so bottled water is often seen as essential for survival. However, even there they are changing their habits. Parker notes, "In June, Kenya announced a ban on single-use plastics at beaches and in national parks, forests, and conservation areas, effective in June 2020, and the South Delhi Municipal Corporation banned disposable water bottles in all city offices."

Since these bans began, we have begun to seek out solutions—replacements for plastic that are biodegradable or compostable, or the use of refilling stations. Parker states, "Last spring, runners in the London Marathon were handed edible seaweed pouches at mile 23 containing a sports drink to slake their thirst. And Selfridges, London's century-old department store, has vanquished plastic beverage bottles from its food court in favor of glass bottles, aluminum cans, and refilling stations."

Parker notes that once bottles have become trash, entrepreneurs around the world are turning them into printer ink cartridges, fence posts, roofing tiles, carpets, flooring, and boats, to name only a few items. Even houses have been constructed from bottles. The latest is a three-story modern on the banks of the Meteghan River in Nova Scotia, promoted as able to withstand a Category 5 hurricane. It only took 612,000 bottles. Biodegradable bottles are also being made out of hemp for those who still prefer to buy their water in a bottle.

However, for these bottles to be viable, a robust recycling plan needs to be implemented, all countries need to charge a recycling fee, and developing countries still need to catch up with recycling efforts. According to Parker, there are three things you can do to be part of the solution:

1. Carry a reusable bottle.
2. Choose aluminum cans over plastic when possible.
3. Recycle all plastic bottles.

Of course, it isn't just plastic bottles that are a problem, as plastic has become deeply entrenched in every aspect of our daily living. Although the U.S. and Japan refused to sign the **G7 pact against plastic pollution** in 2018, many other countries around the world are getting on board to ban plastics. This pact called on countries to reduce single-use plastics, prevent plastic from entering the world's oceans, clean up existing plastic pollution, and invest in technologies for monitoring the impacts of the problem.

Global Citizen reporter Seneo Mwamba states, "Stores such as Iceland and Ikea have introduced plastic bans that will go into effect within the next couple of years. Meanwhile, plastic straws and cutlery have been targeted by numerous retailers, including McDonald's in the UK, and having a plastic bag is enough to land you in jail in some countries. Kenya introduced one of the world's toughest laws against plastic bags in 2017. Now, Kenyans who are caught producing, selling, or even **using plastic bags** will risk **imprisonment of up to four years or fines of $40,000.**"[380]

Mwamba points out that worldwide, about two million plastic bags are used every minute, but one effective solution in Canada and the UK has been implementing a five-cent charge for each plastic bag since 2015. Mwanba tells us this has resulted in an 83 percent reduction in plastic bag use. Still, with our population growing at an alarming rate, plastic use and production is on the rise worldwide. Stronger measures are needed, since we can't simply recycle our way out of this problem. Solutions need to start with government policies, including reducing the availability of the components of plastic. This could be accomplished through eliminating subsidies to gas and oil companies and placing a large tax on that industry for clean up costs. We also need to ban single use plastics worldwide to trigger the use of alternative products.

Rivers—Restoring Our Fresh Water

The sound of water is worth more than all the poets' words.
—Octavio Paz, Mexican poet

World Water Day, held on March 22 every year since 1993, is an annual United Nations observance focusing on the importance of fresh water. In keeping with the aims of this year's World Water Day, the UN is focusing on the connection between water and climate change.

A 2020 UN press release states, "As the impending Climate Crisis draws ever closer, it's increasingly clear how extreme water events are going to make water availability more unpredictable."

The press release notes that with all the impressive activities taking place around the globe for World Water Day 2020, we are reminded to consider how our management of water will help reduce floods, droughts, scarcity, and pollution, and will help fight climate change itself. It's clear that using water more efficiently will reduce greenhouse gases. Furthermore, by adapting water resources management to the effects of climate change, we will protect health and save lives. World Water Day 2020 was about water and climate change—and how the two are inextricably linked.

The IWA states, "The International Water Association (IWA) connects professionals across sectors to deliver water-wise climate solutions. Water is a connector that offers solutions to global challenges. It is everybody's business and climate change affects us all. What we do with water in cities can impact climate change, and climate change can impact water and exacerbate water scarcity in cities where people are already lacking access to safe drinking water. Acting now has never had more meaning. There are a number of things we can be doing to take Climate Action in the Water Sector and the time to act is now."[381]

In 2020, WaterAid and Global Citizen teamed up to make sure that **to combat COVID-19, their project, EVERYONE EVERYWHERE 2030**, assures that every citizen has access to clean

380 Seneo Mwamba, Global Citizen, *10 Plastic Pollution Facts That Show Why We Need To Do More*, June 14, 2018, https://www.globalcitizen.org/en/content/plastic-pollution-facts/
381 International Water Association, *Climate Change and Water – 2020 World Water Day*, March 20, 2020, https://iwa-network.org/news/climate-change-and-water-worldwaterday/

water to wash their hands and stay safe. WaterAid states, "For the first time in history, we have proof that a world where everyone has access to safe water and toilets is an achievable dream. It won't be easy, but we've committed our very existence to making it happen. That's why we're proud to be part of Global Citizen. We simply can't do this alone."[382]

WaterAid adds, "For 34 years and in 26 countries, we've been working with people in the world's poorest communities to build taps and toilets, campaign for change and bolster the skills of local families. Since 1981, we've reached 21 million people with water that is clean and safe to drink and, since 2004, 18 million people with toilets and sanitation."

In Cape Town, South Africa, citizens have proven that they too can avert a water crisis. In 2018, officials in Cape Town announced that the city of four million people was three months away from running out of municipal water, but one year later, the problem had been rectified. Citylab reporter Christian Alexander stated, "Labelled 'Day Zero' by local officials and brought on by three consecutive years of anemic rainfall, April 12, 2018, was to be the date of the largest drought-induced municipal water failure in modern history."[383]

One year later, and this booming city had managed to evade disaster by using a variety of conservation efforts, which could serve as a precedent for cities globally who face this same crisis.

Alexander explains, "One big boost came in February 2018, when the national government throttled allocation of water in the region earmarked for agriculture, allowing more to flow to urban residents. The same month, farmers also agreed to divert additional water stored for agricultural purposes to the city. Cape Town's government ramped up water tariffs and enforcement of prohibitions on heavy users, prohibiting use of municipal water for swimming pools, lawns, and similar non-essential uses. The city's government also implemented a new water-pressure system in January, saving roughly 10 percent of overall municipal water consumption. The effect was stunning, as Cape Town cut its peak water use by more than half in three years."

There was also large-scale cooperation by a wide swath of residents, businesses, and other stakeholders. The municipality's weekly water report became a regular topic at social gatherings and on the radio. Governmental and civic organizations published water-saving techniques, and people traded tips on social media. Mother Nature also helped out with more rain, and so this city is safe for another year.

In Brazil, villagers are facing a threat of a different kind. After plastic waste contributed to deadly floods numerous times each year, one villager took action and started turning plastic pollution into profit. Now people can earn a living by cleaning up the river in a scheme being imitated around the world.

The *Guardian* reporter Sandra Laville shared, "Maria das Gracas started collecting her plastic bottles after she saw the body of her neighbour floating past her house, carried along with the pollution that helped cause the deadly floods. She stores them by the front door of her one-story home, which sits on the litter-strewn banks of the Tejipió River in north-east Brazil. When she has enough

382 WaterAid UK, *Combatting Coronavirus,* 2020, https://www.wateraid.org/uk

383 Christian Alexander, Citylab, *Cape Town's 'Day Zero' Water Crisis, One Year Later,* April 12, 2019, https://www.citylab.com/environment/2019/04/cape-town-water-conservation-south-africa-drought/587011/

she will take them to the local storage skip, where a litter collector will pay her two reals for 50 plastic bottles—about 40 pence. She's not just doing it for the money. She's doing it to stop the tide of plastic drowning this community."[384]

A river that once flowed freely by her house now is stagnant with plastic waste—food wrappers, cola bottles, water bottles, and even plastic chairs. Laville points out, "Globally, some 2 billion people live in communities with no rubbish collection. While international attention has focused recently on the marine plastic litter crisis, the devastating impact of plastic waste on the world's poorest is no less destructive, causing flooding, disease, and hundreds of thousands of premature deaths from toxic fumes caused by the burning of waste."

So residents grew tired of waiting for government action, and organized this project to make a living out of collecting the waste and turning it into products they can sell like handbags, jewelry, and toys, with **young people at the forefront**. Laville adds, "They are supported by Tearfund, the international NGO which is lobbying for global development funding for waste projects to be increased from 0.3% to 3%; a move which would push waste higher up the international agenda, reduce global plastic littering, help cut marine litter and improve the environment and the lives of the world's poorest and most vulnerable."

As the world's plastic crisis becomes more urgent, we are seeing similar projects take root around the world. These projects, often mobilized by schools or community groups, empower people and improve their quality of life. Also, we know that adding more people to the planet only increases the amount of plastic produced, so having fewer children is the best thing we can possibly do for the plastic crisis.

World Wildlife Fund (WWF) Canada's Restoration Fund is working with Coca-Cola Canada to restore Canada's Sockeye, Coho, and Chinook salmon, as well as the endangered Steelhead trout, that have declined drastically in recent years. In 2019, this project focused on the Upper Pitt River and Blue Creek in British Columbia to help restore fish migration, ecosystem connectivity, and water flow.

The *Globe and Mail* points out that the decline of fish stocks was due to climate change, overfishing, and a river blockage due to a landslide that had partially obstructed Blue Creek—a traditional Chinook spawning ground. Sixty years ago, 3,500 Chinook returned to the river system to reproduce, but in 2018, only 75 came back. The Bailey and Katzie First Nations recognized the gravity of the situation and looked to WWF-Canada's Restoration Fund for support.[385]

Since the Upper Pitt River had changed its course because of the landslide, the project included removing the boulders and wood debris to enhance water quality and flow, and protect the lower spawning grounds. On World Rivers Day, September 22, 2019, WWF announced, "Success! The salmon have returned." The strength of projects like this one is their ability to bring together many

384 Sandra Laville, The Guardian, *The Brazilian villagers turning plastic pollution into profit,* Feb 3 2020, https://www.theguardian.com/world/2018/apr/12/the-brazilian-villagers-turning-plastic-pollution-into-profit
385 Globe and Mail Content Studio, *Helping to Restore Canada's Fresh Water, One Partnership at a Time,* October 28, 2019, https://www.theglobeandmail.com/life/adv/article-helping-to-restore-canadas-freshwater-one-partnership-at-a-time/

different partners—WWF-Canada, Coca-Cola, Katzie First Nation, government departments, the Lower Fraser Fisheries Alliance, and other local residents.

The *Globe and Mail* points out that for Coca-Cola Canada, the Blue Creek project is part of a much larger commitment to return 100 percent of the water used in its drinks back to communities and nature. In fact, the company hit that milestone in 2018, two years ahead of its goal.

Making Governments Accountable

When rivers were under threat from the Trump administration, the Center for Biological Diversity filed more than 200 lawsuits against the administration. In 2020, they claimed, "And it's working. We've won 9 out of every 10 cases that have been resolved."[386]

The center explains, "We've fought off some of the most egregious environmental rollbacks and secured protection for imperiled wildlife and plants in the face of unprecedented hostility to endangered species. We've also cast daylight on the most secretive government operations we've ever seen using the Freedom of Information Act, and we've halted destructive projects like the massive Rosemont copper mine in Arizona. The Trump administration serves corporate profits over people—so its attacks on the environment won't stop until its last day. But every single day until then, we'll be battling it ferociously to protect our health and our planet."

Numerous Senate bills have also been introduced by various states to protect rare species and lands. The Grand Canyon Centennial Protection Act, which would make permanent a ban on new uranium mining on about one million acres of public land adjacent to Grand Canyon National Park, has been introduced. These advances were all made possible because of public protest, showing how important it is for citizens to voice their concerns.

Reforming the Meat Industry to Protect Our Water

According to Eco Watch reporter Reynard Loki, the public is also waking up to the meat industry's polluting practices and demanding improvements. We must make polluted rivers and dead zones in our oceans a thing of the past.

Loki states, "The Mighty Earth report recommends that meat producers start employing better farming practices to help curtail the destruction. One way to reduce the need of fertilizers on crops used to feed livestock, for example, is to use cover crops, which involves planting certain species on fields that can suffocate weeds, control pests and diseases, reduce soil erosion, improve soil health, boost water availability and increase biodiversity—all of which would benefit any farm. Mighty Earth also recommends that meat producers employ better fertilizer management, conserve native vegetation and centralize manure processing."[387]

Mighty Earth campaign director Lucia von Reusner asserts, "More sustainable farming practices are urgently needed if we are going to feed a growing population on a planet of finite resources."

---

386 Endangered Earth, Center for Biological Diversity, *More Than 200 Lawsuits Against the Trump Administration,* April 30 2020, https://www.biologicaldiversity.org/publications/earthonline/endangered-earth-online-no1034.html

387 Reynard Loki,Eco Watch, *Corporate Food Brands Drive the Massive Dead Zone in the Gulf of Mexico,* Sep. 01, 2018, https://www.ecowatch.com/corporate-food-dead-zone-gulf-of-mexico-2600841933.html

Their report notes that the federal government is doing little to alleviate the problem, as runoff pollution and greenhouse gas emissions from producing meat are largely unregulated in the U.S. They recommend numerous SOLUTIONS, including stronger regulations to protect the water and climate from the meat industry's ominous seepage of toxins. This pollution makes its way into the Missipi River which flows into the Gulf of Mexico, adding to the catastrophic dead zone which has already killed millions of fish and other sea life.

The environmental damage caused by the meat industry is driving some of the most urgent threats to the future of our food system—from contaminated waters to depleted soils and a destabilized climate.
—Mighty Earth

The report also recommends enforcing the action plan that was agreed upon in 2001 by the Mississippi River/Gulf of Mexico Watershed Nutrient Task Force that was meant to scale down the amount of nutrient pollution in the Gulf by 30 percent. Because this plan lacked teeth, the pollution has instead increased.

Scientists have estimated that the 2018 dead zone measured 5,460 square miles, the size of Connecticut. Unfortunately, the **industry has worked with politicians to prevent regulations to address this issue**.

Mighty Earth adds, "While reforming the meat industry's unsustainable practices is a way to stop the spread of dead zones, change from within isn't coming quickly enough. That's where consumers can play a vital role." The most effective solution would be for consumers to move towards a plant-based diet as recommended by the United Nations. If the demand for meat and dairy decreases, meat production will also decrease and so will the resulting pollution.

The purpose of this chapter has been to highlight the impacts a planet of EIGHT BILLION consumers has on our planet's most precious resource—water. Our escalating numbers and excess have contributed towards piles of plastics in our waterways, toxic industrial chemicals polluting our groundwater and rivers, and overfishing decimating fish stocks in our oceans. However, it is encouraging to see so many innovative groups and citizens volunteering to clean up our oceans, adopt a plant-based diet, promote smaller families, and protect wildlife in our vulnerable water world. After all, unlike other earthly commodities, there is no replacement for water.

PART 3
CRITICAL GLOBAL INFLUENCERS

The most significant characteristic of modern civilization is the sacrifice of the future for the present, and all the power of science has been prostituted to this purpose.
—William James, philosopher and psychologist

It is unfortunate that many of the critical global influencers who have blocked efforts to reduce population growth have purposely misrepresented the many moral and ethical issues involved. Often the debates surrounding **population** are hysterical and full of unfounded accusations that are ethically unsound and illogical. It often provokes name-calling such as authoritarian, sinners, intolerant, racist, chauvinist, and so on. Neither reason, nor truth, nor justice, nor the greater good of society stand a chance in this atmosphere.

Part of the solution would be to compel our governments to bring population out into the open, conduct extensive public consultation, conduct carrying-capacity studies, and develop meaningful population policies for every country. The other part of the solution would be to compel the United Nations to revoke the special veto status that was given to the Vatican at UN conferences, since they have greatly abused this privilege. Then we could finally dispel many of the taboos and myths and move forward to implement voluntary population programs.

So why are so many critical global influencers—those elite growthpushers in positions of power and influence—standing in the way of this process? For one thing, we need to dispel the myth of "green growth" that many policymakers, economists, and influencers are promoting. We now know that neither population growth nor economic growth are realistic in the 21st century.

Foreign Policy reporter Jason Hickel states, "Warnings about ecological breakdown have become ubiquitous. Over the past few years, major newspapers, including the *Guardian* and the *New York Times,* have carried alarming stories on soil depletion, deforestation, and the collapse of fish stocks and insect populations. These crises are being driven by global economic growth, and its accompanying consumption, which is destroying the Earth's biosphere and **blowing past key planetary boundaries** that scientists say must be respected to avoid triggering collapse.... These problems throw the entire concept of green growth into doubt and **necessitate some radical rethinking**."[388]

388 Jason Hickel, Foreign Policy, September 12, 2018, *Why Growth Can't Be Green* https://foreignpolicy.com/2018/09/12/why-growth-cant-be-green/

Study after study has shown that for decades we have been extracting more from our lands and oceans than Earth can replenish or afford to lose, much in the name of "green growth" and increasing our flawed GDP. **Now that we have passed this tipping point, pursuing any degree of growth drives resource extraction up far beyond a sustainable level.**

Yet the **pursuit of constant economic growth** is still being encouraged by most governments and our economic community, and **immigration is still being promoted as a means of accomplishing this in many countries**. Fortunately, some of us have now figured out that ending this addiction to growth would mean living within our means, and a reduction in commuting, family size, single-use products, food waste, meat and dairy consumption, and the use of plastics.

Hickel adds, "Green-growthers trumpet the transformation of European economies in recent decades: higher GDP, falling emissions. But that's mostly because these countries have offshored their emissions: much of their stuff is now made in Asia. Moreover, aviation and shipping aren't counted against national carbon budgets. Once you factor in emissions embedded in imported goods, the EU's carbon emissions are about 19 per cent higher than the bloc's official figures, calculates the Global Carbon Project, a network of scientists; for many big cities, the gap is about 60 per cent."

A long economic depression might be enough to keep the planet habitable. We'd also need to divert money from consumption to family planning, globalize a plant-based diet, grow industrial hemp for plastic alternatives and food, plant trees, and clean up plastic. Kuper adds, **"Economic growth, democracy and CO_2 have always been intertwined."** Degrowthing can only happen if we put the "greater good" first.

Part 3 will demonstrate how ending growth doesn't mean shutting down economic activity or that basic living standards need to take a hit. With a sustainable population and an effective economic system (GPI), we could be living an ideal lifestyle with the best of what the planet has to offer, while still caring for our environment that sustains us. Surviving the 21st century may not be as difficult as we have been told.

CHAPTER 10
POLITICS AND THE ECONOMY

For way too long, the politicians and the people in power have gotten away with not doing anything to fight the climate crisis, but we will make sure that they will not get away with it any longer. We are striking because we have done our homework and they have not.
—Greta Thunberg

2019 was a year to go down in history, first with Greta Thunberg taking centre stage and leading a global climate revolution, then with Australia's horrifying bushfires claiming millions of iconic koalas and other wildlife, and finally COVID-19 seizing centre stage and showing us all that we can make radical changes for the greater good if we are willing to work together. It leaves the entire world stunned, dumbstruck, wondering what the future could possibly hold in store for the 2020s. Anthropologist Neil Turner points out that the disastrous political events occurring in the U.S. during these same uncertain times will also have far-reaching, long-term impacts.

International Politics

No matter how distracted we may be by the number of problems now facing us, one issue remains fundamental: Overpopulation. The crowding of our cities, our nations, underlies all other problems.
—Paul R Ehrlich, president of the Center for Conservation Biology

Perhaps if we had heeded these wise words from the 1960s, children like Greta Thunberg wouldn't have to give up their childhood to fight our adult problems for us in 2020.

Too few voters have heeded Greta Thunberg's warning that we must take back the power politicians and the elite hold over our destiny. Now with social media, many of us have indeed done our homework, and understand that her advice is based on sound science, and crosses all political, religious, racial, and gender barriers! We will no longer be manipulated by deceptive governments or constrained media. As David Suzuki said, "If the people will lead, the leaders will follow." In the 21st century, the youth are now leading, dragging the leaders in their wake. But will the leaders catch up in time to prevent the global crisis that is looming in our near future?

The Myth of Green Growth

Greta Thunberg may not have all the answers, but what she does have is the ability to comprehend the magnitude of the problem and the fact that we have very little time to avoid irreversible disaster. She can see far enough into the future to visualize an unbearable life ahead for both humanity and other animals if we don't make radical change. The problem is that we have run out of time. Neither so-called "green energy" nor "green growth" can save us now. Instead, we must get serious about "degrowth" and put the greater good first. We only have ten years to turn things around, so it is time to look at the driver behind all of these critical world issues—population growth. And our decade of action must start now, not in 2030.

Financial Times reporter Simo Kuper has the same sentiments about the myth of green growth. Kuper states, "Here's the story about climate that we liberals like to tell ourselves: once we get rid of dinosaur politicians like Trump, we'll take on the fossil-fuel lobby and greedy corporations and vote through a '"green new deal.' It will fund clean, fast-growing industries: solar, wind, electric vehicles, sustainable clothes. That will be a win-win: we can green our societies and keep consuming. This story is called 'green growth.'... Except that this is a fairy tale, and green growth doesn't really exist."[389]

> *And when people have money, they convert it into emissions.* ***That's what wealth is.***
> **—Simo Kuper,** *Financial Times*

Anthropologist Neil Turner adds, "While in South America, the countries of Bolivia, Chile, Venezuela, and Ecuador are literally on fire—people are revolting in the streets. In some of these countries, a war against right-wing authoritarian political leadership is taking place while in others famine and starvation are ravishing the country, society, and people. Many people think this is normal for Latin American countries but at this point, we also see political clashes, riots and anti-government protests in Britain, France, Spain, Portugal, Ukraine, and Greece as well."[390] They are also taking extraordinary measures of importing and off-sourcing goods and energy to avoid meeting the UN Sustainable Development Goals, a desperate disappearing act.

Scientists have been telling us for decades that the fuse is running short, and now in 2020, the UN is warning that we have a ten-year window to make radical changes to avert disaster. As we near EIGHT BILLION people to feed, shelter, and clothe, our options are running out. **If world leaders don't make population a priority now and stop growth at EIGHT BILLION, we will have little chance of effectively tackling the other critical world issues.**

> *Human population has pushed these societies far beyond their limits, they are importing many of their goods, off-sourcing most of their dirty energy, importing their low-wage work force, and exporting most of their waste.*
> **—Neil Turner, anthropoligist**

389 Simon Kuper, Financial Times, *The myth of green growth,* October 23 2019, https://www.ft.com/content/47b0917c-f523-11e9-a79c-bc9acae3b654

390 Neil Turner, Perspectives in Anthropology Internet-based magazine, *The Collapse of Societies*, Dec 1, 2019, https://perspectivesinanthropology.com/2019/12/01/the-collapse-of-societies/

International Progress Sadly Lacking

To take us a step closer, in 2019 Kenya hosted the International Conference on Population and Development (ICPD) to further advance the revolutionary program of action developed at the first summit in Cairo, 25 years ago. The summit aimed to examine the progress made since 1994 and to produce pledges of financial support as countries move forward.

VOA News reporter Rael Ombuor states, "Over 6,000 delegates from 160 nations, including heads of state, are attending the three-day forum to discuss reproductive health rights, ending gender-based violence, and sustainable development."[391]

Crown Princess Mary of Denmark, a co-host of the conference, underscored the significance of the summit: "I think we can all agree that ICPD was a turning point, a defining moment in our history. In Cairo, the world articulated a bold new vision about the relationship between population, development and individual well-being and the empowering of women and meeting people's needs for education and health, including sexual and reproductive health, are necessary for both individual advancement and balanced development."

Kenya's President Uhuru Kenyatta pledged to end female genital mutilation (FGM) by 2022, and further stated, "I believe that we can all commit to eliminate child marriages. The percentage of young women between 20 and 24 years of age who are married before their 18th birthday has declined from 34 percent in 1994 to 25 percent in 2019," said Kenyatta. "But the absolute number of girls under 18 who are at risk of child marriage is estimated at 10.3 million in 2019."

U.N. Deputy Secretary-General Amina Mohammed tied women's rights squarely to development, stating, "The power to choose the number, timing and spacing of children is a human right that can bolster well-being, economic and social development. And when people can exercise their rights, they thrive. And they do and, so do societies at large."

UN Women also participated in the Nairobi Summit to commemorate this milestone in the history of gender equality and women's rights. Although most participating countries were very supportive of empowering women, the U.S. joined with a group of authoritarian countries in opposing not just abortion, but the very use of the medical phrase, "**sexual and reproductive health.**" Trump claimed that phrase could undermine the critical role of the family, promote abortion, and be misinterpreted by UN agencies. So the Trump administration discontinued support for the agency, renewed the global gag rule, and made drastic cuts to family planning services.

Foreign Policy reporter Rikha Sharma Rani stated, "One of the signature marks of U.S. President Donald Trump's foreign policy has been deepening cooperation with authoritarian countries. Among the starkest expressions of that trend has been Washington's new alliance with some of the world's worst violators of women's rights (including Saudi Arabia, Libya, Sudan, and the Democratic Republic of the Congo, among others)."[392]

391 Rael Ombuor, VOA News, *Kenya Hosts International Conference on Population and Deveopment,* Nov 12, 2019, https://www.voanews.com/africa/kenya-hosts-international-conference-population-and-development

392 Rikha Sharma Rani, Foreign Policy, Trump's War on the Concept of Women's Health, March 10, 2020, https://foreignpolicy.com/2020/03/10/trump-war-concept-women-health-sexual-reproductive-rights/

This behaviour was mainly orchestrated by the Vatican and the religious anti-abortion groups that formed a core of Trump's supporters. Marie Smith, who leads an international anti-abortion organization and serves as a U.N. observer for the Holy See (Vatican), was particularly vocal.[393] Rani notes, "Trump isn't just going after abortion. He's going after what conservative religious groups decry as a broader feminist takeover of global health.... Services include HIV testing, cancer screening, and education about child marriage and human trafficking. Trump's cuts to sexual and reproductive health are affecting all of these services, too." Although polls suggest that the majority of Americans believe religion has no place in politics, the United Nations, the U.S., and many other countries have been controlled by religious bullying tactics for decades. This issue will be discussed in detail in Chapter 11 on religion. Although Trunp is no longer in power, the Vatican continues to promote his disruptive tactics.

Rani points out that in the developing world, at least ten million unintended pregnancies occur each year among girls and young women aged 15 to 19, and complications from pregnancy and childbirth are the leading cause of death for those in this age group globally. Worldwide, an estimated 15 million adolescent girls and women aged 15 to 19 have experienced forced sex at some point in their life. One in six women alive today were married as children. In 2017, HIV was among the top five causes of death globally for girls aged 10 to 14.

This new Trump alliance is an ominous sign for girls and women globally, and puts a huge burden on other sources of family planning funding like the Thriving Together initiative to fill this gap. Fortunately, the message from other donor countries, including Canada, was clear that they will continue to support sexual and reproductive health rights. In fact, Canada has announced plans to significantly increase funding to this program.

Minister for Foreign Trade and Development of the Netherlands Sigrid Kaag issued a statement that was signed by 58 countries, including the United Kingdom, France, Spain, Germany, Mexico, South Africa, Tunisia, and Australia. It said, "Investing in comprehensive sexual and reproductive health services ... is necessary to address the needs of women, girls, adolescents, and people in the most marginalized situations who need these the most."

According to the Marie Stopes Foundation, contraception allows women to take control of their futures—and in doing so, it drives economic and social development. They point out that $9.4 billion is needed to provide modern contraception worldwide.[394] This is less than what the bankers on Wall Street give themselves in bonuses each year. Perhaps the paltry sum needed for family planning programs could come from a luxury tax imposed on Wall Street. Or funding for this foreign aid could be recouped from military budgets if all countries put war on hold for just *one day*. Not only is our constant state of war depleting our economies, but it is also a losing battle that is unjustified. This will be discussed in detail later in this chapter under "Perpetual War."

393 Kathryn Joyce, Pacific Standard News, *The New War on Birth Control*, Sep 16, 2018 https://psmag.com/magazine/new-war-on-birth-control

394 Marie Stopes Foundation, 2020, https://www.mariestopes.org/the-challenge/

U.S. taxpayers give the military over $2 billion a day ($710 billion per year). According to Wikipedia, the U.S. Department of Defense budget for 2020 is approximately $721.5 billion.[395] A war is ready to start at any time the world military elite decide they need to stock their coffers. A state of constant war is necessary to keep the military employed, the munitions factories operating, and control of resources worldwide. Yet in 2020, Trump pulled most funding for family planning services, having a huge impact since the U.S. is the leading contributor to the UN population fund.

At the third preparatory conference for the International Conference on Population and Development in 1994, the objectives of the UN were defined: the goal would be to stabilize world population at 7.27 billion by 2015. The means would be threefold:

1. Filling unmet demands for contraceptive services
2. Increasing educational opportunity, particularly for girls
3. Providing basic health services, particularly for infants and mothers

United Nations Goals for 2030

Obviously, none of these objectives made it through the International Conference on Population and Development intact or were ever fully implemented. Now, 25 years later, the situation has gotten even more dire and more difficult to remedy. The question remains, will the Vatican be allowed to remain a hindrance to sustainable population worldwide by using their veto power in the UN? The 2020 decade is looking a little more promising on this front, as more women leaders are fighting for science and common sense to take precedence over dogma and hostile religious bullying.

One could argue that the UN population programs need to step up their performance in regard to solving the global population crisis. But are they rich and powerful enough to reverse the trend? We must remember that they depend on participating countries to follow through on their commitments, and this is often lacking. The UN is often underfunded, understaffed, and facing great opposition from the Vatican and others who shape the world in which they operate. The UN is often manipulated by institutions like the International Monetary Fund, World Bank, and World Trade Organization, which are not held accountable and cannot always deliver what they promise. World aid programs all suffer from the way wealthy countries, the U.S. in particular, use their purse strings to manipulate the United Nations.

The UN was limited for years in what it could do by the veto power of the Vatican, all its projects swimming against a hostile tide. In 2019, COVID-19 took centre stage and has diverted our attention and funding to this devastating pandemic. Through all of this, climate change has remained a lead player, with the role of the UN partly to mediate between countries who want to tax their way out of trouble, and those who want to buy and sell permits to pollute.

---

395 Wikipedia, *Military budget of the United States*, 2020, https://en.wikipedia.org/wiki/Military_budget_of_the_United_State

Of course in this public arena, many governments want large populations to enhance their prestige, military strength, and negotiating power. On the international level, many governments feel more secure with a large population so that they can hold their own against domineering neighbours or the vast populations of countries like India or China.

After decades of exponential population growth, there is little that triggers as much panic from the Vatican, political leaders, right-wing populists, and big business as the prospect of population decline. All of these entities have corporate interests they desperately want to protect. Consequently, with the United Nations now strongly supporting a sustainable population, some foresee endless recessions, aging societies, collapsing markets, bankrupt pension funds, and an end to those abhorrent profits they have become accustomed to.

This might be the case if population decline were rapid, but *with the gradual decline* that is likely to occur, the benefits would be numerous—reductions in climate emissions, resource depletion, extinction of species, food shortages, water shortages, forest depletion, fisheries depletion, and plastic pollution. Most importantly, we would also see increased well-being and empowerment of women with a decrease in family size. Now we just need this revelation to reach the echelons of elected politicians—and other critical global influencers—since a population decline is imminent and necessary, according to the UN on their Sustainable Development Goals website.

The UN notes that the 2030 Agenda for Sustainable Development, adopted by all United Nations member states in 2015, provides a shared blueprint for peace and prosperity for people and the planet, now and into the future. At its heart are the 17 Sustainable Development Goals (SDGs), which are an urgent call for action by all countries in a global partnership. They recognize that all goals hinge on Goal #5 for the empowerment of women and access to family planning services. They also recognize that ending poverty and other deprivations must go hand-in-hand with strategies that improve health and education, reduce inequality, and spur economic growth—all while tackling climate change and working to preserve our oceans and forests.[396]

One thing that is not clearly addressed by the SDGs is the impact of international trade. To the contrary, free trade agreements influenced by corporate interests are often encouraged by the international community, when in fact they have a huge negative impact on climate disruption, food security, and overconsumption. For instance, international trade often requires increased air travel for negotiators to secure trade deals and for related social interaction. Also, importing products from other countries, rather than producing them domestically, produces tremendous amounts of climate emissions from transportation. These emissions are not recorded for any particular country, since they are produced offshore and are exempt. So we do not see the true picture of how free trade is negatively impacting climate change and reducing food security.

We should be moving towards each country being more self-sufficient and eliminating the import of goods whenever possible. This really hit home in Canada during the COVID-19 pandemic with the need to import masks, instead of producing our own supplies. It not only required

396 UN Sustainable Development Goals website, *The 2030 Agenda for Sustainable Development*, 2020, https://sustainabledevelopment.un.org/?menu=1300

the transport of masks from China, which created high CO_2 emissions, but also highlighted our dependence on a supply chain that caused severe shortages and quality problems.

International Law and Politics

> *Achieving and maintaining a sustainable relationship between human populations and the natural resource base of the earth is the single most critical long-term issue facing the peoples of the world and this issue will increasingly be the focus of international affairs for the foreseeable future.*
> —Russel E. Train, World Wildlife Fund

The Universal Declaration of Human Rights (UDHR) is a document that acts like a global roadmap for freedom and equality, according to Amnesty International. It was the first time countries agreed on the freedoms and rights that deserve universal protection in order for every individual to live their lives freely, equally, and in dignity.

Amnesty International points out, "The UDHR was adopted by the newly established United Nations on December 10, 1948, in response to the 'barbarous acts which [...] outraged the conscience of mankind' during the Second World War. Its adoption recognized human rights to be the foundation for freedom, justice and peace."[397]

The Declaration outlines 30 rights and freedoms, and Right 25 reads: "Everyone has the right to a standard of living adequate for the health and well-being of himself and of his family, including food, clothing, housing and medical care and necessary social services, and the right to security in the event of unemployment, sickness, disability, widowhood, old age or other lack of livelihood in circumstances beyond his control."

However, this document is not a treaty, so it does not directly create legal obligations for countries. Many argue that without a healthy ecosystem, humanity cannot achieve this health and well-being referred to in the Declaration. This is where the UN international law falls short, and where adopting the crime of ecocide could make up for this shortfall and complete the intent of the Declaration of Human Rights, the most translated document in the world.

Reviving Confidence in the Principle of the Public Trust

The protection and recovery of that which serves us all, the environment, is also the intent of the creation of the ***Public Trust*** concept. The Declaration of Human Rights and the Public Trust Doctrine go hand-in-hand, since both of them identify present and future generations of citizens, you and me, as the beneficiaries of this trust. In 2019, ecologist Dr. Brian Horejsi pointed out, "THE Public Trust, however, refers to something far more critical to each and every one of us; our public land, clean water, clean air, ecologically functional forests and the complex of plants and animals we label biodiversity. The Public Trust Doctrine (PTD) traces its history to Roman

397 Wildlife.org, *The Public Trust Doctrine: Implications for Wildlife Management and Conservation in the United States and Canada*, September 2010, https://wildlife.org/wp-content/uploads/2014/05/ptd_10-1.pdf

times almost 2,000 years ago and is recognized as the **oldest principle of environmental law,** pre-existing all statutory environmental laws across the world. It's been described as *'the law's DNA.'*"[398]

However, Horejsi feels that as our population grows towards EIGHT BILLION, "Public Trust" is slipping away. He states, "Governments across the land have worked, without public consent, and with no or little debate, to steal the public trust from 'we the people,' holding it close to their vest, handing it out to political moneyed and corporate favourites. **We might have avoided the looming collapse of society and environmental integrity, and some of the deep division in the country, had we enacted and defended the public trust."**

In Canada in the 1960s, some citizens and governments began discussing a revival of the Public Trust Doctrine, but the growth lobby managed to fuel immigration with demands for low-paid workers in order to increase consumption and economic growth despite these efforts. Over 100 legal scholars are now working to gain legal support to advance this doctrine in the U.S., despite a worldwide counter-revolution taking place. With the ecocide law that has recently been proposed, they might have a chance at success.

Horejsi notes that PTD speaks to one of the most essential purposes of government: protecting landscapes and ecological function for the continuing survival and welfare of all citizens. He notes, "As a doctrine of common property law, PTD limits privatization, exclusive use, and degradation of trust assets. This means not more growth or consumption, but reform and improvement that serves us all collectively; better government, greater equality in the eyes of the law. The doctrine has often been characterized as an attribute of sovereignty that carries constitutional force. And that is where it should be entrenched!"

The International Court of Justice: Defending the Public Trust

In 1916, Theodore Roosevelt stated, "Defenders of the short-sighted men who in their greed and selfishness will, if permitted, **rob our country of half its charm by their reckless extermination of all useful and beautiful wild things** sometimes seek to champion them by saying that 'the game belongs to the people.' So it does; and not merely to the people now alive, but to the unborn people. The 'greatest good for the greatest number' applies to the number within the womb of time, compared to which those now alive form but an in-significant fraction. Our duty to the whole, including the unborn generations, bids us to restrain an unprincipled present-day minority from wasting the heritage of these unborn generations. The movement for the conservation of wildlife and the larger movement for the conservation of all our natural resources are essentially democratic in spirit, purpose, and method."[399]

The "greatest good" Roosevelt is talking about is meant to apply to the billions living now in misery as well as the unborn generations to come. Is there a universal law to protect them?

398 Dr. Brian Horejsi, eknow.ca, The Public Trust is slipping away, December 29, 2019, https://www.e-know.ca/regions/east-kootenay/the-public-trust-is-slipping-away/

399 Wildlife.org, *The Public Trust Doctrine: Implications for Wildlife Management and Conservation in the United States and Canada*, September 2010, https://wildlife.org/wp-content/uploads/2014/05/ptd_10-1.pdf

According to the International Court of Justice (ICJ) in the Netherlands, "The ICJ is the principal judicial organ of the United Nations. It was established by the UN Charter in 1945 'to bring about by peaceful means, and in conformity with the principles of justice and international law, adjustment or settlement of international disputes or situations which might lead to a breach of the peace.'"[400]

In the field of public international law, the ICJ is the only judicial organ with potentially both general and universal jurisdiction, and it is often called "the World Court." All 193 members of the UN are automatically party to the ICJ. The Court has a twofold role:

1. to settle, in accordance with international law, legal disputes between States
2. to give advisory opinions on legal questions referred to it by duly authorized UN organs and agencies

The Court deliberates on camera. Following the deliberations, it delivers a judgment, which is binding, final, and without appeal for the parties to a case in open session. Cases of non-compliance are extremely rare. In the exceptional case that one of the states involved fails to comply with the decision of the Court, the other state may lay the matter before the UN Security Council, which is empowered to recommend or decide upon the measures to be taken to give effect to the judgment. Since 1946, the ICJ has delivered 113 judgments on disputes concerning various issues, and this is where the crime of **ecocide** is meant to be reckoned with.

In fact, the vulnerable Pacific islands of Vanuatu and Maldives have called for ecocide—wide-scale, long-term environmental damage—to be considered an international atrocity crime on par with genocide.

Ecocide: An International Atrocity

Climate Liability News reporter Isabella Kaminski states, "The International Criminal Court is currently responsible for prosecuting four internationally recognized crimes against peace: genocide, crimes against humanity, war crimes and the crime of aggression. A fifth could be included through an amendment to the Rome Statute."[401] The fifth crime against peace could be the crime of **ecocide**, proposed by British lawyer Polly Higgins.

Kaminsky adds, "Vanuatu, which is particularly vulnerable to sea level rise, has been an advocate of climate justice at international forums for many years, but has been more vocal since 2015, when Cyclone Pam devastated the island, an example of a major storm whose impact was made significantly worse by climate change. Vanuatu's statement is a major victory for the Stop Ecocide campaign, which was launched by Higgins in 2017. The organization wants any agreed-upon criminal definition of ecocide to include the impacts of climate change as well as other forms of environmental harm."

400 UN, The International Court of Justice, http://nvvn.nl/the-international-court-of-justice-icj/

401 Isabella Kaminski, Climate Liability News, *Vulnerable Nations Call for Ecocide to be Recognized as an International Crime*, December 6, 2019, https://www.climateliabilitynews.org/2019/12/06/ecocide-international-criminal-court-vanuatu/

In 2019, the Republic of Maldives announced that it was adding its support as well. Ahmed Saleem, member of the Maldives parliament, stated, "It is time justice for climate change victims be recognized as part and parcel of the international criminal justice system."

Kaminisky pointed out that Jojo Mehta, spokesperson for Stop Ecocide, said she is optimistic that a formal amendment could be submitted to the International Court of Justicce as early as 2021. "This is an idea whose time has not only come, it's long overdue," said Mehta. "It's committed and courageous of Vanuatu to take the step of openly calling for consideration of a crime of ecocide, and it was clear from the response today that they will not be alone. The political climate is changing, in recognition of the changing climate. This initiative is only going to grow—all we are doing is helping to accelerate a much-needed legal inevitability."

Pope Francis has lent his support to the idea of making ecocide a crime, proposing in 2019 that "sins against ecology" be added to the teachings of the Catholic Church. This is ironic, since no entity has played a more sinister role in the overpopulation crisis that is now threatening our planet—unquestionably, a crime against both humanity and the environment. **The dire situation of overpopulation could tick off many of the boxes for international criminal action, including crimes against humanity, ecocide, torture, war crimes, extinction of species, and so on.** Deforestation of the Amazon, which wiped out hundreds of species and millions of acres of forest, would also be a crime that would potentially qualify as a crime of ecocide.

Immigration

> *The accusation that a stand to reduce immigration is racist is music to the ears to those who profit from the cheap labor of immigrants. They are the same people who love to see environmentalists make fools of themselves. And there is no environmentalist more foolish than one who refuses to confront the fact that uncontrolled human population growth is the No. 1 cause of the world's increasing environmental problems.*
>
> —Captain Paul Watson, founder of Sea Shepherd Conservation Society

Immigration is another ethical issue that has people screaming "Racist" if any country protests the huge influx of aliens. Yet, others argue that no country has a moral obligation to admit immigrants if it threatens the country's environment, social well-being, democratic rights, or cultural integrity. This is not racist; this is self-preservation. Economists mistakenly encourage immigration as the solution to our economic woes, but in reality, it only exacerbates the problem. Instead, developed nations should assist developing nations with family planning services, education, and environmental recovery.

Yes, we are a very accommodating species, very good at adapting to almost any situation. However, the present rate of immigration in the world today is often overwhelming and tends to cause disharmony and conflict. It is only amplifying global problems, spreading them out to other countries already reeling from a plethora of their own critical issues.

We often hear the lifeboat scenario used to describe this situation, as the analogy is so fitting. Then the moral question is raised: do we allow unrestricted immigration or protect our own life-support system first, and then help other countries protect theirs? By allowing overpopulated countries to continually encourage their citizens to emigrate rather than provide a healthy existence for them in situ, the country of origin never gets a true picture of the extent of their problem. The present population count of many developing countries does not reflect the huge number of emigrants, who have gone forth to populate greener pastures.

Many countries, like Canada, believe that their present immigration system *is not working*. Every country needs to adopt a population policy with goals for overall population size and annual limits on legal immigration. Since Canada has the highest intake rate in the developed world—more than double that of the U.S. or Australia, many citizens feel that this number should be reduced to equal our immigration rate and then shrunk to a sustainable level. This would be similar to what the Australians Against Further Immigration have proposed in the past.

Is Canada's Global Leadership on Immigration Misplaced?

An October 1990 *Globe and Mail*–CBC News poll indicated that **46 percent of Canadians wanted to see current immigration levels reduced** and only 16 percent wanted them increased. There was clearly no mandate for the government to expand Canada's population and even less to increase 1990 immigration levels (200,000 per year). Yet since that time, immigration levels have increased to over 250,000 in 2019.

In 2020, this trend continues as the Canadian government continues to ignore the wishes of Canadian voters. Canadian Press reporter Teresa Wright states, "New polling numbers suggest a majority of Canadians believe the federal government should limit the number of immigrants it accepts. A recent Leger poll showed that 63 percent said the government should prioritize limiting immigration levels because **the country might be reaching a limit in its ability to integrate them.** Just 37 per cent said the priority should be on growing immigration to meet the demands of Canada's expanding economy."[402]

However, Minister of Immigration Ahmed Hussen is refusing to pay any attention to the wishes of Canadians. Instead he stated, "I think the answer is to continue on an **ambitious program to invest in infrastructure, to invest in housing, to invest in transit**, so that everyone can benefit from those investments, and that we can then use those community services to integrate newcomers, which will also benefit Canadians." So Hussen will be forcing Canadian taxpayers to accept this course of economic and population growth despite continued opposition for over three decades. Will the media then comment on how they don't understand the civil unrest and riots that ensue? Will we leave it to future generations to pay for all this excessive spending?

Hussen states he is concerned about labour shortages, but this only highlights the jobs that Canadians find very distasteful, such as in meat-packing plants and slaughterhouses, as well as those paying minimum wage, indicating that our minimum wage needs to be adjusted to provide a sensible standard of living. It is ironic that most humans refuse to engage in the industries that

402 Teresa Wright, The Canadian Press, *Poll suggests majority of Canadians favour limiting immigration levels*, June 16, 2019, https://www.cbc.ca/news/politics/canadians-favour-limiting-immigration-1.5177814

they vehemently support (like factory farming and meat processing). Something is fundamentally wrong with this picture; we are not acting in line with our core values here. Taking advantage of an alien workforce is not the answer to this dilemma. Perhaps a more rational solution would be to move towards a plant-based diet as suggested by the United Nations, so that we can eliminate many of those undesirable jobs that involve cruelty to animals in the meat industry.

Wright notes, "In 2017, Immigration Department officials **warned of a "tipping point" that could undermine public support for welcoming immigrants** if public discourse was not approached with care, according to internal data prepared by the Immigration Department for a committee of deputy ministers. The data, which was obtained by The Canadian Press through access-to-information law, suggested a majority of Canadians at that time supported the migration levels, but there was also polling data that suggested this **support dropped when Canadians were informed of how many immigrants** actually arrive every year."

Population Institute Canada (PIC) points out that Canada's population will soon surpass 37 million people. In fact, with growth at 1.2 percent per year, it is one of the fastest-growing developed countries in the world. PIC believes there are compelling reasons why this relentless increase in numbers should be critically examined, then reduced and, ultimately, reversed until population size is consistent with Canada achieving an economically and environmentally sustainable future. To ignore facts and the associated imperatives of mushrooming growth is to sleepwalk into a future of almost certain and dire crises.[403]

PIC states, "For many, the very question is absurd. Surely Canada is a huge land of wide-open, seemingly endless spaces with abundant resources and ample room for many more people. This is what most who reflect on Canada's population size think instinctively, and is what informs arguments and policies favouring ever more people. But it is a viewpoint based largely on myths and wishful thinking rather than on rational, scientific observations and assessments."

In reality, there are many indicators that Canada is already in "overshoot" in terms of depleting many of our resources, such as our old-growth forests, top soil, fisheries, and many wildlife species that are now considered endangered. **Yet economic growth is still being encouraged by our government and economic community, and immigration is touted as the best means. Canadians have decided they do not want our population to increase and they are choosing to have smaller families or to be childfree.** Our numbers would begin to decrease if not for our unwarranted high-immigration rate—an enviable position which most Canadians would applaud. Why won't our government heed the wisdom of the majority of Canadians?

It isn't just Canadians that are protesting unreasonably high immigration rates. The *Atlantic* reporter David Frum notes, "According to recent poll numbers, 63 percent of French people believe too many immigrants are living in their country. One-third of the British people who voted in 2016 to leave the European Union cited immigration as their primary reason. In Germany, 38 percent rate immigration as the most important issue facing their country. Thanks in great part to their anti-immigration messages, populist parties now govern Italy, Poland, Hungary, and the Czech

403 Population Institute Canada, *Too Many People in Canada?*, 2020, https://populationinstitutecanada.ca/about-us/our-canada/

Republic. And of course, anti-immigration sentiment was crucial to the election of Donald Trump to the presidency of the United States."[404]

Treading the Politically Fraught Waters of Immigration

Immigration on a very large scale is politically stressful, socially taxing, and environmentally destructive. Yet acknowledging that fact can be hazardous to mainstream politicians, or citizens who dare to voice their concerns. It often results in name-calling, assault, or even riots. This unease related to pressure put on social systems due to immigration is capitalized on by populists who also confuse the subject with overt racism and xenophobia. This is how the voices of common sense and justified trepidation are silenced. This scenario makes little sense for taxpayers, since it is taxpayer dollars that are footing the bill for the spiraling costs of immigration—both legal and illegal. Is it not possible to have a rational conversation about the impacts of immigration?

The fact that Donald Trump got voted in as president demonstrates how many Americans are concerned about the immigration crisis in America, and how much it is costing every taxpayer. Immigration also highlights how those dollars would be much better spent on foreign aid family planning programs that would help those immigrants create a fulfilling lifestyle in their own countries of origin, rather than buying into the illusion of the "American Dream" lifestyle elsewhere.

The Hill contributor Kristin Tate points out, "The costs of illegal immigration are comprehensive. Even after deducting the $19 billion in taxes paid by illegal immigrants, the 12.5 million of them living in the country results in a **$116 billion** burden on the economy and taxpayers each year. About two-thirds of this amount is absorbed by local and state taxpayers, who are often the least able to share the costs. One of the major drivers of the increasing costs is the 4.2 million children of migrants, who automatically become American citizens. Taxpayers are indeed on the hook for over $45 billion in state and federal education spending annually, not to mention the added burden of increased social welfare dollars. Much of the almost $30 billion in medical and assistance funding is sparked by the fact that noncitizen families in the United States are twice as likely to receive welfare payments than native born families. Setting aside the legal and moral questions that shape immigration policy, there is a significant tax burden imposed on citizens and legal immigrants tied to a leaky border."[405]

In 2018, Prime Minister Justin Trudeau announced that he intends to increase immigration by 2021 to 350,000 individuals a year. *National Post* reporter Kathryn Carlson states, "Immigrants to Canada cost the federal government as much as $23-billion annually and 'impose a huge fiscal burden on Canadian taxpayers,' according to a think-tank report released Tuesday. The Fraser Institute report says newcomers pay about half as much in income taxes as other Canadians

---

404 David Frum, The Atlantic, *If Liberals Won't Enforce Borders, Fascists Will*, April 2019 https://www.theatlantic.com/magazine/archive/2019/04/david-frum-how-much-immigration-is-too-much/583252/

405 Kristin Tate, The Hill, *Your taxpayer dollars are footing the spiraling costs of illegal immigration*, April 21 2019 https://thehill.com/opinion/immigration/439930-your-taxpayer-dollars-are-footing-the-spiraling-costs-of-illegal-immigration

but absorb nearly the same value of government services, costing taxpayers roughly $6,051 per immigrant."[406]

Naturally, this is causing some contention regarding immigration policy and how this money could be better used for foreign aid focusing on family planning, which would alleviate social, political, economic, and environmental pressure. Taking this action could help avoid this unnecessarily high immigration rate. It is not simply overpopulation, but also environmental degradation, social and political unrest, as well as war mongering based on resource tension and corporate interests (which are all fueled by population growth) that leads to emigration from home countries. As global population nears EIGHT BILLION, immigration will instigate even more unrest and disharmony in already overburdened host countries in the future.

"It's in the interest of Canada to examine what causes this and to fix it," said Herbert Grubel, co-author of the report *Immigration and the Canadian Welfare State*. "We need a better selection process.... We're not here, as a country, to do charity for the rest of the world."

James Bissett with Immigration Watch International points out, "Our politicians justify their desire for more immigrants by raising the spectre of an **aging population** and telling us immigration is the only answer to this dilemma, and yet **there is not a shred of truth to this argument**. Immigration does not provide the answer to population aging and there is a multiplicity of studies done in Canada and elsewhere that proves this. Moreover, there is no evidence that a larger labour force necessarily leads to economic progress. Many countries whose labour force is shrinking are still enjoying economic buoyancy. Finland, Switzerland, Bhutan, Thailand, and Japan are only a few examples of countries that do not rely on massive immigration to succeed. **Productivity and environmental sustainability are the answers to economic success, not a larger population.**"[407]

Most Canadians assume that our immigrants are selected because they have skills, training, and education that will enhance our labour force, but only about 18 to 20 percent of our immigrants are selected for economic factors, and the bulk of the immigrants we receive come here because they are sponsored by relatives. This is why over 50 percent of recent immigrants are living below the poverty line and costing Canadians billions in government services and benefits.

In 2019, *National Geographic* explained, "The ebb and flow of people across borders has long shaped our world. Data from the past 50 years of international migration help us understand why people make the choice to leave and where they go. Less than 10 percent of these migrants are forced to flee; most are seeking a better life and move only when they can afford to."[408]

In the case of refugees, there remains a social responsibility that governments face which is far more imperative than increasing a cheap labour force. It is time that we reframe the conversation and stop oversimplifying such important issues with name-calling, stop feeling guilty for wanting

---

406 Kathryn Blaze Carlson, *National Post, Immigrants cost $23B a year: Fraser Institute report*, December 12, 2016, https://nationalpost.com/news/canada/immigrants-cost-23b-a-year-fraser-institute-report

407 James Bissett, Immigration Watch International, *Canada: The Truth About Immigration Is That Costs Exceed Benefits*, Sept 30 2008, https://jonjayray.wordpress.com/2008/09/30/canada-the-truth-about-immigration-is-that-costs-exceed-benefits/

408 ALBERTO LUCAS LÓPEZ, RYAN WILLIAMS, AND KAYA BERNE, National Geographic magazine, *Migration Waves*, August 2019, https://www.applyriver.com/newspaper/August/08-august-2019/National_Geographic_USA_-_08_2019.pdf

to protect Canada's environment, stop our government's out-of-control spending, and take the time to do a carrying-capacity study. We need to reduce our immigration level to equal our emigration level until we as a country can decide on what would be a sustainable population and immigration rate. We need to think of the greater good and future generations. We need family planning policies that would alleviate social, political, economic, and environmental pressures worldwide.

Business and Politics

Anyone who believes exponential growth can go on forever in a finite world is either a madman or an economist.
—Kenneth Boulding, Professor of Economics, University of Colorado

New York Times reporter David Gelles tells us that Jane Goodall keeps going with a lot of hope and has a message for business leaders in 2020. Goodall points out, "**One million species are in danger of extinction. So what I say to the business community is: Just think logically.** This planet has finite natural resources. And in some places, we've used them up faster than Mother Nature can replenish them. How can it make sense if we carry on in the way we are now, with business as usual, to have unlimited economic development on a planet with finite natural resources, and a growing population? **But consumers, at least if they're not living in poverty, have an enormous role to play, too. If you don't like the way the business does its business, don't buy their products. This is beginning to create change.** People should think about the consequences of the little choices they make each day. What do you buy? Where did it come from? Where was it made? Did it harm the environment? Did it lead to cruelty to animals? Was it cheap because of child slave labor? And it may cost you a little bit more to buy organic food, but if you pay a little bit more, you waste less. We waste so much. **And eat less meat**. Or no meat. Because the impact on the environment of heavy meat eating is horrible, not to mention the cruelty."[409]

Money = Energy + Intention: Money is a form of energy. It is a human invention that represents the energy we expend through work, investments, or various other means. Like any other form of energy or human behaviour, it can be used to improve Earth's situation or degrade it. The use of money to form our modern economic system is fairly recent, replacing the barter system that prevailed for much of human history.

Throughout history many items have been used to pay for food, including feathers, shells, beads, cattle, and tools. In the South Pacific, at one time society paid for food with dog's teeth. In ancient Rome, soldiers were paid with lumps of salt. That is where the word "salary" came from. Our present form of money-making, often described as a virus of the mind, has reduced humanity into money-making machines.

This "salary" has become part of the present-day economy, based on supply and demand. The problem with this system is captured in the phrase: **supply is limited, demand is not**. When you

---

409 David Gelles, New York Times, *Jane Goodall Keeps Going, With a Lot of Hope (and a Bit of Whiskey)*, Sept. 12, 2019, https://www.nytimes.com/2019/09/12/business/jane-goodall-corner-office.htm

have a variable like demand that knows no limits, the outcome is somewhat unpredictable. Also, this basic formula fails to take into consideration the value of our life-support system. Investopedia points out that economists measure the success of an economy by tracking the gross domestic product (GDP), a measure of the total value of goods and services produced by a country.[410] An increase in the GDP is seen as a good thing. Unfortunately, fighting epidemics, cleaning up oil tanker disasters, and cutting down old-growth forests all help increase the GDP. However, few of us would argue for more disease, bigger oil spills, or the elimination of ancient forests.

When calculating the GDP, economists measure only those activities that enter the monetary economy. They put a value on the timber from a logged forest, but no value on the forest before it is destroyed. So, this system puts very little value on the resource base. It isn't until the resource base starts to run out, as with oil, that the market starts to reflect its scarcity. Then the market goes looking for an alternative commodity.

However, this doesn't work with water. When water is scarce, unlike with oil, there is no replacement. Water is the one resource that every life form depends on, and as our population increases, it becomes increasingly valuable. In the future, it will be the most highly prized commodity that wars will be waged over. As many rivers and aquifers are already running dry, millions of people are migrating toward abundant water supplies.

Refugees are often forced to flee their homes due to droughts, war, natural disasters, and environmental catastrophes. In fact, many of these events that are labelled as natural disasters or wars are actually a result of overpopulation, but this misconception is rarely challenged. Also, if the U.S. stopped perpetuating a constant state of war around the world (as discussed later in this chapter), many of those refugees would be able to remain in their homeland and would not be seeking entrance to other as yet less degraded countries.

The impact is that each person added to this planet in the 21st century is now putting a greater demand on our ecosystems than ever before. As our numbers escalate, the cost of basic necessities increases, resources become more depleted, and more species are on the brink of extinction as we claim more of wildlife's domain for our own. This is a vicious cycle that is now spiraling out of control, since we have passed the "tipping point" and failed to act. The good news is that if we can start reducing our population, this cycle will start to reverse and the global situation will immediately begin to improve.

There will come a day in the near future when the birth of a child may no longer be seen as a cause for celebration, and humanity will clearly recognize that we must stop population growth at EIGHT BILLION. This population growth is a basic economic force that puts unsustainable pressure on the environment. The economic consequences of this pressure are dwindling resources, higher food and housing costs, lower incomes, higher unemployment, and increasing poverty. By casting every problem in an economic light at the expense of the environment, we are ignoring strategies that could help us stabilize or reduce our human population.

---

410 Leslie Kramer, Investopedia, *What Is GDP and Why Is It So Important to Economists and Investors?*, Jun 1, 2020, https://www.investopedia.com/ask/answers/what-is-gdp-why-its-important-to-economists-investors/

Considering a New Way to Calculate Growth

Increasing population may make the growth lobby (cheap labour businesses, media, banking industry, land developers, economists, energy sectors, and governments) happy, but it is creating unprecedented human misery and ecological suicide for future generations. **We need to restructure our economic system to recognize full-cost accounting, based on sustainability not growth. We need to make prices reflect reality, and consumers need to use their spending power to reflect the value they place on a healthy and sustainable environment.** Any country aiming for environmental and economic sustainability must have a population policy that is based on the human carrying-capacity of its land and its natural resources.

So, what is at the root of this myth that we need continuous economic growth spurred on by a constantly growing population? Perhaps Victor Lebo's message in the 1955 *Journal of Retailing* will shed some light on our present situation. He said, "Our enormously productive economy demands that we make consumption our way of life, that we convert the buying and use of goods into rituals, that we seek our spiritual satisfactions, our ego satisfactions, in consumption. The measure of social status, of social acceptance, of prestige, is now to be found in our consumptive patterns. The very meaning and significance of our lives is today expressed in consumptive terms."[411] After World War II, President Eisenhower promoted this ideology and was hugely successful in convincing Americans that buying things made them happy, as their high rate of bankruptcy and credit card debt today testify.

However, if we were as motivated today by our world leaders to reverse this global trend toward growth, we could be just as successful. We have a better-educated and socially connected global population now that could respond with their votes and demand that we protect our home planet. And we also have the **genuine progress indicator (GPI)**, a tried and true economic system that could be put to use immediately to replace the dysfunctional GDP. This will be discussed later in this chapter under Solutions.

Population Institute Canada (PIC) notes, "Indeed all levels of government, strongly encouraged by the economic sector, have consistently pursued growth-promoting policies with varying degrees of urgency. Economic growth—whose negative consequences are never seriously debated, questioned or assessed—is widely seen as a 'good thing,' essential to Canadians' well-being, and with population growth held to be a key economic driver."[412]

> *The "fetish" of pursuing endless economic growth, and the associated folly of following an economic model predicated on continuous population increases, has long been accepted conventional wisdom in Canada.*
> **—Population Institute Canada**

PIC adds that federal and provincial governments, notably Quebec's, have long encouraged population growth through various financial incentives, such as the "baby bonus"; heavily subsidized daycare; income tax breaks; tax exemptions for baby supplies; enticements to rich, prospective

411 Homework UO, *Consumption,* http://homework.uoregon.edu/pub/class/es202/Canvas/consumption.html

412 Population Institute Canada, *Too Many People in Canada?* 2018, https://populationinstitutecanada.ca/about-us/our-canada/

immigrants; and so on. Despite a fertility rate of 1.6 children per female, Canada's current growth rate is 1.2 percent annually, an aggregate increase due solely to immigration. If continued, this rate would result in a doubling of population in 58 years. Absent current immigration rates, Canada's population would have stabilized at well below 30 million people.

The Union of Concerned Scientists tells us, "The current economic model taught universally in business schools world-wide actually has a name, NeoClassical growth economics (NCE). It was installed at the University of Chicago by J.P. Morgan, infamous Wall Street financier and 'robber baron', since it is the system that benefits bankers the most. Soon all business schools were following suit, letting go of professors who taught Adam Smith and the other classical economic theories. Now that the coup is complete, a hundred years later it has taken over the term 'economics', as though NCE was the only way to organize the human economy."[413]

Scientists warn that our present management of money that has shaped Western civilization has captured our minds and controls our behaviors. They advise that we must replace our present economic system as quickly as possible by ecological economics, also known as steady-state-economics, espoused by Dr. Herman Daly and the International Society for Ecological Economics. This system is similar to the genuine progress indicator method of accounting, which tackles the affluent dominance that has only created a greater wealth divide. The 2 percent wealthy elite have built an empire that spends millions of dollars every year on lobbying government to sway policies that protect and enhance their interests further. The most prolific lobbying activities are on budget and tax issues; the military, rifle association, meat industry, pharmaceuticals, finance, insurance, health care, and energy sectors have great influence over budgets, contracts, subsidies, economic accounting, and tax exemptions.

It is our public tax dollars that are going into the pockets of these powerful lobbyists, instead of enhancing the lives of the poor and improving the wretched state of our environment. In fact, Nobel laureate Steven Chu calls our present economic system a pyramid scheme or Ponzi scheme, doomed to failure.

Steven Chu, former U.S. Secretary of Energy and 1997 Nobel Prize winner in physics, recommends finding a new economic model that promotes the education of women and is not based on an ever-increasing population. "The economists know this, but they don't really talk about it in the open, and there's no real discussion in government," Chu said. "Every government says you have to have an increase in population, whether you do it through immigrants or the home population. So, this is a problem."[414]

Everything is not fine, Stiglitz argues, but says economists have been working hard on providing new ways to measure economic health. Embraced more broadly, new economic measures that include accounting for human happiness and environmental well-being could help change the course of humanity. In 2020, we are finding that many countries are responding to our difficult economic times by voluntarily having small families or choosing to be childfree. Japan is one such country, as many young people are having trouble finding steady, well-paid jobs.

---

413 ScientistsWaring.org, Union of Concerned Scientists, *Newsletter June 2019*, Jun 14, 2019, https://www.scientistswarning.org/newsletter/newsletter-june-2019/

414 Jeff McMahon, Forbes, *The World Economy Is A Pyramid Scheme, Steven Chu Says*, Apr 5, 2019, https://www.forbes.com/sites/jeffmcmahon/2019/04/05/the-world-economy-is-a-pyramid-scheme-steven-chu-says/#7a2d8e5b4f17

The Atlantic reporter Alana Semuels points out, "Japan's population is shrinking. For the first time since the government started keeping track more than a century ago, there were fewer than 1 million births last year (2016), as the country's population fell by more than 300,000 people. The blame has long been put on Japan's young people, who are accused of not having enough sex, and on women, who, the narrative goes, put their careers before thoughts of getting married and having a family."[415]

But Semuels adds that there's another, simpler explanation for the country's low birth rate, one that has implications for the U.S.: Japan's birth rate may be falling because there are fewer good opportunities for young people, and especially men, in the country's economy. In a country where men are still widely expected to be breadwinners and support families, a lack of good jobs may be **creating a class of men who don't marry and have children** because they—and their potential partners—know **they can't afford to**.

> *The birth rate is down, even the coupling rate is down. And people will say the number-one reason is economic insecurity.*
> **—Anne Allison, Professor of Anthropology at Duke University**

Semuels notes, "This may seem surprising in Japan, a country where the economy is currently humming along, and the unemployment rate is below 3 percent. But the shrinking economic opportunities stem from a larger trend that is global in nature: the rise of unsteady employment." So the world is watching Japan for a solution to this economic predicament, as many countries are heading down a similar path of fewer couples feeling motivated to marry. As 2020 has come and gone, we are also reeling from the threat of pandemics and civil unrest, further adding to the concerns of bringing an innocent child into a world of chaos.

Millennium Alliance for Humanity and the Biosphere contributor Gregg Miklashek tells us that **overpopulation is not only affecting us economically, but is also causing enormous stress that is making us sick**. He states, "Everywhere women (and men) are **freed from the obligation** in our Capitalist societies to **overproduce offspring to keep the wheels of industry turning** out excessive **profits for the few running these huge corporations**, and women are allowed and encouraged to get higher education and meaningful work, the **fertility rate drops** naturally. The alternative to this voluntary human population reduction movement is increasing disease, increasing medical interventions, decreasing quality of life, unbearable medical costs, and misery, as well as continued environmental exhaustion and endless resource wars. Which do we prefer for ourselves and our offspring?"[416]

---

415 Alana Semuels, The Atlantic, *The Mystery of Why Japanese People Are Having So Few Babies*, July 20, 2017 https://www.theatlantic.com/business/archive/2017/07/japan-mystery-low-birth-rate/534291/

416 Gregg Miklashek, Millennium Alliance for Humanity and the Biosphere, *Population Density Stress is Killing Us Now!*, November 14, 2019, https://mahb.stanford.edu/blog/population-density-stress-killing-us-now/

Perpetual War

Ooo you're slick—you investors in hate
You Saddams and you Bushes; you Bin Ladens and snakes
You billionaire bullies; you're a globalized curse
You put war on the masses while you clean out the purse

Got the world's greatest power and you team up with thugs
Make a fortune on weapons, destruction and drugs…
It's the war racket… … …

The War Racket—a song by Buffy Sainte-Marie on Musixmatch

As global population reaches EIGHT BILLION and resources become increasingly scarce, more wars will be fought over water, land, oil, and food, as society becomes more unstable and desperate. This continual state of war that is being funded and promoted by the United States and other like-minded countries is taking its toll on our environment, economy, and mental well-being.

The Union of Concerned Scientists released a second Warning to Humanity in 2017 that states, "Developing nations must realize that environmental damage is one of the gravest threats they face, and that attempts to blunt it will be overwhelmed if their populations go unchecked. The greatest peril is to become trapped in spirals of environmental decline, poverty, and unrest, leading to social, economic, and environmental collapse. Success in this global endeavor will require a great reduction in violence and war. Resources now devoted to the preparation and conduct of war—amounting to over $1 trillion annually—will be badly needed in the new tasks and should be diverted to the new challenges."[417]

Among those new challenges are achieving a sustainable human population, reducing climate change emissions, dealing effectively with pandemics, restoring biodiversity, empowering women, and ensuring secure food and water supplies. Remaining in a permanent state of war undermines these goals.

Global Research contributor Chris Hedges states, "The embrace by any society of permanent war is a parasite that devours the heart and soul of a nation. Permanent war extinguishes liberal, democratic movements. It degrades and corrupts education and the media, and wrecks the economy. The liberal, democratic forces, tasked with maintaining an open society, become impotent."[418]

Hedges tells us that it was a decline into permanent war, not Islam, that killed the liberal, democratic movements in the Arab world, ones that held great promise in the early part of the 20th century in countries like Egypt, Syria, Lebanon, and Iran. Hedges adds, "It is a state of permanent war that is finishing off the liberal traditions in Israel and the United States. The moral and intellectual trolls—the Dick Cheneys, the Avigdor Liebermans, the Mahmoud Ahmadinejads—personify

---

417 Lorraine Chow, Eco Watch, *20,000 Scientists Have Now Signed 'Warning to Humanity'*, Mar. 09, 2018, https://www.ecowatch.com/warning-to-humanity-scientists-2544973158.html
418 Chris Hedges, Global Research, *The Disease of Permanent War*, May 19 2009, https://www.globalresearch.ca/the-disease-of-permanent-war/13659

the moral nihilism of perpetual war. They abolish civil liberties in the name of national security. They crush legitimate dissent. They bilk state treasuries. **They stoke racism.**"

Hedges warns that massive U.S. military spending, climbing to nearly $1 trillion a year in 2009 and consuming half of all discretionary spending, has a profound social cost. Hedges points out, "Bridges and levees collapse. Schools decay. Domestic manufacturing declines. Environmental protection is virtually ignored. Trillions in debts threaten the viability of the currency and the economy. The poor, the mentally ill, the sick and the unemployed are abandoned. Human suffering, including our own, is the price for victory."

Governments often use the threat of terrorism to keep us in a constant state of fear, even if those "terrorists" are hired for that very purpose. Fear distracts us from real terrorists in our own countries, committing far more murders and rapes under our own noses and in our own families. Fear stops us from objecting to outrageous military spending that guarantees corrupt corporations huge profits. Fear also means that we will be willing to give up our rights and liberties for security.

In *Pentagon Capitalism*, Seymour Melman coined the term *permanent war economy* to characterize the American economy. Melman wrote that since the end of the Second World War, the government has **spent more than half its tax dollars on past, current, and future military operations**. It is the largest, single sustaining activity of the government. The military-industrial establishment is a very lucrative business. It is gilded corporate welfare. And yet this has never been about people. Far more people have died in the U.S. due to COVID-19 and its complications than have died in all U.S.-involved armed conflicts since WW2. And yet look at the spending discrepancy. Something is very wrong with this picture.

Hedges adds, "Foreign aid is given to countries like Egypt, which receives some $3 billion in assistance and is required to buy American weapons with $1.3 billion of the money. The taxpayers fund the research, development, and building of weapons systems and then buy them on behalf of foreign governments. It is a bizarre, circular system. It defies the concept of a free-market economy. These weapons systems are soon in need of being updated or replaced. They are hauled, years later, into junkyards where they are left to rust. It is, in economic terms, a dead end. It sustains nothing but the permanent war economy."

The late U.S. Senator J. William Fulbright described the reach of the military-industrial establishment in his 1970 book *The Pentagon Propaganda Machine*. Fulbright explained how the Pentagon influenced and shaped public opinion through multimillion-dollar public relations campaigns, Defense Department films, close ties with Hollywood producers, and use of the commercial media. The majority of the military analysts on television are former military officials, many employed as consultants to defence industries, a fact they rarely disclose to the public.

Unfortunately, this continual state of war is taking its toll on Gaza and Israeli youth soldiers, where at least 30 percent of the male population is in the 15-to-29 age bracket. *Wall Street Journal* reporter Gunnar Hein Sohn pointed out, "In such 'youth bulge' countries, young men tend to eliminate each other or get killed in aggressive wars.... In Arab nations such as Lebanon (150,000 dead in the civil war between 1975 and 1990) or Algeria (200,000 dead in the Islamists' war against their own people between 1999 and 2006), the slaughter abated only when the fertility rates in these countries fell from seven children per woman to fewer than two."[419]

It takes an abundant supply of soldiers to keep a war operational, especially when they come from a defenseless or vulnerable region.

It also takes the support of the public, so easily manipulated by government deception and media collaboration to maintain a state of perpetual war. As Andrew Blaire posts on his Liberal Education website, the 9/11 event and the global "war on terror" that followed may not have occurred for the reasons we have been led to believe. In fact, it may have been a "false flag" and intentional hoax.[420]

Blaire explained, "Soon after 9/11 the United States demanded that the Taliban government of Afghanistan hand over bin Laden. The Taliban replied that they were ready to cooperate if the U.S. would provide evidence that he was guilty. The U.S. said they did not need to. Then the Taliban said okay, we'll hand him over to a third country, but please do not attack us. The U.S. said it was too late."

Although U.S. Secretary of State Colin Powell and British Prime Minister Tony Blair promised to provide evidence to the public that bin Laden was responsible for 9/11, this never happened. Robert Mueller, director of the FBI, said that they did not have the evidence against him.

Blaire noted, "On October 7, 2001, the U.S., together with the United Kingdom, used the official story to justify launching 'Operation Enduring Freedom' against Afghanistan. Canada, with other NATO countries, joined in. So began the global 'war on terror,' a long series of wars in the Middle East extending the conflict in Afghanistan to Iraq, Pakistan, Syria, Libya, Yemen, Sudan, Somalia, and several other countries. As of August 2019, the U.S. is trying to negotiate a peace with the Taliban in Afghanistan to end the longest war in U.S. history."

In 2003, a propaganda campaign accused Saddam Hussein of having weapons of mass destruction, and a brutal war ensued that involved the murder, torture, and humiliation of Iraqis that were portrayed in the news. This "war on terror" took the lives of millions, according to Body Count, a study done by the Physicians for Social Responsibility. The 9/11 tragedy took the lives of 3,000 people, and the true culprits have not yet been apprehended. Perhaps this is a case to be brought before the UN World Court to judge.

This is just a brief glimpse of the patriarchal influences that are perpetuating war and overpopulation and adding to the pain and misery on this planet. All global citizens who genuinely care about the future of this planet must speak out for civility and justice. We can no longer pretend that we don't know what is happening in this war racket, just because there is less profit in peace. We can no longer stand by and allow developed countries to spread misinformation and make huge profits on others pain and suffering.

419 Gunnar Hein Sohn, Wall Street Journal Europe, *Ending the West's Proxy War Against Israel*, January 12 2009, https://www.wsj.com/articles/SB123171179743471961

420 Andrew Blaire, *Liberal Education and 9/11*, September 3, 2019, https://liberaleducationand911.net/

Assault in the Military: A Microcosm of the Global Rape Epidemic

Assault in the military is also alarmingly common. Wikipedia describes **war rape** as rape committed by soldiers, other combatants, or civilians during armed conflict or war. Rape in the course of war dates back to antiquity, ancient enough to have been mentioned in the Bible. During war and armed conflict, rape is frequently used as a means of psychological warfare in order to humiliate the enemy and undermine their morale. War rape is often systematic and thorough, and military leaders may actually encourage their soldiers to rape civilians.

When part of a widespread and systematic practice, rape and sexual slavery are now recognized as crimes against humanity and war crimes. Rape is also recognized as an element of the crime of genocide when committed with the intent to destroy, in whole or in part, a targeted group. However, rape remains widespread in conflict zones, inflicted by both the developing and overdeveloped countries, such as the U.S.

In 2014, there were at least 20,300 members of the military who were assaulted by military personnel. Reporting assault in the military is also low, with 85 percent of victims not reporting these crimes in 2014 in fear of retaliation. Unfortunately, the military—like the Vatican—is beyond federal law and creates their own legal system for all crimes including sexual abuse. Therefore, these perpetrators are seldom prosecuted either Valnora Edwin, CGG human rights programme manager, warns, "The Silence has to be broken, the closed doors opened. We need to move away from thinking that sexual violence and other discriminatory practices against women should be kept private and not made public. Neither can we hide behind the shroud of customs, or tradition."

Rape of civilians during war is also rampant, underreported, and seldom punished. In 2020, the Global Fund for Women tells us, "In the Democratic Republic of Congo, rape is regularly used as a weapon of war: it is estimated that 48 women were raped each hour in some regions during the height of the conflict. In Iraq and Syria, where ISIS has seized control over major cities, women and girls—especially those in ethnic and religious minorities—have been raped, enslaved, sold, and tortured. The story is the same in armed conflicts around the world as well as in unstable political climates and post-conflict regions: systematic rape is a favorite weapon of war aimed to control, intimidate, and humiliate millions of women and girls."[421]

We need to move away from thinking that sexual violence is acceptable as a form of torture in war. The silence has to be broken, and we must stop hiding behind the shroud of customs or tradition or the military's preferential status of secrecy. We must recognize that **sexual violence is a crime** even in a time of war, and we must open those closed doors to the evil perpetrated against both prisoners and civilians in times of war. For instance, the civil war in Sierra Leone in 2009 directed the world's attention to the problem of sexual violence, as women were raped and sexually violated in shocking and destructive manners. The campaign of terror in this diamond-rich region resulted in burning villages, mutilating civilians, raping women, and recruiting child

421 Global fund for Women, *UNBROKEN: survivors of rape as a weapon of war*, 2020, https://www.globalfundforwomen.org/unbroken/

soldiers. After eight years of violence, 20,000 are dead, and many war criminals still walk free in this devastated country.

AlterNet reporter Naomi Wolf points out, "In Sierra Leone, the soldiers and generals who used rape as an instrument of war have been tried and many convicted. In Bosnia, likewise." Wolf asks, "When will we convict our very own global rapists, the ones who gave the U.S. the hellish distinction of turning us into the superpower of sex crime? ... **Women especially, who understand how sexual abuse and rape can break the spirit in a uniquely anguishing way, should be raising their voices loudly!"**[422]

Wolf pointed out that the *Telegraph* of London broke the news of President Obama refusing to release photos of detainee abuse, which depicted, among other sexual tortures, an American soldier raping a female detainee and a male translator raping a male prisoner. Retired Army Maj. Gen. Antonio Taguba called the images "horrific" and "indecent." Predictably, a few hours later, the Pentagon issued a formal denial.

Wolf stated, "As I wrote last year [2008] in my piece on sex crimes against detainees, 'Sex Crimes in the White House,' highly perverse, systematic sexual torture and sexual humiliation was, original documents reveal, directed from the top:

- President George W. Bush, Vice President Dick Cheney, Defense Secretary Donald Rumsfeld, and Secretary of State Condoleezza Rice were present in meetings where sexual humiliation was discussed as policy.
- The Defense Authorization Act of 2007 was written specifically to allow certain kinds of sexual abuse, such as forced nakedness, which is illegal and understood by domestic and international law to be a form of sexual assault.
- Rumsfeld is in print and on the record consulting with subordinates about the policy and practice of sexual humiliation, in a collection of documents obtained by the ACLU by a Freedom of Information Act filing compiled in Jameel Jaffer's important book *The Torture Administration*."

It is also noted that scores of detainees who have told their stories to rights organizations have told independently confirming accounts of a highly consistent practice of sexual torture at U.S.-held prisons. Wolf contends, "But what is far scarier about these images Obama refuses to release and that the Pentagon is likely to be lying about now, is that it is not the evidence of lower-level soldiers being corrupted by power—it is proof of the fact that the most senior leadership—Bush, Rumsfeld and Cheney, with Rice's collusion—were running a global sex-crime trafficking ring with Guantanamo, Abu Ghraib and Baghram Air Base as the holding sites."

All forms of rape are ugly and inexcusable, but the fact that senior leaders of our governments are condoning this behavior is a travesty. Of course, in order to diminish the horror and possible illegality of torture, many world leaders are avoiding the use of the word "torture." For instance, former Vice President Dick Cheney preferred to use the euphemism "enhanced interrogation" to

422 Naomi Wolf, AlterNet, *Why the Pentagon Is Probably Lying About its Suppressed Sodomy and Rape Photos*, May 30 2009, https://www.themcglynn.com/torturing-democracy/

refer to the torture his administration orchestrated. No wonder so many Americans are demanding a truth commission that will get the facts, put them on the record, and insist on adherence to the law.

In the 21st century, men are starting to speak out about their sexual abuse as well. Supreme Court judge Bozidarka Dodik argued that over the past decade, the Bosnian judiciary has made great progress in the prosecution of war crimes. Dodik states, "Although most of these criminal cases relate to rape and sexual violence committed against women, cases in which men are victims of such violence are also being prosecuted. However, men who are victims of these criminal offences report it even less often than women, mainly because of the stigma and fear that they will be labelled as gay or unmasculine."[423]

The Medica Zenica NGO was founded in 1993 and was the first NGO to help female survivors of sexual violence. Seven years ago (2013), the NGO realized that it needed to start supporting men too. Djelilovic comments, "Medica Zenica's director, Sabiha Husic, said that Bosnia and Herzegovina has no official statistics on the number of men and women who were sexually assaulted and have been officially registered as civilian victims of the war. However, while working on her doctoral thesis about support for survivors of wartime sexual violence, she estimated on the basis of information she gathered from international organisations and reports that about **3,000 men were raped** during the war."

The Hague-based International Tribunal for the Former Yugoslavia classified the sexual abuse and rape of women, men, and children as a war crime for the first time in cases arising from the Bosnian war. The UN-backed tribunal has delivered several judgments that have included allegations of sexual assault and rape of men and boys.

Djelilovic points out, "The first war criminal who was convicted of crimes including sexual violence by the tribunal was Bosnian Serb politician and paramilitary fighter Dusko Tadic, who was **jailed for 20 years** in 1997." The sexual abuse and rape of men during the war has been reported in 23 locations across the country so far, said Bakira Hasecic, president of the Women—Victims of War association. In 2017, the TRIAL International NGO published a report entitled "Rape Myths in Wartime Sexual Violence Trials."

Perhaps it is time that the present war machine was dismantled and the soldiers given jobs that could actually benefit the planet for the common good? Can you imagine the change that could take place if all soldiers were recruited to plant trees, remove plastic from the oceans, restore wildlife habitat, distribute condoms, clean up toxic military sites, and apprehend sex traffickers? Then we would have an army of eco-soldiers we could truly be proud of!

The Environmental Impact of War

Humanity's growing footprint is also playing havoc with the natural world, as ecocide dominates our landscape. A large part of that footprint, the part the military cabal never mentions, is the

---

423 Zinaida Djelilovic, Balkan Insight, *Male Rape Victims Confront the Bosnian War's Last Taboo*, April 24, 2020, https://balkaninsight.com/2020/04/24/male-rape-victims-confront-the-bosnian-wars-last-taboo/

environmental impact of war. The remedy is at hand, as scientists strive to make **harming the environment a crime of war**. Global Citizen tells us that citizens should care because war destroys lives, guts economies, and undermines civil societies. It also has a devastating impact on the environment, and now scientists want the United Nations to develop stronger environmental safeguards in times of conflict.[424]

Global Citizen describes, "Forests burned to the ground. Rivers damaged by broken infrastructure. Animals slaughtered and driven from their habitats. The environmental impacts of war are staggering, yet they're often overshadowed by the societal wreckage created by conflict." Many wars are fought over religious control, and the instigating partners in these wars often work with governments or rebel groups. The very leaders who outwardly speak of peace are often the silent partners of conflict and numerous other violent activities such as rape, ecocide, and human trafficking. In these cases, religious leaders should be held accountable for this travesty.

The scientist's petition states, "We call on governments to incorporate explicit safeguards for biodiversity, and to use the commission's recommendations to finally deliver a Fifth Geneva Convention to uphold environmental protection during such confrontations. Despite calls for a fifth convention two decades ago, military conflict continues to destroy megafauna, push species to extinction, and poison water resources. The uncontrolled circulation of arms exacerbates the situation, for instance by driving unsustainable hunting of wildlife."

Global Citizen warns, "Regardless of where war occurs, it devastates local environments. The United States invasion of Afghanistan in 2001 has led to rampant deforestation, polluted water sources, and widespread air pollution. In addition to the pollution created by bombs, the US military regularly burns garbage in open pits, releasing harmful toxins into the air, and heavy machinery causes more dust to circulate in the atmosphere. When the US attacked Iraq in 1991, bombs containing depleted uranium led to radiation contamination in the soil and water sources, the Guardian reports. The US military also destroyed millions of acres of forest during the Vietnam war with a toxic substance called "agent orange." The environmental effects of that bombing campaign are still felt today."

Global Citizen points out that war has also greatly endangered animal and plant species. During the Congolese civil wars, for example, animals as diverse as antelopes, elephants, and monkeys were killed and forced to flee their destroyed habitats. Even in times of peace, animals regularly step on leftover land mines. The chemicals used to make weapons can irrevocably contaminate water sources, and the lawlessness engendered by war can give rise to destructive activities like illegal mining operations.

But we must not forget the contributions the military makes towards climate disruption. The U.S. military burned more oil in Iraq in 2008 alone than the annual amount that would be used by 1.2 million cars. Overall, the **U.S. military releases more greenhouse gas emissions** by itself than **many countries**. Armies also regularly torch oil wells to thwart their enemies, releasing immense amounts of greenhouse gas emissions into the atmosphere in the process, notes Global Citizen.

---

424 Erica Sanchez, Joe McCarthy and Pia Gralki, Global Citizen, *Scientists want to make harming the environment a crime of war*, July 24, 2019, https://www.globalcitizen.org/en/content/harming-environment-war-crime/

Small steps have already been taken as the UN urges countries to protect the environment during times of conflict through the International Day for Preventing the Exploitation of the Environment in War and Armed Conflict. A UN environment resolution was also adopted in 2016 to promote strong environmental safeguards in war. What is needed now is the recognition of ecocide as a law, and the will of the UN Criminal Court to enforce it.

Global Citizen concludes, "Ultimately, if harming the environment was a war crime, then most acts of modern warfare would essentially be forbidden. After all, there's no way to drop a bomb without harming the ground it falls on."

An even more destructive form of military power is assault using hypersonic weapons that are now operational. *Axios* reporter Rashaan Ayesh tells us, "Russian President Vladimir Putin has emphasized that Russia is the **only country armed with hypersonic weapons.** He also has compared the success of the Avangard's development to the Soviet Union's first satellite launch in 1957."[425]

The Russian military said the Avangard is unique because it can fly 27 times faster than the speed of sound. Ayesh noted that this weapon also has the ability to make sharp maneuvers in flight, which will render traditional missile defense systems useless. Not to be outclassed, the Pentagon said that it is currently working on developing hypersonic weapons, but Defense Secretary Mark Esper said in August that "it's probably a matter of a couple of years" before the U.S. can obtain one. And so the arms race continues, while what the world desperately needs is healing and goodwill.

On www.BetterWorld.net, Isaac Asimov asks, "Which is the greater danger—nuclear warfare or the population explosion? The latter absolutely! To bring about nuclear war, someone has to DO something; someone has to press a button. To bring about destruction by overcrowding, mass starvation, anarchy, the destruction of our most cherished values—there is no need to do anything. We need only do nothing except what comes naturally—and breed. *And how easy it is to do nothing."*

Environmental Politics

> *It is vitally important to recognize that some of the major participants in the population debate have no interest in rationally solving our problems, but are using this issue to advance their own agendas.*
>
> —Madeline Weld, Population Institute Canada

Speaking of government agendas, most are based on economic gains and give little importance to environmental health or human well-being. Unfortunately, most political leaders have been deluded into believing that the only way to promote economic progress is through consistently increasing population growth. The other delusion that has long been a basis for governance is that a healthy economy is in conflict with a healthy environment. Genuine environmental defenders are regarded as a hindrance to government agendas, an obstacle to overcome. This is certainly the

425 Rashaan Ayesh, Axios, *Russia's new hypersonic weapon becomes operational*, Dec 27 2019, https://www.axios.com/russia-hypersonic-weapon-avangard-missile-defense-247883d2-d987-4766-87a2 ff70930a9e15.html

case in Mexico, where the government has waged a war on environmental activists.

In fact, Grist EcoWatch reporter Rachel Ramirez emphasizes that for Indigenous protesters, defending the environment has been fatal in the last decade. Ramirez states, "Adán Vez Lira, a prominent defender of an ecological reserve in Mexico, was shot while riding his motorcycle in April. Four years earlier, the renowned activist Berta Cáceres was shot dead in her home in Honduras by assailants taking direction from executives responsible for a dam she had opposed. Four years before that, Cambodian forest and land activist Chut Wutty was killed during a brawl with the country's military police while investigating illegal logging."[426]

A 2020 report from the Universitat Autònoma de Barcelona notes that these acts of violence against environmental activists are not unusual. Ramirez states, "As police crack down on protests demanding justice and equity in the wake of the police killing of George Floyd in the U.S., it's clear that activism in general comes at a heavy price. Environmental activists specifically—particularly indigenous activists—have for years faced high rates of criminalization, physical violence, and even murder for their efforts to protect the planet."

These efforts to protect their homelands include protesting the construction of pipelines on tribal lands, construction of fossil fuel refiners, illegal mining and logging in the Amazon rainforest, oil extraction in the Arctic, construction of a border wall in the U.S. that blocks wildlife migration routes, and lack of protection for wetlands and wildlife habitat worldwide.

Ramirez notes, "The analysis draws on a 2019 report from the human rights and environmental watchdog organization *Global Witness,* which found that at least 164 environmental activists were killed in 2018 alone. The Philippines was named the deadliest country in the world for environmental defenders, who have been **called terrorists** by President Rodrigo Duterte. In fact, not long after these findings, 37-year-old Brandon Lee, an American environmental activist who was in the Philippines on a volunteer mission, was shot four times in Ifugao province by unknown assailants after his group, the Ifugao Peasant Movement—a farmers group opposing a hydropower project — had been labeled an 'enemy of the state' across social media by propagandists. As of April, Lee was recovering in his hometown of San Francisco, but he remains paralyzed from the chest down."

The lead author of this study, Arnim Scheidel, said, "Being aware of these connections may help to connect struggles against various forms of racism worldwide. Protest is key for the success of such struggles, particularly when using diverse channels and building on broad alliances."

Perhaps such studies will give lawmakers and the public a better understanding of the causes of the violence that protesters still face around the world. These senseless killings will highlight the need for a law of **ecocide** to be recognized and enforced by the United Nations World Court. In reality, most of these protests to protect the environment are protests against overpopulation, the root cause of the environmental destruction.

During President Trump's presidency, he also waged a war against the environment by promoting population growth, increased fossil fuel production, weaker wildlife protection laws, and

426 Rachel Ramirez, EcoWatch, *For Indigenous Protesters, Defending the Environment Can Be Fatal,* Jun. 11, 2020, https://www.ecowatch.com/environmental-activist-violence-2646168966.html?rebelltitem=3#rebelltitem3

reduced protection for our water system. However, President Biden promises he has his own plans to unravel this mess and restore order in 2021.

Grist reporter Zoya Teirstein explains, "A number of Democratic candidates for president have released ambitious environmental plans that make the environmental platforms of yore look like yesterday's lunch. And many of them include proposals aimed at correcting environmental injustices—protecting vulnerable communities that are often exposed to pollution or are on the frontline of climate change. Elizabeth Warren just became the latest candidate to unveil such a plan. It will direct at least $1 trillion to low-income communities on the frontlines of climate change, and contains similar themes to justice-centered proposals put out by the likes of Bernie Sanders, Kamala Harris, and Cory Booker. In at least one respect, however, the plan stands out: It contains a section on how Warren aims to rein in the rampant wildfires burning in the American West."[427]

Teirstein points out that Senator Warren planned to collaborate with tribes to incorporate "tradition ecological practices" to prevent wildfires and return public resources to Indigenous protection wherever possible. This would create an opportunity for tribes and government agencies to collaborate on addressing the climate crisis, something that did not occur under the Trump administration. Of course, this all depends on whom the voters elect.

Across the globe in Australia, increased heat waves, wildfires, respiratory illnesses, and suicides plague the country, and many citizens believe the government is to blame for most of their woes. *The Guardian* reporter Melissa Davey states, "The federal government's lack of engagement on health and climate change has left Australians at significant risk of illness through heat, fire and extreme weather events, and urgent national action is required to prevent harm and deaths, a global scientific collaboration has found."[428]

In 2019, an international *Lancet* "Countdown" report on climate change and health revealed, "There continues to be no engagement on health and climate change in the Australian federal parliament, and Australia performs poorly across many of the indicators in comparison to other developed countries; for example, it is one of the world's largest net exporters of coal.... As a direct result of this failure, we conclude that Australia remains at significant risk of declines in health due to climate change, and that substantial and sustained national action is urgently required in order to prevent this.... **This work is urgent.**"

The first two *Lancet* assessments were published in 2017 and 2018, with annual assessments continuing until 2030, consistent with the near-term timeline of the Paris climate agreement. Spokeswoman for Doctors for the Environment Australia, Dr. Arnagretta Hunter, agreed Australia was poorly prepared for the health challenge of climate change. Doctors describe Australia as the developed country with the most serious vulnerability to climate change, and warn that the governance system must recognize that a **sustainable population** level is the key.

---

427 Zoya Teirstein, Grist, *Elizabeth Warren's new climate plan uses wildfire wisdom from tribes,* Oct 11 2019, https://grist.org/article/warrens-new-climate-plan-uses-wildfire-wisdom-from-tribes/

428 Melissa Davey, The Guardian, Coalition Inaction On Climate Change And Health Is Risking Australian Lives, Nov 16 2019, https://www.theguardian.com/environment/2019/nov/14/coalition-inaction-on-climate-change-and-health-is-risking-australian-lives-global-report-finds

Doctors for the Environment state, "Population numbers in Australia should be based on what science tells us is the ecological carrying capacity of Australia taking into account projected water and fertile land resources, and the need to reduce greenhouse gas emissions in concert with expected international contraction and convergence policy. Having defined the constraints on population growth, all major projects should have a population health impact statement in order to move away from the 'given' that a new project must be developed immediately if economically advantageous."[429]

> *Population stabilisation will require progressive modification of the economic ethos of continued growth and expanding consumption.*
> **—Doctors for the Environment**

In 2019, the Australian Medical Association, Doctors for the Environment Australia, and the World Medical Association recognized climate change as a health emergency. There has been a loud chorus of protestors in Australia calling for Prime Minister Scott Morrison and the federal government to acknowledge the link between population, climate change, and health. Should this purposely neglectful government be held responsible for placing the fossil fuel industry and the growth lobby above the health and well-being of citizens, when Australians are crying out for smart leadership? Should there be some accountability in the World Court?

Business As Usual - In Disguise

Of course, Australia is not the only country that is favouring the growth industry and fossil fuel companies over the well-being of humanity and wildlife. In a world of almost EIGHT BILLION humans, some industries are now seeing the writing on the wall. Shell is one of those companies. The world knows that pressure to abandon fossil fuels is in force. *Time* reporter Justin Worland states, "How should a company that generates most of its profits by serving the world's enormous appetite for oil navigate a long-term future in which shifting political and economic tides threaten to make fossil fuels obsolete? The pressure to abandon oil and gas is already in force. In recent years, protesters have swarmed Shell's headquarters; advocates representing 17,000 Dutch citizens have sued the company; and powerful investors successfully coerced executives to say they will reduce emissions. In 2015, countries around the world promised to aggressively tackle greenhouse-gas emissions, in order to meet the target laid out by the Paris Agreement: goals that require buying and burning significantly less oil and gas."[430]

As climate change increasingly defines the way companies operate, they are expanding their investment in renewable energy, which still depends on natural gas, and in plastics, which still depend on oil, in hopes that this strategy will appease shareholders. However, as governments and the public become more informed about the enormous negative impact of plastics and the true

429 Doctors for the Environment Australia, *Population and Australia Position Paper,* 2020, https://www.dea.org.au/images/general/Doctors_for_the_Environment_Australia_population_policy_with_endnotes.pdf

430 Justin Worland, Time, *The Reason Fossil Fuel Companies Are Finally Reckoning With Climate Change,* Jan 16 2020, https://time.com/5766188/shell-oil-companies-fossil-fuels-climate-change/

consequences of alternative energy, these solutions may not be the magic bullet they had hoped for. Some economists warn that oil could peak as early as 2025.

Worland points out, "Markets are already jittery about the industry: energy was the worst-performing sector on the S&P 500 index in 2019. In 1980, the energy industry represented 28% of the index's value, according to the Institute for Energy Economics and Financial Analysis (IEEFA). Last year, it represented less than 5%."

While some companies like ExxonMobil plan to squeeze the last out of the oil economy, others like Shell are positioning themselves to create a more eco-friendly image. Worland explains, "By the 2030s, the 112-year-old fossil-fuel giant wants to become the world's largest power company." However, some scientists and activists are beginning to see beyond this **smokescreen**, and recognize that this strategy will not achieve any significant reductions in emissions. All too often the façade of "clean electricity" only **gives dirty energy a whitewash**. Unfortunately, most environmentalists and governments have bought into this hoax, backing mega wind and solar projects that still rely on fossil fuel backups and are killing off millions of birds, bats, and insects every year. Although better solutions exist, they are not as economically lucrative as these taxpayer-funded "renewable energy" schemes. "Shell is doing a lot of the right things," says a senior energy official, who asked to remain anonymous to speak freely. "The question is: **What award do you get for being the best-painted deck chair on the Titanic?"**

While Shell has committed to reducing emissions by 3 percent by 2021, Worland notes that the UN Intergovernmental Panel on Climate Change concluded in 2018 that to keep temperatures from rising to levels that would bring a wide range of catastrophes, countries must **halve** their greenhouse-gas emissions by 2030 and hit net-zero emissions by 2050. That would mean more than incrementally reducing emissions; it means keeping vast reserves of oil already discovered in the ground. Worland adds, "Executives at Shell knew decades ago that burning fossil fuels would cause the planet to warm." Shell executive Van Beurden stated, "Yeah, we knew. Everybody knew," he said. "And somehow we all ignored it."

In fact, it may be the most recent political and economic pressure brought on by climate activists that has compelled Shell to make a leap forward. Worland states, "In 2018, Climate Action 100+, a powerful group of global investors that now represents $41 trillion in assets, delivered an ultimatum: either Shell committed to short-term emissions-reduction targets, or it risked losing the support of some of its largest shareholders."

Shell, a Netherlands-based company headquartered in the Hague, has also received pressure from protesters in Amsterdam who in 2019 took to the streets by the thousands to demand change. The U.K. declared a climate emergency and prominent members of the U.S. Congress called to eliminate the country's fossil-fuel emissions by 2030. Unfortunately, Shell's proposal for change has come in the form of increased electricity production, expanded investment in natural gas, and the construction of additional facilities to produce an expanded line of plastics. Their new 400-acre Pittsburgh chemical facility is what activists describe as an environmental nightmare, emblematic of its future business model, notes Worland. Studies have found plastic in tap water, in food products, and in the bellies of sea birds and whales. And plastic production is a significant driver of climate change.

"Their own company is built on the death and destruction of nature and of people all around the world," says Farhana Yamin, a lawyer turned activist who glued herself to the cement outside Shell's London headquarters in 2019 while Extinction Rebellion was protesting there. Meanwhile, the Dutch branch of Friends of the Earth is suing Shell for reneging on its "duty of care" obligation under Dutch law. "All of these initiatives add to the pressure," says Freek Bersch, a campaigner at Friends of the Earth Netherlands.

So are politicians listening to the chorus of protesters? Since the gas and oil industry powers the relentless GDP growth and modern capitalism, they hold enormous political power which they have used to block climate legislation and acquire funding. Worland notes, "Globally, fossil fuels receive roughly $5 trillion annually in government subsidies, a figure that includes the cost of environmental damage caused by industry that's left to everyone else to clean up, according to a 2019 International Monetary Fund paper."

However, the industry's untouchable status may end soon as they consider the uncertainty of the next decade: souring public perception, shifting consumer behaviour, reduced support from investors, and politicians' bold promises to dramatically reduce emissions. **The primary factor now driving gas and oil consumption is our constantly growing human population**, which somehow remains the elephant in the room.

National Policies

Remember, population growth in America is not an act of God—it is an act of Congress.
—Edward C. Hartman, author, *The Population Fix*

It is not that world governments are unaware of the urgency of the population crisis. It is just that most of them have their own agendas, which do not include reducing population as a priority. In 2020, Canada still has no meaningful population policy. We have Ministers of Environment, Health, Human Resources, and Immigration, but they all act as if these are independent portfolios. We have no Minister of Population, as Australia does, to tie these issues together. Politicians refuses to acknowledge the links between environmental degradation and population increase in Canada—only identifying developing nations as overpopulated and in need of reducing their numbers. High Canadian consumption rates do not seem to register as problematic, but rather as an advantage according to the GDP. We refuse to consider that since we consume and pollute far more per capita, the planet might also benefit from Canadians reducing our numbers.

For decades now, warnings about overpopulation have been issued from various governments, scientists, and progressive thinkers, but most often these warnings have been systematically ignored.

The so-called "global gag rule" on family planning has been used as a political wedge to further alienate voters, changing position every time a new government is elected in the U.S. Even advice from within a government's own administration is often ignored if lobbying by industry and the religious right is persuasive enough, revealing how shamefully derelict in their duty governments can be.

In 1972, following two years of research and public hearings, the final report of the President's Commission on Population Growth and the American Future (the Rockefeller Commission Report) was released. The commission was established by Congress in response to a proposal by then President Nixon. Its mission was to assess the social, economic, and environmental impacts of continued population growth in the United States. What was its major conclusion? The commission stated, "We have looked for, and have not found, any convincing economic argument for continued national population growth. The health of our economy does not depend on it. The vitality of business does not depend on it. The welfare of the average person certainly does not depend on it."[431]

In fact, it was found that the U.S. economy could actually benefit from a move toward a stable population. Quality of life would improve, demand for resources would decrease, and stress on the environment would ease. So, whatever became of these recommendations, and why was this not the impetus for a drastic reduction in population? Well, in his book *The Population Fix,* Edward Hartman explains, "Six months prior to seeking reelection, reportedly under immense pressure from religious leaders, President Nixon simultaneously received and renounced the report."

History was repeated when the U.S. House-Senate Concurrent Resolution 17 was submitted to the House of Representatives in 1999 by Rep. Tom Sawyer. It read: "Resolved by the House of Representatives (the Senate concurring), that it is the sense of the Congress that **the United States should develop, promote, and implement, at the earliest possible time and by voluntary means consistent with human rights and individual conscience, the policies necessary to slow the population growth** of the United States, and thereby promote the future well-being of the people of this Nation and of the world."[432]

This advice was also ignored. In fact, we have a situation in most countries where government policies are subsidizing and abetting the trend to overpopulation. Disregarding the writing on the wall, governments and their traditional economic advisors espouse their usual philosophy on the glories and benefits of increasing human numbers while being blinded to the enormous negatives of such advice and policies. But it isn't just warnings overpopulation that some leaders are choosing to ignore, fearing that panic will constrain the economy.

> *The costs of inaction are so steep, to the point that economists struggle to account for them.*
> **—Justin Worland,** *Time*

Time reporter Justin Worland states, "Report after report has documented the ample warning President Trump received about the growing threat the new coronavirus posed to the U.S. But Trump remained defiant, reports say, fearing that a reaction might rattle markets and disrupt the three years of economic growth that had underpinned his reelection

431 David Simcox, Floridians for a Sustainable Population, *Nixon and American Population Policy*, http://www.flsuspop.org/NixRockefeller.html

432 Rep. Tom Sawyer, CONGRESS.GOV, 1999, https://www.congress.gov/bill/106th-congress/house-concurrent-resolution/17?s=1&r=8

campaign. Even after recommending social distancing and the closing of the U.S. economy, Trump insisted that the measures to protect human life couldn't last too long because he did not want to 'let the cure be worse than the problem itself.'"[433]

This familiar fear has also paralyzed many leaders when dealing with climate change, overpopulation, biodiversity loss, and water scarcity. With both climate change and the coronavirus pandemic, simple economics based on the cost-benefit analysis show that the benefits of early action far outweigh the costs. In fact, Trump's failure to act has done far more harm than good in the long-term. Of course, this depends on priorities. The Obama administration calculated the social cost of carbon to be around $50 per tonne of carbon dioxide emitted; the Trump administration said it was as low as $1.

Worland adds, "Calculating the monetary value of addressing climate change is similarly complicated, but the evidence is firmly on the side of urgent action. If handled right, policies to reduce emissions will actually deliver huge economic benefits, from clean air to green jobs."

Isaac Asimov reflects on inaction on population to say, "I use what I call my bathroom metaphor. If two people live in an apartment, and there are two bathrooms, then both have what I call freedom of the bathroom, go to the bathroom any time you want, and stay as long as you want to for whatever you need. And this to my way is ideal. And everyone believes in the freedom of the bathroom. It should be right there in the Constitution. But if you have 20 people in the apartment and two bathrooms, no matter how much every person believes in freedom of the bathroom, there is no such thing. You have to set up times for each person, you have to bang at the door, aren't you through yet, and so on. And in the same way, democracy cannot survive overpopulation. Human dignity cannot survive it. Convenience and decency cannot survive it. As you put more and more people into the world, the value of life not only declines, it disappears. It doesn't matter if someone dies."

China has also been criticized for political decisions in the past, even though in this case those decisions have proven to be for the greater good. China's family planning effort initiated in 1980 remains perhaps the most controversial, complex, and misunderstood on our planet. In their desperate attempt to deal with their burgeoning population, which was nearing one billion at the time, the government finally took action with a one-child policy. The official government statement of justification was that population growth interferes with economic development. They have since proven this to be correct, and the sweeping changes that the world judged as coercive have actually paid off in many ways. Perhaps in light of the state of the world in 2020 due to overpopulation, global citizens should reconsider their harsh judgment of China's handling of their overpopulation crisis. Now in hindsight, if we are to meet the UN Sustainable Development Goals, the rest of the world may have to consider this same one-child scenario, but with voluntary methods based on educating couples about the impact each child has on the planet. We now know that the 50 percent of unintended pregnancies could be prevented through education and access to family planning services. So we can accomplish the same goal, but without using coercive measures, and avoid the

433 Justin Worland, Time Newsletter, 2020, *Ignore cost-benefit analyses at your own peril*

tragedy that took place around this policy (as it relates to female infanticide) due to cultural beliefs and patriarchal traditions.

India, on the other hand, lagged in introducing fundamental economic and social reforms to match China's, and they will soon surpass China as the most populous nation in the world. Now India is **encouraging high emigration levels**, as recipient countries struggle to manage the influx of immigrants.

So how can leaders decide on a wise and rational vision for their country? Fortunately, since 1984, the Worldwatch Institute, founded by Lester Brown, has been producing an annual report on progress toward a sustainable society. Each year, this report is presented in the form of a book entitled *State of the World,* which is translated into all major languages. In the absence of a comprehensive annual assessment by the United Nations or any national government, this book is now accorded semi-official status by national governments, UN agencies, and the international development community. More than 1,100 U.S. college and university courses—ranging from biology to geography to political science—use the volume. As the *New York Review of Books* points out, *State of the World* "deals with calamitous events rationally and constructively, and always offers logical solutions."[434]

Successes and Solutions

I hope that Greta's message is a wake-up call to world leaders everywhere that the time for inaction is over. It is because of Greta, and young activists everywhere that I am optimistic about what the future holds.

—Leonardo DeCaprio

Author Julian Cribb points out, "Thunberg is very much a child of the 21st Century: she was born in 2003. She has lived all her short life under the shadow of a world progressively being rendered uninhabitable by a climate out-of-control, weapons of mass destruction, by extinction, resource depletion, global poisoning, **overpopulation** and the untrammeled destruction of soil, water, life, atmosphere and oceans. A world being trashed by that figment of the human imagination called **money.**"[435]

"You say you love your children above all else and yet you are stealing their future in front of their very eyes," she upbraided world leaders at the climate summit in 2018. Bluntly dismissing the almost universal cowardice that does not want to face up to the realities she speaks of, Thunberg states, "I don't want you to be hopeful. I want you to panic. I want you to feel the fear I feel every day. I want you to act. I want you to act as you would in a crisis. I want you to act as if the house is on fire, because it is."

434 Deutsche Welle Media (DW), *Worldwatch publishes 'State of World' report,* Nov 04 2012, https://www.dw.com/en/worldwatch-publishes-state-of-world-report/a-15874848

435 Julian Cribb, John Menadue, *The Rise of Woman. Greta Thunberg.* March 11 2019, https://johnmenadue.com/julian-cribb-the-rise-of-woman-greta-thunberg/

Cribb declares, "Thunberg is a new stamp of world leader. She contrasts refreshingly with people like Trump, a dangerous ning-nong by any leaderly yardstick; with Putin who thinks he is Peter the Great reincarnated and wants to prove it by re-establishing an C18th Empire on the back of the oil, gas and coal that will be its ruin; with Xi-Jinping who, though better than most at reading the wind still wants China to be a great nation in a world where nations are history; with Brazilian President Bolsonaro who wants to destroy his country's chief asset, the Amazonian forest; with May, who seems to have lost sight of everything but the debilitating local gangrene of Brexit.... She has fired the youth of the world, crushed beneath the weight of their parents' greed and smug self-assurance, with a soaring ambition to make a difference. But Greta Thunberg is also another kind of new world leader. She's a woman."

In Cribb's compelling book *Surviving the 21st Century*, he argues that female leadership is the only way that humanity and Earth can be saved. If men are left in charge, we're doomed. He believes female leadership is essential if we wish to survive the global crisis our testosterone-driven legislators have created. Cribb adds, "Greta Thunberg is a small candle shining in a dark world. She symbolises new hope to young people, to women and to enlightened men alike."

We also must have the wisdom to hold the very wealthy accountable for being the world's most prolific consumers. Perhaps along with reducing the number of births in wealthy countries, we should be preventing the creation of extremely wealthy people. Reducing **global income inequality** would require changing our tax system, taking white-collar crime more seriously, implementing a luxury tax and a tiered carbon tax, eliminating government subsidies to the wealthy and corporate elite, and making changes to legislation regarding off-shore accounts. As it stands, the wealthy are allowed to write off most of their expenses as part of doing business, and this travesty must change. By reducing population and the number of extremely wealthy, we may yet have a chance at tackling our most critical world issues.

In fact, this century has seen a growing tide of wise women rise to positions of political power in countries like Finland, Bolivia, Germany, Ethiopia, New Zealand, Iceland, Slovakia, Switzerland, Taiwan, Norway, Nepal, Serbia, Bangladesh, Belgium, Barbados, Singapore, Myanmar, Denmark, and many others. In 2020, the Greek Parliament also elected a high-court judge, Katerina Sakellaropoulou, as their first female president with an overwhelming majority.

ABC News reported, "In her first comments as president-elect, the judge noted the 'difficult conditions and challenges of the 21st century, including the financial crisis, climate change, the mass movement of populations and the consequent humanitarian crisis, the erosion of the rule of law and all manner of inequalities and exclusions.'"[436]

Greece has a historically low number of women in senior positions in politics—in the current Greek Cabinet, all but one of the 18 senior positions are held by men. ABC News points out that Ms. Sakellaropoulou enjoys broad support from across the political spectrum. She has written numerous papers on environmental protection and chairs a society on environmental law, so perhaps now Greece will move ahead into a new era of equality and environmental integrity.

436 ABC News, *Greek parliament elects top judge Katerina Sakellaropoulou as first female president*, Jan 23, 2020, https://www.abc.net.au/news/2020-01-23/greece-parliament-elects-first-female-president/11892432

Incentives for Change

This new era of equality could start with all countries revamping their present system of economics, and those based on the gross domestic product (GDP), replacing it with a system like the genuine progress indicator (GPI) that would result in sustainable economics and support efforts towards a sustainable population.[437]

A solution to our presently flawed tax system in developed countries—which encourages large families through subsidies like child tax credits, child care benefits, and income tax deductions—would be to create incentives to encourage fewer children. It should be replaced with a system where parents are responsible for most costs for children after the first two, as we have seen in the successes of countries like Thailand, Iran, Tunisia, and so on, as discussed in Chapter 13. This new system would not discriminate against childfree citizens, punishing them with a higher tax load for not producing offspring, as the present system does. It would also encourage parents to plan and prepare for having a family, as almost half of pregnancies today are unintended. Humans clearly need economic and social incentives to think more practically for the planet.

Penalizing small and childfree families through our tax system is not the answer. To the contrary, we must reward those individuals who consider the environmental and social impacts of their child-bearing decisions, both morally and financially. In fact, some countries have already **proposed policies** that would reward those with small families: "In an effort to check the country's population boom, India plans to bar couples who marry too young from state jobs. A new national population policy would ban men married before 21 and women who married before 18 from government jobs.... The draft of the new policy also recommends that only couples with no more than two children should be eligible for tax benefits and for promotion in government jobs. There will be laws banning those who do not adopt the small-family norm from entering politics."[438]

These kinds of measures are apparently necessary as we have yet to take the population crisis seriously enough to cap ourselves. Although India's proposed policy was not fully implemented, just imagine the difference it could have made if ultimately put into action. Instead of overtaking China as the most populous nation, India could be closer to enjoying a utopian lifestyle while restoring their environment and wildlife habitat.

437 Mark Anielsk, Pembina Institute, *The Genuine Progress Indicator - A Principled Approach to Economics*, Oct. 1, 1999, https://www.pembina.org/reports/gpi_economics.pdf

438 Rahul Bedi, The Daily Telegraph, *No State Jobs for Couples Who Marry Too Young, Jan 2, 1997, https://www.newspapers.com/newspage/496145564/*

International Successes

To begin the process of healing and restoring our damaged and most beautiful earth, we need fewer people, considerably fewer. Population reduction is the highest and most moral of all human goals.... Efforts should be made to improve the quality of life in areas of human concentration, and not on spreading people over parts of the earth that are not yet savagely degraded."

—Dr. Ted Mosquin, ecologist, Tropical Conservancy

The United Nations was established in 1945 and is currently made up of 193 member states. Their most remarkable accomplishment to date is the development of the Sustainable Development Goals in 2015, following the groundbreaking International Conference on Population and Development (ICPD), which took place in Cairo in 1994.

The United Nations reports that The 2030 Agenda for Sustainable Development, adopted by all United Nations member states in 2015, provides a shared blueprint for peace and prosperity for people and the planet, now and into the future. At its heart are the 17 Sustainable Development Goals (SDGs), which are an urgent call for action by all countries—developed and developing—in a global partnership. They recognize that ending poverty, environmental degradation and extinction of species must go hand-in-hand with strategies that improve health and education, reduce inequality, and provide universal access to sexual and reproductive health—all while tackling climate change and working to preserve our oceans and forests.

The SDGs build on decades of work by countries and the UN, and it is urgent that we achieve them all by 2030 if we wish to avoid catastrophe and create a better future for all.[439]

Goal #5, Empowering Women, is seen as pivotal to the accomplishment of the SDGs, and that is why the 2019 UN International Conference on Population and Development (ICPD), on the 25th anniversary the 1994 Cairo conference, was so momentous. Leadership from WHO participated in ICPD +25 in Nairobi, Kenya, alongside heads of state, ministers, parliamentarians, thought-leaders, technical experts, civil society organizations, grassroots organizations, young people, business and community leaders, faith-based organizations, Indigenous peoples, international financial institutions, people with disabilities, academics, and many others interested in the pursuit of sexual and reproductive health and reproductive rights.

Together, we will work to make the next 10 years, years of action and results for women and girls, in keeping with the decade of delivery on the SDGs. Together, we will make sure that promises made are promises kept.

—Dr. Natalia Kanem, UNFPA

Co-convened by the governments of Kenya and Denmark and UNFPA, the ICPD+25 Summit aimed to be a watershed moment for women and girls everywhere. At the 1994 conference, diverse views on human rights, population, sexual and reproductive health, gender equality, and sustainable development merged into a remarkable global consensus that placed individual dignity and

439 United Nations, Sustainable Development Goals, 2018, https://sustainabledevelopment.un.org/?menu=1300

human rights, including the right to plan one's family, at the very heart of development. A quarter of a century later, the world has seen significant progress. There has been a 25 percent increase in global contraceptive prevalence rate around the world. Adolescent births have declined steeply, and the global maternal mortality ratio has fallen. Despite intense efforts by the Vatican to derail the progress of the UN, the majority of individual countries stepped up to pledge meaningful measures towards reproductive health and gender equality.

The United Nations notes that as the Nairobi Summit ends, it is time to focus on accountability. There were more than 1,200 concrete commitments to help ensure sexual and reproductive health and end sexual violence by 2030. Donor countries pledged around $1 billion in support to sexual and reproductive health and gender equality programs. Private-sector firms agreed to mobilize a combined $8 billion.[440]

But the pledges made were much more than financial. Organizations, governments, and businesses also put forward bold, visionary plans to spark change and meet unmet needs. For instance, Kenya, the host country and co-convenor of the Summit, pushed forward its deadline to end female genital mutilation, now calling to end the practice by 2022, eight years earlier than its previous target. They walk away with a clear roadmap of actions, including a statement that calls for achieving "three zeros"—zero maternal deaths, zero unmet need for family planning, and zero gender-based violence and harmful practices against women and girls—by 2030, the deadline for achieving the Sustainable Development Goals (SDGs).

"I am pleased to announce that UNFPA will **create a new high-level commission to drive this agenda** and our commitments forward," UNFPA Executive Director Dr. Natalia Kanem announced in her closing remarks. "We will draw from the full spectrum of stakeholders—government and the private sector, young people and activists, civil society and philanthropy. The commission will propose ways to monitor progress on the commitments made here this week, while accounting for all existing global, regional, and national follow-up mechanisms."

The United Nations also offers an Ambassador program to train youth to participate in these programs. In 2019, the UN Model Youth Summit was held in New York. This one-day event featured workshops and plenaries led by inspiring student leaders and United Nations experts who shared their experiences and ideas on how to transform Model UN into a force for positive change. Participants left with a toolkit of ideas, a network of support and a new action-oriented vision for their Model UN clubs and conferences. More than 400 students from 30 countries took part.[441]

Other alliances have been formed as a result of past UN population conferences, include the South-South initiatives. The goal of the partners was to foster cooperation and exchange of technical information between developing countries, rather than following the development strategies sent out from the industrial nations.

The Population and Development South-South Center of Excellence in Beijing hosted its first training workshop on "Population Data Collection, Analysis and Utilization" in 2018. This

440 United Nations ICPD, *As Nairobi Summit ends, time to focus on accountability*, Nov 20 2019, https://www.nairobisummiticpd.org/news/nairobi-summit-ends-time-focus-accountability
441 United Nations, *Model United Nations Program*, 2020, https://www.un.org/en/mun/page/events

landmark partnership brought together 19 population experts from over 17 countries in China and Africa, enhanced their capacity to collect and harness data, and provided population projection software and skills, for use in their home countries. Partner countries included China, Burundi, Cambodia, Comoros, Côte d'Ivoire, Eritrea, Gambia, Kenya, Maldives, Mauritania, Papua New Guinea, Sierra Leone, South Sudan, Sri Lanka, and Tanzania.[442]

The World Court Investigates War Crimes

In its capacity as International Court of Justice, the UN is also authorized to investigate war crimes. A 2020 decision by the International Criminal Court is the **first time** the prosecutor has been authorized to **investigate U.S. forces**. *New York Times* reporters Elian Peltier and Fatima Faizi state, "The International Criminal Court ruled on Thursday that its chief prosecutor could open an investigation into allegations of war crimes in Afghanistan including any that may have been committed by Americans, a step that **infuriated the Trump** administration. The ruling by an appeals chamber of the court in The Hague reversed a lower chamber's decision that had halted an inquiry into the behavior of forces from the United States, which does not recognize the court's jurisdiction. Washington revoked the visa of the court's chief prosecutor, Fatou Bensouda, last year after she had signaled her intentions to pursue the case."[443]

"The ICC Appeals Chamber's decision to greenlight an investigation of brutal crimes in Afghanistan despite extreme pressure reaffirms the court's essential role for victims when all other doors to justice are closed," said Param-Preet Singh, the associate international justice director at Human Rights Watch.

Peltier and Faizi point out, "The prosecutor has said that the court had enough information to prove that U.S. forces had 'committed acts of torture, cruel treatment, outrages upon personal dignity, rape and sexual violence' in Afghanistan in 2003 and 2004, and later in clandestine CIA facilities in Poland, Romania, and Lithuania. The wide-ranging investigation would also look into allegations against the Afghan government forces, which are accused of torturing prisoners, as well as those against the Taliban and antigovernment forces. The United Nations' mission in Afghanistan has documented the killings of more than 17,000 civilians by the Taliban since 2009, including nearly 7,000 targeted killings. Yet, last April, a U.N. report found that **U.S. and Afghan forces had killed more civilians** in the first three months of 2019 than the Taliban did."

The ruling came days after the United States signed a deal with the Taliban to withdraw U.S. troops from Afghanistan after nearly two decades of conflict:

"This decision vindicates the rule of law and gives hope to the thousands of victims seeking accountability when domestic courts and authorities have failed them," said Jamil Dakwar, director of the American Civil Liberties Union's Human Rights Project. He added, "Countries must

---

442 United Nations, 5 *Results achieved through South-South and triangular cooperation*, 2019 https://www.unfpa.org/sites/default/files/board-documents/Annex_5-Results_achieved_through_South-South_and_triangular_cooperation.pdf

443 Elian Peltier and Fatima Faizi, New York Times, *I.C.C. Allows Afghanistan War Crimes Inquiry to Proceed, Angering U.S.*, March 5, 2020, https://www.nytimes.com/2020/03/05/world/europe/afghanistan-war-crimes-icc.html

fully cooperate with this investigation and not submit to any authoritarian tactics by the Trump administration to sabotage it."

Accountability is also being pursued for former Sudan president al-Bashir. In 2020, Associated Press reporter Nadir Ahmed states, "A court in Sudan convicted former President Omar al-Bashir of money laundering and corruption on Saturday, sentencing him to two years in a minimum security lockup. That's the first verdict in a series of legal proceedings against former Sudan president al-Bashir, who is also wanted by the International Criminal Court on charges of war crimes and genocide linked to the Darfur conflict in the 2000s."[444]

"The trial for these charges of financial crimes does not address the human rights violations that so many Sudanese have experienced," said Jehanne Henry, a Human Rights Watch associate director who focuses on Sudan. "So the sentence will not likely satisfy the many thousands of victims of abuses under al-Bashir's 30 year rule."

With this new shift in power, Sudanese women are demanding equality and decision-making power. However, months after al-Bashir's removal in 2019, male-dominated politics returned to Sudan.

Independent reporter Justin Lynch notes, "It was the iconic image of a woman protester that came to define Sudan's revolution, and women made up the majority of demonstrators—but activists complain that they have been almost entirely excluded from the new system. For three decades, Omar al-Bashir enforced a raft of oppressive laws aimed at subduing women, apparently with the objective of satisfying the country's ultra-conservative Islamic forces, which propped up his regime. Child marriage was allowed, marital rape was permitted, and women were not allowed to wear trousers in public."[445]

It is not surprising that women hoped for change with the new leadership, but this was not to be. This lack of female leaders in Sudan's democratic movement is not just a question of equality for the sake of equality, say women's activists, but will affect the quality of the transition and, ultimately, the success of the revolution. But perhaps with the latest commitment at the UN Population and Development Conference in 2019 there will be incentive for change in the near future, as countries will be held accountable.

The International Criminal Court is also indispensable in the global fight against impunity.

In 2019, *Washington Post* contributor and minister of foreign affairs of the Netherlands, Stef Blok states, "The ICC is the only institution that can deliver justice to the victims of the most serious crimes when all other venues fail. The value of this was on display earlier this month, when the court sentenced the former Congolese rebel leader Bosco Ntaganda to 30 years in prison for crimes against humanity and war crimes, such as murder, sexual slavery and conscripting child soldiers. The court's recent decision to authorize an official investigation into the deportation of the

444 Nadir Ahmed, The Associated Press, *Ex-Sudan strongman al-Bashir gets 2 years for corruption*, Dec 14, 2019, https://www.vancourier.com/ex-sudan-strongman-al-bashir-gets-2-years-for-corruption-1.24035331

445 Justin Lynch, Independent, *Women fueled Sudan's revolution, but then they were pushed aside*, Aug 04 2019, https://www.independent.co.uk/news/world/africa/sudan-revolution-women-uprising-democratic-transition-army-bashir-a9038786.html

Rohingya from Myanmar to Bangladesh is **another signal of hope** for the victims of this humanitarian crisis and others."[446]

Although many of the world's worst criminals continue to escape justice, a review is underway to identify ways to achieve stronger support from member nations. Countries appointing more qualified judges for the court, increasing financial contributions, and stopping the obstructive attitude of some nonstate parties—like the Vatican—would help to make the ICC more efficient and effective. Blok notes, "Sanctions against court officials are unacceptable, and it's worrying to see such measures come from the United States, a long-standing friend and ally. If we truly believe that international justice is an attainable goal, we must practice what we preach and help the ICC to function in the way it was intended."

Turning Crisis into Opportunity

While some people may feel that these efforts to move forward are being hindered by the COVID-19 pandemic, others have found that this is a perfect opportunity to make the world a better place. Live Kindly staff writer Audrey Enjoli tells us, "Pakistan is working to help laborers who have lost their jobs amid the pandemic. The country has hired more than 63,000 people to help plant billions of trees. The new tree-planting initiative is part of the country's 10 Billion Tree Tsunami program. The prime minister of Pakistan Imran Khan founded the five-year project in 2018. It aims to help reduce rising temperatures and the risk of natural disasters like floods and droughts caused by climate change."[447]

According to the Global Climate Risk Index 2020, Pakistan ranks number five on a list of countries most heavily impacted by global warming over the past 20 years. "This tragic crisis provided an opportunity and we grabbed it," Malik Amin Aslam, climate change advisor to the prime minister, told the Thomson Reuters Foundation. "Nurturing nature has come to the economic rescue of thousands of people."

When it comes to "economic rescue," putting pressure on the finance world could be one of the most effective ways to fight climate change. *Time* reporter Bill McKibben stated, "Time for something like panic: last week [2019] the UN climate talks in Madrid essentially collapsed, even as scientists were reporting that the 2010s had been by far the hottest decade since records began. Most of the blame fell on countries like the U.S., Brazil, and Saudi Arabia, but around the world political systems simply aren't responding to the **greatest crisis they've ever faced**—they're so corrupted by fossil fuel money, so overcome by inertia, so preoccupied with the next election or coup."[448]

But increasingly, amid the climate crisis and pandemic, protesters are going after not only governments but high finance as well. In 2019, McKibben notes, Goldman Sachs announced that it

446 Stef Blok, Washington Post, *The International Criminal Court must do better. Reforms are urgently needed.*, Dec. 2, 2019, https://www.washingtonpost.com/opinions/2019/12/02/international-criminal-court-must-do-better-reforms-are-urgently-needed/

447 Audrey Enjoli, LIVE KINDLY, *Pakistan Just Hired 63,000 People to Plant 10 Billion Trees*, May 7 2020, https://www.livekindly.co/pakistan-plant-10-billion-trees/

448 Bill McKibben, Time Magazine, *Putting Pressure on the Finance World Could Be One of the Most Effective Ways to Fight Climate Change, Dec* 18, 2019, https://time.com/5752188/financial-world-pressure-climate-change/

would restrict lending to the coal industry and no longer fund drilling in the Arctic. Liberty Mutual stopped funding coal companies due to pressure from a coalition of environmental groups called Insure Our Future.

McKibben points out, "Washington and Wall Street are deeply linked, but they're also distinct, and both need to shift dramatically. Even if we can get politicians to make real change, it will come slowly, and one national capitol at a time. But if these **financial giants** begin to move, the effects will be both quick and global—and those are the two things most required for effective progress on the climate."

In 2020, activists will need to go after the worst of the power centres, like Chase Bank—lending billions for the most extreme fossil fuel projects on Earth—and the American money giants who dominate global finance. A daunting task? McKibben explains, "But look at them another way: unlike Exxon or Shell, they're able to make plenty of money from things other than fossil fuel. Destroying the planet is only a side business for them—heck, they could make money financing the transition to sun and wind. And they're potentially vulnerable to citizen anger. **Anyone with a pair of scissors can cut up a credit card.**"

> *Addressing the financial risks of the climate crisis is an international issue. We will not defeat the climate crisis if we have to wait for the financial industry to self-regulate.*
> **—U.S. Senator Elizabeth Warren**

When looking for solutions to our climate crisis, Senator Elizabeth Warren recommends taking aggressive steps to rein in Wall Street and avoid financial collapse by using a number of levers at a president's disposal.[449]

Grist reporter Zoya Teirstein tells us about Warren's suggestions, "She'd ask the Securities and Exchange Commission and Department of Labor, the two agencies in charge of regulating pensions, to identify carbon-intensive investments. Current pension systems, she writes, are 'leaving all the risk of fossil fuel investments in hard working Americans' retirement accounts.' She'd join other world powers in making climate change a factor in monetary policymaking, and prompt the Federal Reserve to join the Network on Greening the Financial System, a global coalition of central banks. And she'd make implementation of the Paris Agreement a prerequisite for future trade agreements with the U.S."

More importantly, we will not defeat the climate crisis or threatening pandemics if we wait any longer to tackle our alarming population emergency. Russell Train with the World Wildlife Fund put it this way, "Achieving and maintaining a sustainable relationship between human populations and the natural resource base of the earth is the single most critical long-term issue facing the peoples of the world and this issue will increasingly be the focus of international affairs for the foreseeable future."

When it comes to dealing with these pandemics, celebrities have stepped up and are urging world leaders and citizens to take environmental action. They say going "back to normal" after COVID-19 is not an option.

---

449 Zoya Teirstein, Grist, *Eliabeth Warren's new cliate plan can go the distance, even if her ampaign can't*, Mar 1, 2020, https://grist.org/politics/elizabeth-warrens-new-climate-plan-can-go-the-distance-even-if-her-campaign-cant/

Live Kindly reporter Audrey Enjoli notes that more than 200 prominent public figures—including Hollywood celebrities Penélope Cruz, Adam Driver, Robert De Niro, and Cate Blanchett—have signed an op-ed calling for a "radical transformation" of society amid the coronavirus pandemic. An editorial entitled "**No to a Return to Normal**" was published in the French paper *Le Monde* last week. Oscar-winner Juliette Binoche and astrophysicist Aurélien Barrau penned the letter.

Enjoli states, "A lengthy list of notable artists, musicians, directors, scientists, athletes, and actors lent their signatures to the op-ed. High-profile celebrities Barbra Streisand, Joaquin Phoenix, Wim Wenders, and Madonna signed the piece. Other signees include the former president of the National Museum of Natural History, Gilles Bœuf; Buddhist monk Matthieu Ricard; pianist Khatia Buniatishvili; and French philosopher Jean-Luc Nancy."[450]

The letter reads: *"The COVID-19 pandemic is a tragedy. This crisis is, however, inviting us to examine what is essential. And what we see is simple: "adjustments" are not enough. The problem is systemic."* It goes on to warn of how the *"ongoing ecological catastrophe"* will *"have immeasurable consequences."* The letter explains one of these consequences would be *"the massive extinction of life on Earth. The pursuit of consumerism and an obsession with productivity have led us to deny the value of life itself. The radical transformation we need–at all levels–demands boldness and courage."*

Enjoli adds, "The group of signees urges world leaders and citizens '*to leave behind the unsustainable logic that still prevails,*' and says a complete overhaul of the world's '*goals, values, and economies*' is needed in order to prevent mass extinction. They declare that there is no going 'back to normal.'"

In the U.S. in 2020, the Republicans and Democrats also seem to agree that there is no going back to business as usual. Grist reporter Emily Pontecorvo states, "It's not often Republicans and Democrats in Congress come together to pass legislation that benefits the environment, especially in the Republican-controlled Senate. But last week, they did just that: The upper chamber voted overwhelmingly in favor, 73-25, of a new bill that will permanently funnel money into the Land and Water Conservation Fund. The Great American Outdoors Act authorizes $900 million in oil and gas revenues to be directed into the fund annually, along with an additional $9.5 billion over five years to address maintenance backlogs on public lands, including the National Park system."[451]

Where to from Here: Solutions to Shape Our Future

By now it is also obvious that the present operation of the United Nations, which gives the Vatican veto power on population-related issues, needs an overhaul. The **solution** is to revoke this undeserved privilege, so that greater progress can be made toward formulating sustainable population goals and initiatives in the UN.

Our present political system, greatly prejudiced by the Vatican's overpowering influence, does not truly reflect the wishes of the majority of citizens regarding population issues. A **solution** would be to require politicians to make population and immigration issues part of their election

450 Audrey Enjoli, LIVEKINDLY, *More Than 200 Celebrities Call for Urgent Environmental Action Post–COVID*, May 11, 2020, https://www.livekindly.co/celebrities-urgent-environmental-action/

451 Emily Pontecorvo, The Grist, *Senate unites for the outdoors*, June 22 2020, https://grist.org/beacon/senate-unites-for-the-outdoors/

campaign and a focus of their platform. Elections should include a referendum on population choices, encouraging aroused and informed public debate on the issue. Polls in many countries indicate that the majority of people do not want population or immigration to continue growing, and they do not want religion to be reflected in the political arena.

The task before our leaders is to recognize the consequences if they continue to ignore their responsibility regarding overpopulation and to take effective and timely measures to achieve a sustainable population to benefit the common good.

The task before our citizens is to recognize that we can no longer sit back and allow big power governments, big corporations, and international institutions to determine our global destiny. We must no longer allow global decision-making to be controlled by a handful of tyrants with a questionable agenda. It would benefit humankind to insist that global decision-making be more accountable to the public, especially when it comes to reproductive health care.

For instance, small Canadian families have shown that we want a declining birth rate, and if not for immigration our population would be decreasing. It is the politicians, media, and government-funded advisory groups like the Economic Council of Canada that refuse to draw the link between increasing population and decreasing quality of life. We must let Prime Minister Trudeau know that his plan to increase immigration to 350,000 a year is unacceptable.

The Carrying Capacity Network (CCN) in 1994 issued this statement: "On the local, state, national, and global level, every problem of the environment such as pollution, loss of biodiversity, etc.; every problem of social justice, such as racism, sexism, unequal distribution of wealth, etc.; every economic problem such as unemployment, inflation, etc.; every problem related to resource scarcity and conflict, such as war, famine, etc.; is made worse by increases in population. Conversely, **all these problems would be easier to solve if populations were not growing or were smaller**." This statement still holds true in 2020, and we must make our governments understand this.

The persistence of war related to resource conflict that CCN speaks of is especially outdated and unnecessary. This temporary respite taking place during the COVID-19 pandemic should become the "new normal" for the 2020 decade.

The Pentagon is in a position to lead a global campaign to withdraw all forces from foreign countries in an effort to reevaluate the "permanent state of war" mentality, which clearly is not working. A new strategy of cooperation, relationship building, and conflict resolution could be designed to end the senseless continuation of war many of our leaders have become obsessed with. If the population were to begin to decline, there would also be a substantial decline in conflict over natural resources, bringing peace to many regions.

The "war machine," strongly promoting and greatly benefiting from perpetual war, would be dismantled as demand for its services waned and its methods were recognized as unacceptable. Public outrage is needed to bring about major change. The needless ethnic and religious dimensions to the population issue, often exploited by politicians and religious leaders for personal gain, winning elections and appropriating resources, would be recognized as senseless and be rejected.

The United Nations must lead the world in this endeavour and insist that every country have a population policy in place to guide people to sustainable lifestyles. These policies would reflect a country's carrying capacity, a meaningful public consultation process, as well as cumulative and long-term impacts of population levels. Most certainly they would include guidelines for appropriate immigration numbers.

Of course there have been many enlightened thinkers who have broken the mould and helped advance the population movement. These people provide hopeful examples in an otherwise grim political landscape of mass denial, lobbyist-controlled agendas, and power-hungry politicians. Notably, the pioneering spirit of Ireland's first female president, Mary Robinson, helped guide humanity into the 21st century, and she turned 75 in 2019.

Irish Central News reporter Niall O'Dowd states, "Robinson was elected president of the Irish Republic in 1990, creating perhaps the biggest shock wave in the lifetime of Irish politics. Her victory was achieved **against all odds**—at a time when women were decidedly second class in Ireland."[452]

In 1990 Ireland, family planning was forbidden, with condoms and birth control pills illegal, women could not serve on juries, women had to retire from teaching and other government positions when they married, and women were paid about half the rate men were for similar jobs.

O'Dowd adds, "The all-powerful church found her a juicy target and honed in on her advocacy of family planning, not surprising given their obsession with female bodies. They became aware of this uppity woman early on when she proposed a family planning bill. The local bishop denounced her from the pulpit in her own hometown of Ballina for advocating contraception.

Despite all that, Robinson made it to the presidency with an approval rating of 93 percent. She left in 1997 to become UN High Commissioner for Human Rights and is still active on the world stage. Hopefully, she has set an example for many more women and enlightened thinkers to come, even in the face of archaic religious bullying tactics. Other progressive thinkers have dared to challenge prevailing economic myths. Contrary to popular opinion, a stable or declining population is nothing to fear and economically is now most often seen as a boon, not a threat.

> *A growing population no longer has any economic benefit.*
> **Adair Turner, Market Watch**

Marketwatch reporter Adair Turner advises, "Our expanding ability to automate human work across all sectors—agriculture, industry, and services—makes an ever-growing workforce increasingly irrelevant to improvements in human welfare.... In all countries that have achieved middle-income status, and where women are well educated and have reproductive freedom, fertility rates are at or below replacement levels. We should be wary of declaring a universal rule of human behavior, but it seems this might be one.... Conversely, automation makes it impossible to achieve full employment in countries still facing rapid population growth."[453]

Unfortunately, the U.N. projects that Africa's population will soar from 1.34 billion to 4.28 billion by 2100 if steps aren't taken imediately. Therefore, this is where our foreign aid should be

---

452 Niall O'Dowd, Irish Central News, *The woman who changed Ireland - Mary Robinson turns 75*, May 23 2019, https://www.irishcentral.com/news/woman-changed-ireland-mary-robinson?utm_campaign

453 Adair Turner, Market Watch, *Opinion: A growing population no longer has any economic benefit*, July 5, 2019, https://www.marketwatch.com/story/a-growing-population-no-longer-has-any-economic-benefit-2019-07-05

focused in an effort to raise Africa out of poverty and provide access to family planning services. The solution is for all countries to unite and make this a priority for the 2020 decade of change. Stopping population growth at EIGHT BILLION must be the defining issue of the 21st century. Turner notes, "Automation has turned conventional economic wisdom on its head: there is greater prosperity in fewer numbers."

CHAPTER 11
RELIGIOUS INFLUENCES

We think there are many problems. No, there are not. There is only one problem—population. If you don't cap that consciously, Nature will do it to you in a very cruel manner.
—SADHGURU 2019—Indian yogi, author. and public speaker at UN Millennium World Peace Summit

Ancient traditional beliefs of pagan animism assumed that every rock, tree, animal, and mountain had its own guardian spirit and deserved respect and reverence. This ecocentric ideology reigned for centuries. If we had continued on this path, I wouldn't need to write this book at all. So what changed this paradigm?

Christianity: Ecological Villain?

Counter Punch contributor Christopher Ketham explains that in 1966, a professor of medieval history presented a lecture on this topic that would go on to live in infamy, titled "The Historical Roots of Our Ecological Crisis." The author was Lynn White Jr., who singled out Judeo-Christian religion as the historical villain, calling it "the most anthropocentric religion the world has seen." Ending the ecological crisis was White's primary concern.[454]

As Ketham notes, "White's argument was that the Judeo-Christian conception of a planet made solely for man's exploitation, as laid out in the book of Genesis, freed humankind to lay waste to the environment." And indeed we have, with serious consequences. The verse in Genesis giving humankind "dominance" over everything on Earth—which is then relegated to soulless matter—has literally destroyed the planet.

When this Christian worldview replaced **pagan animism**, respect and inhibitions crumbled, we became disenchanted with nature, and our lands soon became unrecognizable. Ketham adds, "The fact that most people do not think of these attitudes as Christian is irrelevant. No new set of basic values has been accepted in our society to displace those of Christianity. Hence we shall continue to have a worsening ecologic crisis until we reject the Christian axiom that nature has no reason for existence save to serve man."

454 Christopher Ketham, Counter Punch, *The American West as Judeo-Christian Artifact*, July 23, 2019, https://www.counterpunch.org/2019/07/23/the-american-west-as-judeo-christian-artifact/

This self-centred humanism now holds sway over every corner of the planet, and has seeped into our political system, education system, and economic system—contaminating our honour and respect for nature. However, as we throw off this blanket of delusion, our mastery over nature will diminish. Over the centuries, there have been many who have seen past this fallacy.

Chinese political theorist, statesman, and scholar, **Hong Liangji**, was most famous for his critical essay to the Emperor, which resulted in his banishment. Wikipedia tells us that today he is best remembered for his 1793 essay *Zhi Ping Pian*, ("On Governance and Well-being of the Empire") on population growth and its socio-political consequence, in which he raised many of the same issues that were raised by Malthus, writing during the same period in England. Liangji was especially concerned with population control and government corruption. He critically re-evaluated the common Chinese assumption that a growing population was the sign of a good government.[455]

Hong's time experienced one of the fastest expansions of population in Chinese history. With the promotion of New World crops such as corn, Chinese population tripled from 100 million (1651–1661) to 300 million (1790). The population boom resulted in a series of socio-economic problems and caused concerns among the Mandarins. In 1791, Qianlong Emperor expressed his worry to the court officials that the resources might not be able to support the growing population. Of course, **like Malthus, Liangji was correct in his anticipated hardships**, and millions of people have starved to death mainly due to overpopulation since their predictions.

Malthus: A Pioneer of Evolutionary Biology

When talking about population, one would be amiss to not discuss the **half-truths** surrounding **Thomas Robert Malthus, an Anglican pastor** who famously observed that human population, if unchecked, would grow faster than its food supply. He argued that education in "moral restraint" might prevent starvation from being the operative check on population growth. His warning went unheeded, partly due to strong opposition from the Vatican.

Malthus was rash enough to predict that the crash would come by the second half of the 19th century. Because it didn't, economists in general tend to believe that he was wrong, when actually his prediction was simply premature because he hadn't foreseen the misguided "Green Revolution" that was to take place in agriculture. The fact remains that Malthus formulated a principle of population growth that no one has been able to falsify and that forms a foundation stone of the theory of evolution by natural selection.[456]

Can anyone logically dispute his view that if society relied on human misery to limit population growth, then sources of misery (e.g., hunger, disease, and war) would inevitably afflict society, as would volatile economic cycles? On the other hand, "preventive checks" to population that limited birth rates, such as later marriages and birth control, could ensure a higher standard of living for all, while also increasing economic stability. His views became influential, and controversial, across

455 Wikipedia, *Hong Liangji*, https://en.wikipedia.org/wiki/Hong_Liangji

456 Ward Chesworth, Michael R. Moss, Vernon G. Thomas, *Malthus and The Third Millennium*, 2000, https://www.abebooks.com/book-search/title/malthus-third-millennium/author/chesworth-moss-thomas/

economic, political, social, and scientific thought. Pioneers of evolutionary biology read him, notably Charles Darwin, and he remains a much-debated writer today.

In the 20th century, we are seeing the latter scenario of "preventive checks" play out in countries like Thailand, Japan, and Bhutan with exactly the positive effects Malthus and Liangji predicted. **Perhaps now we can put to bed the notion that we need a never-ending supply of people to create a healthy economy and prosperous society.**

Intellegent Economist reporter Prateek Agarwal notes, "The Malthusian Trap (or 'Malthusian Population Trap') is the idea that higher levels of food production created by more advanced agricultural techniques create higher population levels, which then lead to food shortages because the higher population needs to live on land that would have previously been used to grow crops."[457] Of the relationship between population and economics, Malthus wrote that when the population of labourers grows faster than the production of food, real wages fall because the growing population causes the cost of living (i.e., the cost of food) to go up.

The rapid increase in the global population of the past century exemplifies Malthus's predicted population patterns. On the whole it may be said that Malthus's revolutionary ideas in the sphere of population growth remain relevant to economic thought even today and continue to make economists ponder about the future.

Church History on Birth Control

> *It can be said that the family planning movement is one of the important social changes of this century, and especially of the last half-century.... No country except the Holy See(Vatican) objected to the services provided by the family planning movement.*
> —Roderic Beaujot, 1995, Canadian National Advisory Council for the Cairo International Conference on Population and Development[458]

On the other end of the spectrum from Malthus, in 2020 we have Pope Francis denouncing women who decide not to produce offspring. The *Independent* reporter Loulla-Mae Eleftheriou-Smith states, "Pope Francis has condemned couples who deliberately choose not to have children, labelling their decision as 'selfish,' just weeks after insisting that Catholics do not need to 'breed like rabbits.'"[459] Despite these contradictory statements, Pope Francis maintains that the Catholic Church does not in any way support artificial birth control.

It is not only Pope Francis that has presented a contradictory view, but the Vatican itself has presented contradictory views throughout history. Notre Dame Law contributor John T. Noonan Jr. tells us, "In the

457 Prateek Agarwal, Intellegent Economist, *Malthusian Theory of Population,* Nov 21, 2020, https://www.intelligenteconomist.com/malthusian-theory/

458 Research Gate, *Currently Married Women with an Unmet Need for Contraception in Eritrea,* March 2011, https://www.researchgate.net/publication/267997407_Currently_Married_Women_with_an_Unmet_Need_for_Contraception_in_Eritrea_Profile_and_Determinants

459 Loulla-Mae Eleftheriou-Smith, The Independent, Feb12 2015, *Pope Frances: The Choice not to have chidren is selfish,* https://www.independent.co.uk/news/people/pope-francis-the-choice-not-to-have-children-is-selfish-10041224.html#comments

Mediterranean world in which Christianity appeared, abortion was a familiar art. The most learned of Greco-Roman gynecologists, Soranos of Ephesus (c. 98–138 CE), discussed abortion in terms of expelling what has been conceived. He recommended such methods as walking about vigorously, carrying things beyond one's strength, or using particular drugs composed of plant mixtures."

Noonan notes, "In Plato's *Republic,* abortion is proposed as a solution to prevent endangering the optimum population of the state.... Aristotle also proposes abortion if a couple has too many children for the good of the state, but he does so with remarkable caution, saying it is to be done before there is 'sensation and life.'"[460] Over the centuries, the morality of practicing abortion has been debated by physicians, philosophers, and religious teachers. For almost 2,000 years, the Catholic Church's stance on birth control has been one of constant change depending on which Pope held power, since **nothing in scripture explicitly prohibits contraception.**

Acacia was one plant used by ancient Egyptian women for contraception, which proved to have a powerful spermicidal effect. Both the Bible and the Koran refer to coitus interruptus (the withdrawal method). Catholic theology, which now regards the early fetus as a person, did not always do so. The Church first adopted the belief of Aristotle, St. Jerome, St. Augustine, and St. Thomas Aquinas that ensoulment occurs several weeks after conception. Pope Innocent III, who ruled at the turn of the 13th century, made that belief part of church doctrine, allowing abortion until fetal animation. It was not until 1869 that the church prohibited abortion at any time and for any reason.

The Conversation reporter Lisa McClain states, "The church, however, had little to say about contraception for many centuries. For example, after the decline of the Roman Empire, the church did little to explicitly prohibit contraception, teach against it, or stop it, though people undoubtedly practiced it. Most penitence manuals from the Middle Ages, which directed priests what types of sins to ask parishioners about, did not even mention contraception. It was only in 1588 that Pope Sixtus V took the strongest conservative stance against contraception in Catholic history. With his papal bull 'Effraenatam,' he ordered all church and civil penalties for homicide to be brought against those who practiced contraception."

Plato and Aristotle thought of abortion as a way of preventing excess population.
—John T. Noonan Jr., Nottre Dame Law

McClain emphasizes, "However, both church and civil authorities refused to enforce his orders, and laypeople virtually ignored them. In fact, three years after Sixtus's death, the next pope repealed most of the sanctions and told Christians to treat 'Effraenatam' 'as if it had never been issued.' By the mid-17th century, some church leaders even admitted couples might have legitimate reasons to limit family size to better provide for the children they already had.... When an 1886 penitential manual instructed confessors to ask parishioners explicitly whether they practiced contraception and to refuse absolution for sins unless they stopped, 'the order was virtually ignored.'"[461]

---

460 John T. Noonan Jr., Notre Dame Law School, *Abortion and the Catholic Church: A Summary History,* Jan 1 1967, https://scholarship.law.nd.edu/cgi/viewcontent.cgi?article=1125&context=nd_naturallaw_forum

461 Lisa McClain, The Conversation, *How the Catholic Church came to oppose birth control,* July 9, 2018 https://theconversation.com/how-the-catholic-church-came-to-oppose-birth-control-95694

During the 19th century, reproductive knowledge improved, and in the 20th century, contraceptive technology made a giant leap forward with the birth control pill in 1960, which was an instant success by 1962. But not for the Vatican, which petitioned the UN and was granted observer status in 1964 in order to oppose use of the pill. Thus the Pope and his appointed supporters began their war against contraception with unlimited power in the UN to veto proposed legislation.

McClain points out, "By the 20th century, Christians in some of the most heavily Catholic countries in the world, such as France and Brazil, were among the most prodigious users of artificial contraception, leading to dramatic decline in family size. As a consequence of this increasing availability and use of contraceptives by Catholics, church teaching on birth control—which had always been there—began to become a visible priority."

With no scripture to guide them, devout Catholics wanted solid permission from their priests to use the pill. McClain notes, "Church leaders confronted the issue head-on. Some leaders believed the church could not know God's will on this issue and **should stop pretending that it did**, as Dutch Bishop William Bekkers said outright on national television in 1963. The question was left for consideration by the Pontifical Commission on Birth Control, held between 1963 to 1966. This commission by an overwhelming majority—**a reported 80 percent**—recommended the church expand its teaching **to accept artificial contraception.** That was not at all unusual. The Catholic Church had changed its stance on many controversial issues over the centuries, such as slavery, usury (lending money at an exorbitant interest) and Galileo's theory that the Earth revolves around the sun. Minority opinion, however, feared that to suggest the church had been wrong these last decades would be to **admit the church had been lacking in direction by the Holy Spirit.** Paul VI eventually sided with this minority view and in 1968 issued 'Humanae Vitae,' prohibiting all forms of artificial birth control."

Pope Paul VI felt the need to preserve church authority, but an outcry ensued from both priests and laypeople. They could not sway the pope. Eventually, millions of Catholics worldwide decided to simply ignore this outdated ruling. The year 2018 marked the 50th anniversary of the landmark "Humanae Vitae," and Pope Francis continues to uphold this strict and tenuous prohibition against artificial contraception.

Vatican in the United Nations

If humanity expects to flourish into the 21st century, it must take action with its intellect rather than its emotional and religious paradigms that prove outdated, outmoded and irrelevant.

—Bob Woodruff, ABC journalist, Earth 2100 (www.earth2100.tv)

Vatican City is the smallest country in the world—smaller than an 18-hole golf course. Yet, it functions as one of the most powerful political entities within the UN system. Since 1964, the Holy See (Vatican City) has held permanent observer status at the United Nations, the only world religion allowed this special privilege to influence policy. However, according to Catholics for Human Rights (CHR), the Vatican DOES NOT qualify as a state, and therefore should not legally be allowed to participate as a member of the UN.

The CHR states, "According to the criteria codified at the 1933 Montevideo Convention on the Rights and Duties of States, Article 1, a state must have a defined territory, a government, the ability to enter into relations with other states and a permanent population. None of the entities involved, the Holy See, Vatican City or the Roman Catholic church, possess all four of these attributes that define a state."[462] The population of Vatican City is about 800 people, and only 450 live there permanently with citizenship. For years, organizations have argued that the Holy See should not hold the status of permanent observer and that the UN should be a site of secular decision-making. Alternatively, some have asserted that Christianity and Catholicism should not be given preferred religious status to the exclusion of other religious entities. They claim that this privilege is **discriminatory.**

The UN also enjoys a special status, free of accountability for condoning these male-dominated crimes and behaviours, where women have never had a voice in governance or authority. The CHR explains, "Few governments are willing to comprehensively challenge the Holy See or to hold it accountable for not adequately complying with the human rights obligations that come with UN participation. Nor have they held the Holy See to account for human rights violations it—and the individuals who represent it—have perpetrated or suppressed information about. This, of course, is of particular interest given **charges against it of torture and crimes against humanity** in terms of sexual violence and the detail that has finally surfaced in recent days (and months and years) about clergy and sexual abuse of children, as well as sexual assault of nuns."

Yet, the Holy See is particularly active in the arenas of women's rights and rights of children, rights related to sexual orientation and gender identity and expression, and to contraception and sexuality education. In fact, the CHR notes that overall, it has tenacious interest in areas of sexual and reproductive rights and health. In addition, every Pope who has reigned since the Holy See received its permanent observer status has been allowed to address the U.N General Assembly to influence policy regarding these issues, and is allowed full diplomatic relations with 177 countries in the UN. Its reach is broad, and its command, given its global influence, sometimes coercive.

Its presence and its ideological grounding, the CHR adds, affect the development of and discourse about human rights standards. In certain areas, and particularly in relation to gender and sexuality, this influence serves to limit rather than enhance protections; in many instances, the Holy See's positions further entrench discrimination, allow violence, and serve to deny information and services. Women and young people often bear the brunt of the Holy See's efforts to integrate its ideological commitments into the UN human rights system.

> *The Roman Catholic church pursues disastrous policies worldwide. At the recent world conferences on women and population development, the Catholic Church successfully led the effort to block the inclusion of safe, legal abortion on the list of basic reproductive rights for women.*
> **—National Secular Society, Britain**

462 Catholics for Human Rights, *Challenging the Holy See at the United Nations*, March 14, 2019, https://www.womensordination.org/blog/wp-content/uploads/2019/03/Catholics-for-Human-Rights-Report-revised.pdf

The National Secular Society (NSS), Britain's only organization working exclusively towards a secular society, maintains that the Catholic Church uses and abuses its unique position at the UN as the only religion allowed to vote, and deserves a voice which is no louder than any other NGO. The NSS states, "Each year 5.8 million people become HIV positive and 2.5 million die from AIDS. Within the UN, the Roman Catholic church attempts to block international policy decisions that would make condom education and use a major tool in the prevention of HIV/AIDS."[463]

The role of the Vatican in the UN has been under scrutiny for decades, with many questioning how the Catholic Church wangled its way into the UN and why millions of both people and animals are dying as a result. Should the Vatican be held accountable for ecocide as well as crimes against humanity? Why was the Holy See granted the veto power to undermine the overwhelming consensus of member states?

The Holy See owes its participation in the UN since 1964 to an accident of history, due to confusion regarding use of the terms Holy See and Vatican City. Contrary to some claims, the Holy See was not invited to participate in the UN as a permanent observer. Pope John Paul II confirmed that the Holy See invited itself into the UN. The NSS emphasizes, "No vote has ever been taken on the Holy See's presence at the UN by the General Assembly. The Holy See's membership in the UN agencies such as the International Atomic Energy Agency—which allowed it to qualify as a Non-member Permanent Observer—was also not subject to vote by the general conference."

In fact, NSS notes that in October 1944, the Pope inquired of U.S. Secretary of State Cordell Hull what the conditions for membership would be for the future United Nations. Hull replied that, "the Vatican would not be capable of fulfilling all the responsibilities of membership." The Holy See also requested to be admitted to the League of Nations, but according to NSS was denied due to concerns about its statehood status and the possibility it would **have undue influence on the votes of Catholic member states.**

The NSS adds, "Because UN conferences operate on consensus, the ability to disagree with the majority consensus has significant power. The official documents of the recent UN conferences on women and population and development are replete with 'objections' by the Vatican to the majority consensus." This misuse of power has a name—**tyranny**.

Given its role at the UN, these official objections, entered formally into the final report of the conference, serve to weaken support for the conclusions of the majority. Moreover, they represent sectarian religious positions, not governmental public policy positions. The NSS notes that this is exactly what the Vatican and its handful of allies—nations such as Libya, the United States, and the Sudan that do not support full human rights for women—intend.

The activities of the Roman Catholic Church have been detrimental to women throughout its history. NSS explains, "From decrying emergency contraception for women who had been raped in Kosovo to burning boxes of condoms as AIDS ravages Africa, the hierarchy of the Roman Catholic church has allowed outdated doctrinal concerns to take priority over the lives of real people. Nowhere is that more evident than in the UN, where the Holy See insists on foisting its limited and

---

463 National Secular Society, *Catholic Church abuses its position at the UN*, Feb 4 2004, https://www.secularism.org.uk/32964.html

largely rejected view of gender, sexuality and reproductive health on a world intent on creating a more progressive personal ethic that is respectful of the common good."

Although many world religious leaders continue to discriminate against women and promote sexual abuse, this is not the intent of the religion itself, but misguided leaders. In addition, no other religion is given veto status at the UN, and allowed to undermine political policy. Although the Catholic religion itself (and many priests, nuns, and practicing Catholics) often support family planning, the Vatican does not. As a representative of the Roman Catholic Church, the pope refuses to give up this outdated and unpopular stance. Therefore, in this chapter, I will be primarily addressing this unyielding position of the Vatican's dominance over international relations.

Why the Long Silence on Population?

Dr. Martha M. Campbell from Berkeley School of Public Health tells us why the tripling of the world's population growth since 1960 has received little public attention the past decade. Numerous reasons for the silence around this subject constitute a **"perfect storm"**: visibility of actual fertility decline in the developed countries, as well as a number of the developing ones; well-justified attention to the impact of high levels of consumption on the environment; the tragedy of AIDS dominating international health concerns; and the 1994 Cairo conference's focus on examples of coercive family planning while nearly ignoring the coercion of women forced into unwanted child-bearing.[464]

At this 1994 UN conference, the Vatican insisted that the **population subject be removed** from the global policy discourse, which served to break the population/environment connection and alienate environment groups. In an attempt to mitigate the strategy of abandoning population, the UN decided to focus on the health and education of girls instead. This was a major victory for the Vatican in their campaign against family planning, and a major disaster for the planet. It was not until the 2019 UN conference that "population and family planning" were again considered acceptable terms, after **decades of opportunities to achieve UN targets had been squandered**.

The Catholic position not only thwarts a woman's right to control her own reproduction, but also blocks the most important weapon in protecting the earth from overpopulation. We need realistic public policy that promotes family planning and safe sex, not one designed to capture votes from the religious right. It is absurd to build a population policy based upon the notion that older teens and adults will abstain from sex, especially in many countries where the choice is not theirs. Why is the UN allowing a religious tyrant to steer global policy?

Our children will live in a much better world if human population growth is checked by the rational decision to reduce family size, rather than by famine, epidemics, and war. So where is our international and national leadership on this issue? For four years, the Trump administration worked to dismantle both national and global population control efforts, cutting funding to these programs over the issues of abortion and birth control in deference to the Catholic Church.

---

464 Dr. Martha M. Campbell, Research Gate, *Why the silence on population?*, May 2007, https://www.researchgate.net/publication/226701482_Why_the_silence_on_population

Similarly, when President Bush ran and won with a pro-life Christian platform, he actually declared war against the poor and working women all over the world, who would suffer most from lack of family planning programs largely funded by the U.S. By now it is ominously clear that the Vatican's position is simply criminal and this destructive influence must be stopped. It is only the UN that is in the position to do this by revoking the Vatican's unjustified observer status privilege.

Support for reducing population to a sustainable level is almost unanimous from the science community. Every credible scientist from Jane Goodall to David Suzuki, and including the Union of 1,600 Concerned Scientists worldwide, has come to the realization that we must bring environmentally damaging activities under control to restore and protect the integrity of our life-support systems. By defying this tenet, the **Vatican is committing the crime of ecocide**.

Church Crime

> *At the root of #ChurchToo stories are patriarchy, male leadership coupled with female submission, purity culture, evangelical personality cult culture, lack of sex-positive and medically accurate sex education, homophobia, and white supremacy. Commit to dismantling these things and addressing these root causes in your faith community today.*
> —Emily Joy, leader of the #ChurchToo movement

It is not only the Vatican's stance on contraception that is a crime, but the list goes on to include sexual abuse, slavery, corruption, forced religious conversion, immigration manipulation, use of religious bias in foreign aid, torture, ecocide, and crimes against humanity.

In the 21st century, the #MeToo and #ChurchToo movements have forced a recognition that power imbalances such as those in spiritual relationships can breed abuse. Christian patriarchy in particular is riddled with harmful views about gender roles, and encourages power dynamics that contribute to sexual abuse of women and children. With the rising tide of woman empowerment, men often fear they are losing control and turn to the church to justify this male dominance as God's will. However, this toxic thinking is inconsistent with the justice system the rest of society must abide by. By joining the church, these men receive impunity, and by going to confession instead of court they are instantly forgiven. Church is the perfect hide-out and social network for sexual offenders.

On the website BishopAccountability.Org, Bill Frogameni notes, "What happens if a Catholic priest molests children? Usually, he's protected by the Church hierarchy. Maybe he'll eventually have his parish or diocese taken away, or be switched to another one—often after years of serial abuse. But there's a good chance he'll stay in the Church."[465]

"Nearly 5000 Catholic priests (in the US) have sexually abused over 12,000 Catholic children ... but they were not excommunicated," says Father Roy Borgeois. Grim tales have also emerged regarding the Irish Catholic Church, making news headlines globally. For years, they ran hellish

465 Bill Frogameni, BishopAccountability.Org, *The Vatican's Perverted Sense of Justice*, April 2009 http://www.bishop-accountability.org/news2009/03_04/2009_03_04_Walsh_CourtTells.htm

work camps under the guise of reform schools, where sadistic priests were given free rein over the children. In 2020, **is it finally time to stop equating religion and morality?**[466]

In light of the #ChurchToo movement, many women have accused high-profile church leaders of sexual abuse, causing them to resign, but usually without criminal process. Until we can dismantle the systems that gave male church leadership impunity and unjustified power over women and children, this sexual violence will continue. Fortunately, groups like Catholics for Choice and Catholics for Human Rights are supporting legal action against church officials who have enjoyed absolute impunity in the past.

In 2019, Catholics for Human Rights commented, "The scope of the abuse was so massive that in September 2011, the U.S.-based Center for Constitutional Rights (CCR) and the Survivors Network of those Abused by Priests (SNAP) filed an 80 page complaint, accompanied by 22,000 pages of supporting material, to the International Criminal Court (ICC) in The Hague asking for the investigation and prosecution of Pope Benedict XVI and three top Vatican officials for abetting and covering up the rape and sexual assault of children by priests. The language of impunity is important here: challenging impunity is one of the main principles of human rights. The International Criminal Court has jurisdiction over war crimes, crimes against humanity and genocide committed after July 1, 2002, the year the court opened. It is independent of the United Nations and has jurisdiction in the 123 countries that have ratified the statute that created the court. To this day, neither the **U.S. nor Vatican City** have signed or ratified the Rome Statute that created the Court. **These are not innocent omissions**. In 2013, a prosecutor of an international court opted not to pursue the case, claiming that the matter did not fall within the jurisdiction of the ICC. The Holy See benefitted from this either as a politically motivated decision, or one which failed to recognize the relationship between widespread sexual assault and abuse and crimes against humanity. Between those two rationales, the Holy See was "let off the hook."[467]

The UN provides a platform for the Vatican's coalition building, including with governments in Egypt, Saudi Arabia, and the U.S. Numerous other government and religious entities have also united with the Vatican in a worldwide counter-revolution, designed to paralyze progress in women gaining control over their bodies and fertility. With the Vatican's vast wealth and influential UN position, attention is now focused on how humanity will rein in this ruthless world power. Or perhaps will that collapse come from within the church itself?

Besides Catholic churches losing worshippers due to their unpopular worldviews, they are losing priests at a remarkable rate as well. *TC Times*

> *It [the Vatican] advocates against language that allows rights related to abortion, it condemns the terms "reproductive rights" or "sexual rights, and argues against support of rights related to sexual orientation and gender identity.*
> **—Catholics for Human Rights**

466 The Conversation, *The Catholic Church's grim history of ignoring priestly pedophilia – and silencing would-be whistleblowers,* Oct 9 2018, https://theconversation.com/the-catholic-churchs-grim-history-of-ignoring-priestly-pedophilia-and-silencing-would-be-whistleblowers-102387

467 Catholics for Human Rights, *Challenging the Holy See at the United Nations,* March 14, 2019, https://www.womensordination.org/blog/wp-content/uploads/2019/03/Catholics-for-Human-Rights-Report-revised.pdf

reporter Vera Hogan states, "Catholic priests have been on the decline for a number of years as an aging priest population isn't being replaced by younger priests. Sociologists have pointed to several factors, which Pope Francis also pointed out, like the celibacy requirement, smaller family size, and the sexual abuse scandals that plagued the church for so many years. Another issue is women not being allowed to become Catholic priests."[468]

According to economist.com, Pope Francis suggested to a German magazine in 2018 that he would be open to the idea of allowing married men to become priests. This would not actually be a break from tradition, since *Future Church* notes that throughout the history of Christianity numerous Popes have married and had families: Apostle St. Peter; St. Felix III, 483–492 (two children); St. Hormidas 514–523 (one son); St. Silverus (Antonia), 536–537; Hadrian II, 867–872 (one daughter); Clement IV, 1265–1268 (two daughters); and Felix V, 1439–1449 (one son). In fact, the Eastern Catholic Church still has married priests.[469]

Nowhere does the New Testament explicitly require priests to be celibate. However, Catholic canon law maintains that if a priest were to have a family there would be the potential for conflict between his spiritual and familial duties.

The Church Manipulating Migration

But it isn't just this spate of outdated decrees that are of concern, as the Vatican's power over migration in the United Nations and within the Christian community also allows for manipulation of migrants through global channels of the Vatican's choice. As the world watches the refugee crisis unfold on global news, we witness record numbers of people fleeing violence, a growing proportion of the world's poor facing political and economic turmoil, and climate change creating increasing chaos and international migration. This crisis sets the stage for the power elite to set the agenda for immigration to wherever is seen as most advantageous to them. If Catholics in developing countries can be convinced to produce a never-ending supply of Catholic migrants, the unemployed and unwelcomed in their homelands can then be distributed to areas where worshippers are waning.

For example, Mexico with the second-largest number of Catholics in the world and an unsustainably high population, has seen a mass wave of emigration to the U.S. With an unsupportive political system and a shortage of jobs, emigration was an escape valve for the teeming millions.

Migration Policy reporter David Fitzgerald comments, "The desire to cooperate across borders has also come from the Vatican, particularly following the Second Vatican Council in 1962. Both the Mexican and U.S. branches have followed the Vatican in enthusiastically taking up the global rights revolution demanding an expanding list of protections and privileges based on universal rights of personhood.... Most importantly, the **Catholic Church** has become one of the major

468 Vera Hogan, TCTimes, *Worldwide shortage of priests is growing,* Nov 5, 2018, https://www.tctimes.com/news/worldwide-shortage-of-priests-is-growing/article_181a0f9a-dec9-11e8-b8aa-2fc9e6fc645a.html

469 Future Church, *A Brief History of Celibacy in the Catholic Church,* https://www.futurechurch.org/brief-history-of-celibacy-in-catholic-church

interest groups in the United States and around the world promoting a more accommodating stance toward immigration even when such positions have been politically unpopular."[470]

At first, afraid the Mexican emigrants would lose faith once in U.S., the Catholic Church provided a system in the US to ease their transition and provide places of worship. Fitzgerald notes, "The Church's decision to help those who decide to migrate has arguably eased the path for travelers. Along with travel agents, human smugglers, immigration lawyers, and labor brokers, Catholic organizations offering refuge along the border are important nongovernmental actors that enable international migration." This strategy also increases the number of Catholics in the most powerful country in the world—a perfect position for the Vatican.

How the church confronts migration leads to two questions, Fitzgerald explains: how its policies have shaped migration between Mexico and the United States and how its policies have affected government policy. Although church and government don't always share the same goals, they use similar techniques to govern a mobile population. They both tried to document the amount of collective remittances emigrants had and how they were spent. The Catholic Church continues to be one of the major voices for comprehensive immigration reform in the United States.

Fitzgerald states, "An end to remittances would be a financial disaster for many parishes. One of the proposals to the Mexican government in a meeting between Mexican Church officials and then President Vicente Fox in 2003 was to reduce remittance transfer fees, a goal the Mexican government already shared." The Vatican continues to influence governments globally, especially through the United Nations. They also encourage Catholics to continue worship in their receiving country and ensure that payments to families left behind would include a tithe to the church. Instead, governments should examine the root causes of migration—such as overpopulation, endless wars, climate change, poverty, and loss of traditional habitats—and seek long-term solutions.

Conversion, Coercion, and Terror

Of course, in many countries this loyalty to the church may not be guaranteed, since the emigrants may have been victims of forced conversion and may return to their original religion once they have settled abroad. Wikipedia defines forced conversion as "adoption of a different religion or irreligion under duress. Someone who has been forced to convert may continue, covertly, with the beliefs and practices originally held, while outwardly behaving as a convert."[471] Anthropologists report of missionaries often using coercive methods to convert those they were supposedly helping.

Throughout history, rapport between religion and politics has varied greatly, supporting or opposing one another to various degrees. They have battled for power, cooperated on dubious schemes to generate financial gain, and joined forces to fight plagues or disasters.

The battles for power often involved forced conversion, as Wikipedia notes, "Pope Innocent III pronounced in 1201 that if one agreed to be baptized to avoid torture and intimidation, one

470 David Fitzgerald, Migration Policy, *Uncovering emigration policies of the Catholic Church in Mexico*, May 21, 2009, https://www.migrationpolicy.org/article/uncovering-emigration-policies-catholic-church-mexico
471 https://en.wikipedia.org/wiki/Forced_conversion

nevertheless could be compelled to outwardly observe Christianity. During the European colonization of the Americas, forced conversion of the continent's indigenous, non-Christian population was common, especially in South America and Mesoamerica, where the conquest of large indigenous polities like the Inca and Aztec Empires placed colonizers in control of large non-Christian populations. Historians broadly agree that most native populations that converted did so under the threat of violence, often because they were compelled to after being conquered, and that the Catholic Church cooperated with civil authority to achieve this end."

These conversions continue today in many countries, often using foreign aid in the form of food, water wells, or other necessities as a bargaining chip. Other times it is just by brute force. India has experienced an ongoing plague of terrorism in the name of Christianity, with the terrorists often finding support from missionary groups. In 2009, the *Assam Times* reported that a group of Christian militants calling themselves the Manmasi National Christian Army tried to force Hindu residents of Assam to convert to Christianity.

Swarajyamag reporter Kampanna states, "There has never been a clear focus on the terrorism by adherents of Christianity. Indeed, there is almost a clear all-encompassing silence about terrorism perpetrated in the name of Christianity or terrorist outfits that have found support from Christian groups. Whether it is due to the alleged closeness between Church groups and our mainstream media or it is due to other reasons is a point for people to ponder over."[472]

Kampanna added that the most telling statement in the above link is the remark of the Archbishop of Delhi in a church event, which states: "Inaugurating the two-day event, Most Rev Anil Couto, Archbishop of Delhi, spoke of the indispensability of the media in our evangelizing mission, for carrying out the command of the Lord to proclaim the good news to the ends of the earth in season and out of season."

The Global Security website concedes that in India, the National Democratic Front for Bodoland (NDFB) is mostly Christian. NDFB, being strongly backed by Christian missionaries of the West, carried out large-scale religious conversion of the Bodos, to the extent that the rich culture of the Bodos was threatened. Atrocities by the various factions of the NDFB that perpetrate terror in the name of Christianity can be found in the South Asia Terrorism Portal (SATP) records.

Kampanna adds, "The Hindu Bodos were isolated and some even forcefully converted under the influence of the missionaries-backed NDFB. Due to such activities, NDFB was able to attract significant funds from the west and pleased missionaries that they had M-16 guns with them. Even the ULFA who raised much more funds from Assam never had M-16 guns till very recent days. Under influence of the Christian West, they have tried to de-Indianise their language by forcing people to adopt the Roman script for their language. On 19 August 2000, **NDFB killed Bineshwar Brahma, the president of the Bodo Sahitya Academy**, for using Devanagari script and refusing to adopt the Roman script for the Bodo language."

When it comes to forced conversion, the gripping headlines of the hundreds of unmarked graves of indigenous students found at church-run residential schools in Canada in 2021 have caused

472 Kampanna, Swarajyamag.com, *Terror in the Name of Christ in Northeast*, January 7, 2015, http://blogs.swarajyamag.com/2015/01/07/terror-in-the-name-of-christ-in-northeast/

outrage around the world. When similar Christian schools in other countries are investigated, these numbers are sure to run into the thousands. This cultural genocide prompted the hunt for similar finds in the United States in 2021, but the families of these abused and murdered students, some as recent as the 1990s, may never get justice for these atrocities.[473]

Political Ties to Catholic Aid

But it isn't just the media supporting the Christian abuses abroad, as the U.S. government is also greatly invested. In 2019, *ProPublica* reporter Yegeneh Torbati questioned how then Vice President Mike Pence's U.S. office meddled in foreign aid to reroute money to favoured Christian groups. Officials at USAID warned that favouring Christian groups in Iraq with funds totalling more than $265 million since 2015 could be unconstitutional and inflame religious tensions. After all, it is common knowledge that faith groups don't just deliver meals or heal the sick, but try to convert people to their religion. Also, Bashar Warda, a powerful archbishop based in Iraq, was a key figure in this effort.

Torbati notes, "U.S. officials urged the U.N. in the summer of 2017 to pay special attention to the Nineveh Plains, an ethnically and religiously diverse region of northern Iraq where many of the country's Christians live.... Pressure from Washington built.... Influential religious groups like the Knights of Columbus and current and former Republican members of Congress advocated throughout 2017 for direct U.S. aid to religious minorities, including Christians and Yazidis."[474] The UN was compliant.

However, USAID regulations state that awards "must be free from political interference or even the appearance of such interference and must be made on the basis of merit, not on the basis of the religious affiliation of a recipient organization, or lack thereof." Although USAID's inspector general is now investigating U.S. activity in Iraq, Torbati points out that additional USAID religious programming is also planned for Lebanon, Morocco, and Tunisia. Of course, increasing Christian foreign aid was promoted throughout Trump's presidency.

Origins of Church Wealth

It is clear that the Vatican is now in a panic to preserve their Catholic dominance, especially since the percentage of Catholics worldwide has been declining over the last 20 years—even in the two countries with the largest number of Catholics, Brazil and Mexico. In fact, one in ten adults in the United States is a former Catholic, according to the Pew Research Center's 2009 report "Faith in Flux."[475]

---

473 Ian Austen and Dan Bilefsky, New York Times, *Hundreds More Unmarked Graves Found at Former Residential School in Canada*, July 30, 2021, https://www.nytimes.com/2021/06/24/world/canada/indigenous-children-graves-saskatchewan-canada.html

474 Yegeneh Torbati, ProPublica, How Mike Pence's Office Meddled in Foreign Aid to Reroute Money to Favored Christian Groups, Nov. 6, 2019, https://www.propublica.org/article/how-mike-pences-office-meddled-in-foreign-aid-to-reroute-money-to-favored-christian-groups

475 Pew Research Center, *"Nones" on the Rise*, October 9, 2012, https://www.pewforum.org/2012/10/09/nones-on-the-rise/

Is this change of heart partly due to online media coverage of atrocities committed by the Catholic Church throughout history? It isn't only the sex scandals that have tarnished the Vatican's image. According to Wikipedia, the first extensive shipment of black Africans in what would later become known as the transatlantic slave trade, was initiated at the request of Catholic Bishop Las Casas and authorized by Charles V, Holy Roman Emperor in 1517. Later the Episcopal Church—which describes itself as Protestant, yet Catholic—was deeply involved in the slave trade in the 18th century. Tiny Rhode Island played an outsized role in the trade, thanks to the state's financiers, a seafaring work force, and officials who turned a blind eye to antislavery laws. This criminal behaviour by the Christian community was never reckoned with.

New York Times reporter Katharine Seelye notes, "One of the darkest chapters of Rhode Island history involved the state's pre-eminence in the slave trade, beginning in the 1700s. More than half of the slaving voyages from the United States left from ports in Providence, Newport and Bristol—so many, and so contrary to the popular image of slavery as primarily a scourge of the South, that Rhode Island has been called 'the Deep North.' That history will soon become more prominent as the Episcopal diocese here, which was steeped in the trans-Atlantic slave trade, establishes a museum dedicated to telling that story."[476]

"I want to tell the story," Bishop Knisely said, "of how the Episcopal Church and religious voices participated in supporting the institution of slavery and how they worked to abolish it. It's a mixed bag." Bishop Knisely said his research had revealed shameful episodes in church history. For example, he said, when Quakers and Baptists in Newport began turning against slavery, some slave owners in those churches switched to the Episcopal Church, where they were welcomed and their slaveholding was not challenged. "We were happy to receive their financial support," Knisely added.

Numerous churches supported slavery, and some profited from it even after slavery had been banned in the state. Seelye states, "Among the most notable Episcopalian slaveholders were Thomas Jefferson, who was active for some time in the church, and George Washington." Because of dwindling membership, the Rhode Island Cathedral closed in 2012, and funds were raised to convert it into a museum and reconciliation centre to improve racial relations. These horrendous crimes of slavery by Catholics and other Christians were never accounted for.

Was it this centuries-old blood money that helped build the Vatican's significant wealth? Among the Vatican's many assets are the network of Catholic cathedrals and churches throughout the world, and a valuable artwork collection from artists like Michelangelo and Raphael. *All Things Finance* reporter Tamila McDonald states, "Along with those assets, the Vatican's net worth also includes a large investment portfolio. They have billions of shares in numerous corporations. Some of the most recognizable ones

> *More than half of Americans say they often feel a deep connection with nature and the earth (58%), while more than a third classify themselves as "spiritual" but not "religious" (37%).*
> **—Pew Research Center**

476 Katharine Q. Seelye, NY Times, *Rhode Island Church Taking Unusual Step to Illuminate Its Slavery Role*, Aug. 23, 2015, https://www.nytimes.com/2015/08/24/us/rhode-island-church-taking-unusual-step-to-illuminate-its-slavery-role.html

include General Motors, General Electric, Chase-Manhattan, Shell, and Gulf Oil."[477] It appears that climate change doesn't concern these jet-setters.

Yet, exactly who funds and influences the big business of the Vatican remains an enigma. How does the wealthy and powerful Vatican in turn manage to influence world governments even outside of the UN arena? We do know that the Vatican has big investments in banking, insurance, chemicals, steel, construction, and real estate. Unlike ordinary stockholders, the Vatican pays no taxes on this income, even though the Vatican's net worth has been estimated at $10 to $15 billion. In any other country, this tax evasion would be a crime.

CNN News reporter Ahiza Garia points out, "The Vatican Bank, which has about $8 billion in assets, has often been at the center of scandal and corruption since it was founded in 1942. Pope Benedict began the process of cleaning the bank up, and Francis has continued that work. Vatican Bank accounts are only supposed to be held by residents of Vatican City and church personnel. But according to Gerald Posner, a Vatican bank scholar and the author of *God's Bankers*, these accounts were often awarded to powerful Italian officials looking to stash money without paying taxes. The bank closed over 4,000 accounts to weed out corruption and currently has a total of 33,400 accounts." It is little wonder that whole books can be written about crimes of the Vatican.

In 2020, crime in the Vatican continues. Vatican officials said an investigation into suspicious transactions at the Vatican bank could be widened to include "additional individuals." *US News* reporter Philip Pullella states, "Vatican police on Tuesday raided the department in charge of maintenance and restoration at St. Peter's Basilica, seizing documents and computers for an investigation into suspected corruption. The raid was similar to one last October that involved another investigation into a separate department over the purchase of a building in a posh area of London."[478]

Pope Francis last month (2020) set up a commission, including Harvard law professor Mary Ann Glendon, to review the Vatican bank and propose reforms, where necessary. The Vatican bank serves thousands of Catholic charities, religious orders, and dioceses around the world. Perhaps this is how the Vatican has become the most powerful corporate force in the United Nations working against achieving a sustainable global population. The Vatican and the Trump administration have also been working closely with international anti-abortion groups like the Chris and Marie Smith Gospel of Life Ministries, C-Fam, and U.K.'s Society for the Protection of Unborn Children (which had been active in the International Family and Pro-Life Conference in Nigeria) to thwart efforts for effective family planning policies.[479]

Yet, this diabolical **hindrance** would be the easiest removed. All it would take to stop the Vatican from having veto power in the UN would be for the UN officials to revoke this special privilege given the Vatican in 1964, a privilege which was not approved by member states and is being sadly abused. This in turn would reduce the Vatican's influence on the World Bank, World

---

477 Tamila McDonald, All Things Finance, *The Vatican and Pope Francis' Net Worth*, March 2018, https://allthingsfinance.net/the-vatican-and-pope-francis-net-worth/

478 Philip Pullella, US News, *Vatican Police Carry Out New Raid Over Suspected Corruption*, June 30, 2020, https://www.usnews.com/news/world/articles/2020-06-30/vatican-police-carry-out-new-raid-over-suspected-corruption

479 Kathryn Joyce, Pacific Standard, *The New War on Birth Control*, Aug 17, 2017, https://psmag.com/magazine/new-war-on-birth-control

Health Organization, countless other UN agencies, and individual governments. It would also contribute to a decreased opportunity for religious terrorism to occur in poor countries depending on religious-dominated agencies for aid. Fortunately in 2020, Catholicism lost more members than it gained at a higher rate than any other denomination.

Postchristianity, according to Wikipedia, is the loss of the primacy of the Christian worldview in public affairs, especially in the Western world, in favour of alternative worldviews. These would include other religious ideologies, secularism, nationalism, environmentalism, militant atheism, and the newly recognized religious movement of ethical veganism.[480] The fastest growing numbers are for the "no affiliation" faction, especially among millennials, with more than a third of all U.S. millennials claiming to be "nones."

This postchristianity can be witnessed globally as churches close at an unprecedented rate. Even rising Catholic migrant numbers haven't been enough to reverse the trend. Decreasing attendance, dwindling money in the coffers, and rising maintenance costs are causing thousands of churches to be abandoned or repurposed around the world.

Solutions and Successes

> *Religious leaders of all dominations must embrace the future, and the challenges it presents, not merely cling to the past—or risk being seen by society as redundant. They must focus on the evidence and what it says about our situation, as well as the moral good.*
> —Julian Cribb, author of *Surviving the 21st Century*

It can be said that the **rapidly shifting view on world religion is one of the most important social changes of the 21st century.** Today, a rising number of citizens are speaking out to demand that religion be kept out of politics, a growing force of protesters are demanding that the Vatican's observer status at UN conferences be revoked, and powerful entities have united to provide funding for contraception services. People are speaking out, and changes are being made.

A large majority of Americans feel that religion is losing influence in public life, according to a 2019 Pew Research Center survey. Nearly two-thirds of Americans in the new survey are resoundingly clear in their belief that religious institutions should stay out of politics. Three-quarters of the public expresses the view that churches should *not* come out in favour of one candidate over another during elections, in contrast with efforts made by President Trump during his time in office to roll back existing legal limits on houses of worship endorsing candidates.[481]

---

480 Wikipedia, *Postchristianity*, https://en.wikipedia.org/wiki/Postchristianity

481 Pew Research Center, *Americans Have Positive Views About Religion's Role in Society, but Want It Out of Politics*, November 15, 2019, https://www.pewforum.org/2019/11/15/americans-have-positive-views-about-religions-role-in-society-but-want-it-out-of-politics/

Religious Networks—Taking Action

Delivering family planning services and education to those who most need them may seem a daunting task, but a powerful infrastructure is already in place that can help us increase access to contraception: the global faith community.
—Rev. Canon Grace Kaiso and Dr. Ahmed RA Ragab

In Africa, social change has gone a step further, as the clergy has committed to champion quality sexual and reproductive health across the continent. **Faith to Action Network** supports faith actors to empower people to live healthy, peaceful, quality lives. They declare, "As a global interfaith network of more than 100 Bahai, Buddhist, Christian, Confucian, Hindu and Muslim faith organisations, we focus on issues that faith actors are grappling with, including sexual and reproductive health and family planning; gender equality and women's rights."[482]

They promote partnership building to create bridges, and provide faith organizations with access to small grants to finance innovative family planning projects while still respecting the core tenets of their faith. Examining their successes provides a valuable blueprint for faith-based solutions.

Perhaps this powerful interfaith network will have the courage to stand up to the elite growth pushers—the economic, religious, and cultural forces that oppose reproductive services and population stabilization. Clearly, we must rally all forces to raise the modest sum to provide universal family planning access to all. *The Standard* reports, "At a convention held in Nairobi, the faith leaders highlighted achievements saying **they were no longer excluded from key health decisions affecting the continent**."[483]

"**Before**, we were regarded as **obstacles** to realising the aspirations of family planning among other sexual and reproductive health programs, the notion has changed," Reverend Grace Canon Kaiso said. "We **now** are **at the centre** of most development blueprints for Africa and the world."

The Standard notes that at the convention, the clergy also shared advocacy successes and interventions they use to manage the rising number of teenage pregnancy and school dropout cases in the country. In Zimbabwe, GRACE Foundation offers the homeless information on family planning. The Evangelical Association of Malawi takes *family planning messages* to pulpits in churches with supportive scriptures from the Bible aiming to **end the firmly held misconception about family planning.** Elsewhere, Muslim Family Counseling Services (MFCS) in Ghana is celebrating their advocacy effort that led to clinical family planning methods included in the **national health insurance** scheme. While in Rwanda, besides the Anglican church of Rwanda working to combat (gender-based violence), they too work with the government and CSOs to advocate for family planning.

In Germany, Bread for the World is part of the global ecumenical movement that is committed to enhancing peace, environmental protection, and engagement against gender-based violence.

---

482 Faith to Action Network, *Who we are*, https://www.faithtoactionnetwork.org/who-we-are/

483 The Standard, *Clergy commit to champion quality sexual and reproductive health*, Aug 14 2019, https://www.standardmedia.co.ke/article/2001338129/clergy-commit-to-champion-quality-sexual-and-reproductive-health

These organizations are proving that a lot of progress can be made when the Vatican is not allowed to undermine their efforts. Developed countries could also learn from these successes.

International organizations are increasingly looking to develop partnerships with faith-based organizations to promote development objectives, according to a special issue of *The Ecumenical Review* presented at a World Council of Churches (WCC) meeting in Geneva.[484]

"This marks a significant milestone in global conversations around the nexus between religion and international affairs," writes Azza Karam, coordinator of the UN Inter-Agency Task Force on Religion and Development, in an article on **"The Role of Religious Actors in Implementing the UN's Sustainable Development Goals."**

There is a growing realization today that we need to get all sectors on board to realize the UN's 17 development goals. In 2020, almost all religious leaders understand that they have a moral obligation to promote family planning as a primary means of accomplishing these goals, as all of the goals hinge on Goal #5—providing gender equality and contraception services.

The Guardian contributors Rev. Canon Grace Kaiso and Dr. Ahmed RA Ragab point out, "As men of faith, we share a deeply held conviction that all families—regardless of their religious beliefs—are entitled to lead healthy lives free from suffering and deprivation. But, tragically, an absence of basic family planning services deprives millions of people of this fundamental right every year. More than 200 million women worldwide lack access to modern contraceptives."[485] Kaiso and Ragab add that an estimated 40 percent of health services in sub-Saharan Africa are provided by faith-based organizations, ensuring that condoms and contraceptive pills reach rural communities.

> *Because of this influence, faith leaders worldwide have an unparalleled opportunity – indeed, a moral obligation – to prioritise conversations about family planning and close the contraception gap.*
>
> **—Rev. Canon Grace Kaiso and Dr. Ahmed RA Ragab**

According to the UN Development Programme, for every **$1 spent on family planning, governments can save up to $6 for other development priorities**. There is also wide agreement among global experts that people who have access to family planning information, services, and supplies are likelier to complete their education, live more prosperous lives, and raise healthier children.

In the Muslim world, many imams, such as the grand sheikh of Al-Azhar, spread information about family planning to their followers, encouraging the use of long-acting reversible contraception methods such as IUDs and contraceptive implants. Moreover, in 2012, 200 Indonesian imams came together to approve **vasectomy**, which was previously prohibited. These religious leaders have declared that **no text in the Qur'an specifically prohibits these family planning methods.**

484 World Council of Churches, *Ecumenical Review focuses on role of religion in development*, April 27 2017, https://www.oikoumene.org/en/press-centre/news/ecumenical-review-focuses-on-role-of-religion-in-development

485 Rev Canon Grace Kaiso and Dr. Ahmed RA Ragab, The Guardian, *Leap of faith: why religious leaders have a moral duty to promote family planning*, Mar 4 2019, https://www.theguardian.com/global-development/2015/oct/15/family-planning-religious-leaders-leap-of-faith

"Many faith-based organisations are already taking a leading role in the promotion of family planning in developing countries, while respecting the core tenets of their faith. Examining their successes provides a valuable blueprint for faith-based solutions," state Kaiso and Ragab. This is in stark contrast to the inexcusable role of the Vatican.

Holding the Vatican Accountable

Outrage and protests have finally influenced the Vatican's decision to end the "secrecy" rule on child sexual abuse in the Catholic Church. In 2020, Pope Francis declared that pontifical secrecy will no longer be an excuse for church officials to refuse to share information, as he used his authority to rewrite specific articles of canon law or parts of previous papal documents. "This is an epochal decision," Archbishop Charles Scicluna of Malta, the Vatican's most experienced sexual abuse investigator, told Vatican Radio.[486]

> *The veil of secrecy which surrounded these abominable crimes and which prevented victims obtaining justice and reparation has been lifted.*
> **—UN Human Rights Council**

The Guardian reports that at a summit on sexual abuse held at the Vatican in 2019, church leaders demanded zero tolerance for sexual abusers and that victims not be obligated to silence. They argued that secrecy in cases of child sexual abuse was outdated and that some church officials were hiding behind it instead of cooperating with authorities. This is also true of the Vatican's law against family planning, and the ultimate victory will come when this law too has been rewritten to reflect the beliefs of the majority of humanity, including Catholics.

Survivors of abuse want Francis to do more and make bishops who allegedly cover up abuse accountable as well. Ms. De Boer-Buquicchio, the UN Special Rapporteur on the sale and sexual exploitation of children stated, "We have **victims to thank** for the courage to speak out on this devastating issue, but the burden of addressing this evil should not fall on them alone. The world is waiting for States and the Church to live up to their duty to end this scourge. Actions must follow words."[487]

So, it would appear that the Vatican, as well as the more religiously compliant countries it controls, is holding out the strongest religious opposition to achieving a sustainable population. The residents of these countries, often living in poverty and being less educated, are more easily controlled and terrorized than those in industrialized countries where family planning programs are available and more widely accepted.

Efforts to Strip the Vatican of its Power to Block Consensus in the UN

- In 1995, Catholics for a Free Choice (CFFC) initiated a petition asking the United Nations to reconsider the status of the Holy See. Hundreds of NGOs from around the world signed

486 The Guardian, Vatican City, *Pope ends 'secrecy' rule on child sexual abuse in Catholic church* , Dec 17, 2019, https://www.theguardian.com/world/2019/dec/17/pope-francis-ends-pontifical-secrecy-rule-child-sexual-abuse-catholic-church

487 UN News, *Rights expert welcomes Pope's decision to lift 'veil of secrecy' on child abuse cases,* Dec 19 2019, https://news.un.org/en/story/2019/12/1054031

the petition, along with approximately 2,000 individuals who signed the petition at the Fourth World Conference on Women and the NGO Forum.

- In 1996, Mouvement Ecologie-Demographie in Belgium forwarded a petition, or "Call to the UN organization to reconsider the status of the Holy See," sponsored by USPDA (case postale CH-3052, Zollikofen, Switzerland) along with various women's organizations, including (surprisingly) the National Coalition of American Nuns.
- In 1999, CFFC launched the "See Change" campaign to change the Holy See's status at the UN. This initiative included an international postcard campaign to the secretary-general of the UN and was endorsed by a large coalition of women's, religious, and reproductive rights organizations. It stated in part, "Demands of the Vatican that the edicts of "Humanae Vitae" be followed are made to ensure survival of the principle of infallibility and the institution of the papacy. The price of these demands in human terms—in suffering, illness and premature death—**is so great as to make such demands immoral**. On these grounds, we believe the 'non-member state permanent observer status' of the Holy See in the United Nations, as well as its similar relationships with specialized agencies in the UN system, is inappropriate and should be terminated as soon as possible."
- In 1999, the editor for the Women's International Network wrote, "The Vatican opposed the very organization of UNFPA (the United Nations Fund for Population Activities) at the 1974 Population Conference in Bucharest, Romania, where members of the Vatican disguised as representatives of various institutions, lobbied against every positive move to make contraception available. The day before the Conference opened the Vatican delegation called a special press conference to announce that the meeting should be canceled as the subject was not fit to discuss at an international meeting. Since then the Vatican has continuously opposed UNFPA and its programs for women."
- In 2000, the government of Brazil, the world's largest Catholic country, blasted the local Roman Catholic Church for opposing condom use to prevent the spread of AIDS. The National Bishop's Association reiterated its stance against Catholics using condoms, despite internal opposition from some clergymen.

As of 2020, the Vatican's dogma of papal infallibility has precluded any effective efforts to rescind or modify the ban on contraception imposed in 1968.

Atheists Rise with Millennials and Gen Z

Why believe in a god? Just be good for goodness' sake.
—American Humanist Association, Christmas time bus ad in Washington, D.C.

The 21st century has seen a lot of doors opening, and a lot of suppressed people coming out of the closet. The LGBTQ2 community is exercising more freedom of expression and receiving greater acceptance than ever before. Vegans and environmentalists are less often labelled as "granola nut bars" or "tree huggers." Many women around the world are beginning to demand their rightful status and equality in the workplace. Many survivors of abuse by priests, celebrities, coaches, medical staff, and so on are now speaking out with the #MeToo and #ChurchToo movements. And finally, atheists are coming out of the closet to shrug off the outdated cloak of secrecy.

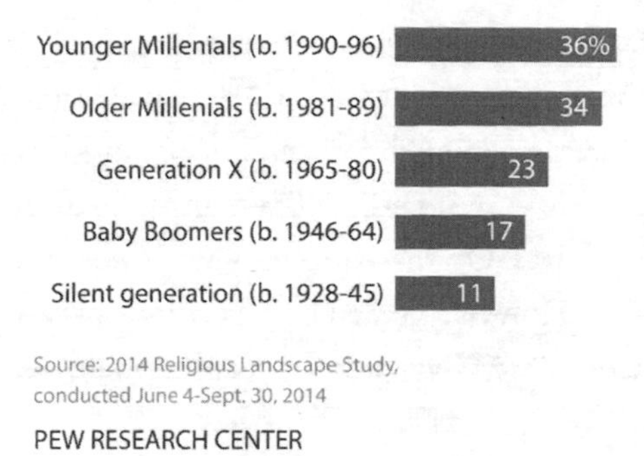

Figure 11. Pew Research Center

A new secular confidence is transpiring, and this remarkable trend reflects **today's most significant change—that of atheists emerging** as the portents of a science-based ideology.

Could religion be going the way of the dodo? Consider this: Faith-based surveys are finding that the rise in the "unaffiliated" category is occurring remarkably fast, about the same speed as population and climate change are accelerating. What if the response just shows that young people are realizing that it is science-based solutions that are needed here? No amount of prayer is going to solve these critical world issues for us—it is in our hands to take urgent immediate action to tackle these imminent threats. The onus is on humanity to change the course of history.

For half a century, there has been a rising chorus from women, elders, the scientific community, and more recently from youth to draw our attention to a spate of disregarded concerns. Since the gods have not proven to remedy any of these problems, it has been predicted that religion will lose significance in the scheme of things. This prophecy is now coming to pass.

National Geographic reporter Gabe Bullard tells us that the world's newest major religion is: **No Religion.** He states, "The religiously unaffiliated, called 'nones,' are growing significantly. They're the second largest religious group in North America and most of Europe. In the United States, nones make up almost a quarter of the population."[488]

> *In the past decade, U.S. nones have overtaken Catholics, mainline protestants, and all followers of non-Christian faiths.*
> **—Gabe Bullare,** *National Geographic*

As secularism grows, we are thinking differently about what we teach our kids, how we vote, how we treat Earth, the role of religion in politics, and how we think about death. Bullard notes, "France will have a majority secular population soon. So will the Netherlands and New Zealand. The United Kingdom and Australia will soon lose Christian majorities. Religion is rapidly becoming less important than it's ever been, even to people who live in countries where faith has affected everything from rulers to borders to architecture."

However, in some developing regions like Africa, where people often depend on missionaries and religious aid groups to survive, the poor often succumb to forced Christian conversion and

488 Gabe Bullard, National Geographic, *The World's Newest Major Religion: No Religion*, April 22, 2016, https://www.nationalgeographic.com/news/2016/04/160422-atheism-agnostic-secular-nones-rising-religion/

religion is on the rise. As this population becomes more independent and better educated, this trend will likely reverse.

Michael Lipka with the Pew Research Center points out that their massive 2014 Religious Landscape Study makes clear just how quickly this trend is happening—across genders, generations, and racial and ethnic groups. Religious "nones" now make up **roughly 23 percent of the U.S. adult population.** Lipka states, "Overall, religiously unaffiliated people are more concentrated among young adults than other age groups—35% of Millennials (those born 1981-1996) are *Nones*."[489]

World Population Review notes, "While there is a trend in fewer people believing in God around the world, only a few countries have more than 20% of citizens who are **atheists**. The six countries with the highest percentage of their populations identifying as atheists are China, Japan, Czech Republic, France, Australia, and Iceland.[490]

Approximately 40 to 49 percent of China's population says that they're atheists. This is the highest population of atheists in the world. Confucianism, which is one of China's oldest philosophical systems, is notable for its lack of a belief in a deity.

The rise of atheists in America has been attributed to an increased exposure to diverse perspectives in multicultural societies, which challenges the claim of any one particular worldview. Millennials and Gen Z are the most racially, ethnically, and religiously diverse generation in the U.S. We desperately need their ingenuity and logic to mend our overpopulated, politically fractured, environmentally wounded planet. Also, if we could stop the Vatican from derailing progress on family planning, this would truly bring the United Nations into the 21st century—and align the decision-making process with the environment and the greater good, not religious affiliation.

489 Michael Lipka, Pew Research Center, May 13, 2015, https://www.pewresearch.org/fact-tank/2015/05/13/a-closer-look-at-americas-rapidly-growing-religious-nones/

490 World Population Review, *Most Atheist Countries*, 2020, https://worldpopulationreview.com/country-rankings/most-atheist-countries

PART 4

POPULATORS—THE NEXT GENERATION

You can recycle everything that you can get your hands on for the rest of your life.... but you can't possible recycle as much as an unintended human will consume.
—Dr. Doug Stein, Florida urologist, in the 2013 documentary *The Vasectomist*[491]

At this fork in the road, the next generation of populators will need to confront the difficult ethical and environmental questions surrounding family planning. But more than that, the questions they will be facing about the future of humankind will be critical to our collective well-being. The 21st-century populators will be on a seemingly impossible mission to lower population numbers, and with only a window of ten years to show results. If they should succeed, they will not only give other species a better chance at life, but also rescue humanity. Additionally, we are at a crossroads of another kind. Although the exploitation of children has occurred throughout history, we have now **reached a pivotal time** that demands that humanity take action to end these atrocities we are inflicting on today's children by 2030—if Earth's inhabitants are going to survive and prosper.

The Sustainable World Coalition makes this **dedication**: "Indigenous human cultures around the world celebrate the sacredness of nature. They have developed richly diverse spiritual ways of life that honor the laws of nature, which rule this beautiful world. We, the divine humans upon sacred Earth, now find ourselves at a **prophetic crossroads** with our profound planetary destiny. Each of us brings to this immense moment all of our **beloved ancestors, and seven generations of those not yet born.** We focus our lives and resources to foster and perpetuate a nature-friendly planetary community in the arenas of religion, commerce and politics. All women—the life-givers—are Mother Earth's feminine presence. Feminine leadership and values are essential now to balance the masculine. Thus we renew and celebrate our sacred relationships with mother nature, our most precious gift. We do so for all children of all creatures—to ensure that they will inherit a healthy, happy and peaceful world from us. We can't overestimate the importance of understanding the big issues confronting humanity."[492]

491 Screen Australia, *The Vasectomist*, 2013, https://www.screenaustralia.gov.au/the-screen-guide/t/the-vasectomist-2013/31724/
492 Sustainable World Coalition, *Sustainable World Sourcebook*, 2014, https://olyclimate.files.wordpress.com/2013/09/sustainable-world-sourcebook-2014.pdf

Today we face unprecedented global challenges in a world reeling from environmental, social, political, and economic crises—a result of unsustainable population growth. We have seen why, after decades of dire warnings and alarming scientific findings, overpopulation remains the most urgent, yet most neglected problem on the planet. Part 4 explores who the next generation is, what motivates them, and what challenges they face. Then it goes a step further, and looks at how they are adapting and working to change global priorities. It shines a light on the many solutions, and points a way to prosperity.

CHAPTER 12
AT A GENERATIONAL CROSSROADS

The problems in the world today are so enormous they cannot be solved with the level of thinking that created them.
—Albert Einstein

The majority of world scientists, the United Nations. and numerous enlightened thinkers agree that the 21st century, and especially the coming decade, is crucial for the planet and all species that make Earth their home. British astronomer Sir Martin Rees told the *Globe and Mail*, "For the first time, a single species—our own—carries the future of the planet in its hands. The choices we make in the next few decades could lead to dramatically different outcomes for our descendants and the rest of life on Earth.... The world's population is higher than ever and we are, each of us, more demanding of energy and resources and possibly pushing the planetary environment and climate over tipping points."[493]

Rees advises that the problem is to motivate politicians to use some of their goodwill and capital to make hard trade-offs on issues which are long term and where the main beneficiaries may be people who live decades in the future and in distant parts of the world. He believes that the youth activism we are seeing on climate change is important because "politicians care about what's in their inbox and what's in the press."

The youth are the future populators of this planet, and our future relies on them to make better decisions in terms of family planning than previous generations have. In response to this momentous defining 2020 decade, the United Nations has created the 17 Sustainable Development Goals to guide us. Goal #5 for gender equality and access to contraception is a pivotal goal for the others, and the basis for making meaningful progress.

The Rumah Kitab site tells us, "It [goal #5] asks that governments, civil society organizations, religious and community leaders, and families work together to essentially re-evaluate the worth of their girls. The economic case for change is powerful. By limiting girls' education and lifetime earning potential, child marriage may be costing trillions of dollars, according to a **report by the World Bank and the International Center on Research for Women**, which was released in 2017.

493 Ivan Semeniuk, Globe and Mail, *21st century crucial for humanity and the planet: Sir Martin Rees*, Sept 29, 2019, https://www.theglobeandmail.com/canada/article-21st-century-crucial-for-humanity-and-the-planet-sir-martin-rees/

The report looked at child marriage in 25 developing countries where at least a third of women marry before age 18."[494]

Quentin Wodon, a World Bank lead economist and the report's co-author, emphasizes the importance of keeping a girl in school and delaying marriage to reduce population, and therefore reduce poverty. "Investments in adolescent girls tend to have very high economic returns," he says. That's "not the most important reason to end child marriage—the moral argument is—but these economic returns are very useful to convince various policymakers to invest in ending the practice."

If you have high population growth, it's very difficult to provide quality services for everybody, whether it's for school ... or whether it's for health services or even basic infrastructure.
—Quentin Wodon, World Bank economist

Now, at this youth-focused crossroads, the fate of millions of girls may depend on a family transaction that includes their marriage, or sale into slavery. In this 2020 decade of change, we must decide to empower our girl children and raise them out of this dismal despair. At the same time, our young men must be protected from slavery, whether for work or as child soldiers, and together our youth will pave the way for a better future.

Can Generations X, Y, and Z Save Us?

About half the modern world doesn't have the same basic amenities the ancient Romans took for granted.
—Peter Gleick, world-renowned scientist, Pacific Institute in California

Sir Martin Rees believes that young people should be focusing their attention on science: "Science is part of our culture and it's the one truly global culture which straddles all boundaries of nationality and faith. But also, so many of the decisions that have to be made have a scientific element, whether it's in the area of health, climate, energy ... all these things. In order for citizens to be able to vote responsibly, they have to have an awareness of science, and enough of a feeling for numbers not to be bamboozled."

Tom Brokaw coined the term the Greatest Generation as a tribute to Americans who lived through the Great Depression and then fought in WWII. The silent generation refers to seniors 70 years of age and older in 2020 who as a rule were not as vocal as following generations, as their focus was finding a job and working hard. The Baby Boomers are the hippie generation who were born during the Baby Boom after WWII and grew up during great social upheaval, the Beatles, and the Vietnam War. They symbolized the love children idealists, demonstrators, and the first vegans, and they started the environmental movement. I am happy to be part of that generation, which kept me both physically and mentally active, and was not yet totally dependent on technology.[495]

494 Rumah Kitab, *The Worth of a Girl*, April 30, 2020, https://rumahkitab.com/category/news/page/2/

495 Iberdrola website, From the baby boomer to the post-millennial generations: 50 years of change https://www.iberdrola.com/talent/generation-x-y-z

Characteristics of Generations X Y Z, Alpha and Gen C (born 1965 - 2020)

The Generations reporter Michael Robinson notes that it is important to note that the last three generations each brought one billion additional people with them, an alarming population growth rate that has had dire consequences on the planet.[496]

Gen Xers tend to be independent and self-reliant from growing up with two working parents; they were left to fend for themselves and became great problem-solvers. They also witnessed a period of radical technological evolution and the rise of online media.

Generation Y was labelled the Millennials, and they were digital natives. The Iberdrola website states, "Technology is part of their everyday lives: all their activities are mediated by a screen. The concept of on and off is completely integrated into their lives." The economic crisis forces them to be better trained, as job competition is increasing and they become more goal oriented.

Next came Generation Z, labelled as Centennials for being born at the turn of the century. Iberdrola notes, "They arrived with a tablet and a smartphone under their arms." A world of chaos and seven billion competitive consumers awaited them, most not able to cope with the fast rate of change.

Alpha Gen, into which Greta Thunberg fits, is the next generation. They are immersed in the digital world and give a strong voice to social causes. However, they may be inclined to neglect interpersonal relationships to a greater extent, as they take the world by storm with their social movements. In this sense, they are indeed martyrs for a just cause, but not always by choice. The latest generation is known as the Coronials (the Covid generation) or Gen C.

Despite today's social diversity, **Generations Y and Z predominate**. According to the study *New Kids On The Block*, today there are 2 billion millennials and 2.4 billion centennials, representing 27 percent and 32 percent of the world population, respectively. **They are the ones that will be voting in future governments.**

Since the new breed of Gen Z parents, the multi-taskers, have a short attention span, they are in the greatest danger of being distracted parents. They are the generation most shaped by the Internet, and have YouTube DNA in their blood. They lead the way on social activism both online and on the streets, but the 21st century will not make finding a job or raising children easy for them. *The Atlantic* reporter Erika Christakis points out, "Smartphones have by now been implicated in so many crummy outcomes—car fatalities, sleep disturbances, empathy loss, relationship problems, failure to notice a clown on a unicycle. Our society may be reaching peak criticism of digital devices ... yet **more than screen-obsessed young children**, we should be **concerned about tuned-out parents.**"[497]

The mind-numbing impact of being plugged into a screen is taking a toll on both children and parents in the form of low-quality relationships. We also know that fast-paced and violent imagery are damaging to anyone's brain. We Baby Boomers and Millennials have left the young parents of

496 Michael T. Robinson, Career Planner, *The Generations*, https://www.careerplanner.com/Career-Articles/Generations.cfm

497 Erika Christakis, The Atlantic, *The Dangers of Distracted Parenting*, July/August 2018, https://www.theatlantic.com/magazine/archive/2018/07/the-dangers-of-distracted-parenting/561752/

today a tough world to navigate, and a scary place to raise a family. In 2019, *The Economist* reports that Generation Z is stressed, depressed, and exam-obsessed. They find getting good grades is a bigger worry than drinking or unplanned pregnancies. They are more lonely than ever before and say YouTube is the number-one platform they turn to when they want to relax or cheer up.[498]

These challenges that our young people are facing should be a wake-up call, a warning that we need some bold changes if we are to create a bearable planet for future generations. It is not only Greta Thunberg and her generation of youth activists that are making a colossal change, but also the new social media who have the foresight to imagine the possibilities **factual journalism** could make—and the tenacity to expose the shortcomings of our traditional media.

> *Time spent on devices is time not spent actively exploring the world and relating to other human beings. The new parental-interaction style can interrupt an ancient emotional cueing system, whose hallmark is responsive communication, the basis of most human learning.*
> **—Erika Christakis,** *The Atlantic*

Younger Generations Are Procreating Less

Is it really such a bad thing that young parents are having fewer children? Dr. Charles A. Hall, a systems ecologist, noted, "Overpopulation is the only problem. If we had 100 million people on Earth—or better, 10 million—no others would be a problem." We are now nearing EIGHT BILLION. New research in 2020 from the University of Southampton has shown that large ecosystems that have existed for thousands of years, such as rainforests and coral reefs, can collapse at a significantly faster rate than previously understood. The findings suggest that ecosystems the size of the Amazon forests could collapse in only 49 years and the Caribbean coral reefs in just 15 years.[499]

Some scientists argue that many ecosystems are currently teetering on the edge of this precipice, with the fires and destruction both in the Amazon and in Australia. Lead author Dr. Simon Willcock of Bangor University emphasizes, "These rapid changes to the world's largest and most iconic ecosystems would impact the benefits which they provide us with, including everything from food and materials, to the oxygen and water we need for life."[500]

We are not just tourists here. We **must** take action to conserve and restore biodiversity. We also must accept that the era of growth is over—economic growth, population growth, and the myth of "green growth" can no longer benefit humanity, and we must start looking out of the box for the answers.

498 The Economist, *Generation Z is stressed, depressed and exam-obsessed,* Feb 27 2019, https://www.economist.com/graphic-detail/2019/02/27/generation-z-is-stressed-depressed-and-exam-obsessed

499 University of Southampton, EurekAlert, *Planet's largest ecosystems collapse faster than previously forecast,* Mar 10 2020, https://www.eurekalert.org/pub_releases/2020-03/uos-ple030620.php

500 Science Daily, March 10, 2020, *Amazon rainforest could be gone within a lifetime,* https://www.sciencedaily.com/releases/2020/03/200310124713.htm

Gen Z Embraces Science

This amazing planet is in fact finite. Its ability to provide food, absorb wastes, and accommodate humans is finite. It is becoming exceedingly clear that we have already surpassed many of nature's limits, and our future is uncertain. Our massive meddling with ecosystems that we know little about has released an ominous reign of terror over fellow species, many of whom may never recover. To rectify this situation, we must create a symbiotic alliance with nature and a deep appreciation for all other species that call this planet home. How does one make up for decades of abuse, and help recover the endangered South China tiger, Hawksbill sea turtle, or Asian elephant? We do know that Generation Z embraces science, and their protests and activism are starting to be felt around the world. From COVID-19, we know that people are able to make radical change if need be, and Gen Z is using this discovery to propel humanity into action. At this historic turning point, science will support them all the way.

In 2017, the Union of Concerned Scientists (UCS), over 1,600 of the world's most distinguished scientists, renewed their 1992 "Warning to Humanity." The UCS joins many other scientists, such as David Suzuki, Jacques Cousteau, and world science academies, in lamenting the irreversible loss of wildlife. They recognize that human beings and the natural world are on a collision course, and fundamental changes are urgently needed. The Baby Boomers and following generations have all ignored these warnings, but Gen Z is a different breed, embracing science and common sense, making up 32 percent of the world's population and the largest percentage of religious "nones." They are Earth's best hope.

Safeguarding Planet Earth

There are promising projects in the works today which Generations X, Y, and Z will want to monitor and safeguard with their lives if we are to rescue the planet in the next **ten years**. Whether putting aside half the planet for wildlife, assuring that the banking giants stop funding destructive energy projects, or advancing the initiative for billionaires to donate half their net worth to charity—the new generations will become sentinels of this rock we call Earth.

If we are to engage today's youth in re-inventing this troubled world, the Baby Boomers and Millennials will have to get on board as well and become role models. After all, since we created most of this mess, we have no right to just bail out now when things are heating up. We are all in this together, and we must unite in the fight of our lives.

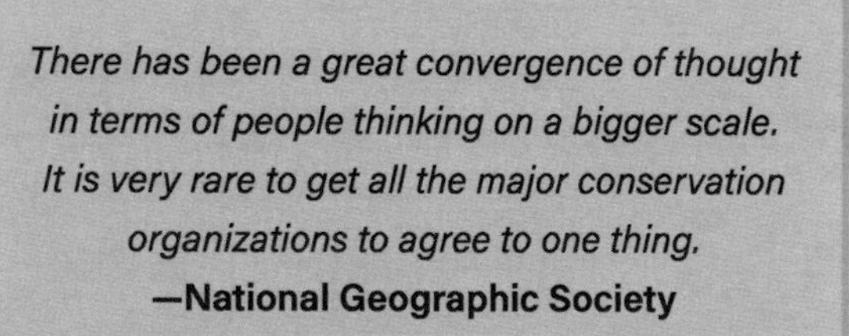

In 2009, Nature Needs Half was formally launched by the WILD Foundation at the 9th World Wilderness Congress in Mexico, boldly encouraging the conservation community to act publicly on what everyone was saying privately, **nature needs half.**[501]

501 Nature Needs Half, https://natureneedshalf.org/who-we-are/history/

E.O. Wilson's 2016 book *Half Earth* echoed the same sentiment, and in 2019, a joint statement was released by some of the world's largest conservation organizations to protect 30 percent of the planet for wildlife by 2030 and 50 percent by 2050. *National Geographic* reports that signatories of this statement posted in 2019 include BirdLife International, Conservation International, the National Geographic Society, the Natural Resources Defense Council, the Nature Conservancy, and nine other NGOs.[502]

This would be a radical change from today's 12 percent of earth protected for wildlife, and **will require a radical reduction in human population.** Youth-powered groups like Fair Start Movement, Youth for Wildlife Conservation, and Environmental Youth Alliance will be watching closely to see that these organizations and governments follow through on their targets, and will be playing a major role as part of the solution.

Supporters say that having an ambitious and clear target may help the crisis of biodiversity loss get the attention it deserves from governments and private institutions. In recent years, concern over climate change has captured more attention, although extinction of species and loss of habitat are clearly as devastating. Similarly, by setting a **clear population target** of stopping growth at EIGHT BILLION and then reducing towards sustainability, we can better provide space for recovering wildlife and their habitat. And we know that this number is attainable and must be met if we are to curtail this mass extinction. Studies have also shown that reducing human population woud produce greater reductions in climate change than moving to alternative energy—or any other solution, for that matter. To clearly illustrate this has been the work of this book.

In 2020, the *Guardian* is one of those vehicles for change, reporting with tenacity, rigour, and authority on the most critical events of our lifetimes. The *Guardian* notes that their editorial independence allows them to set their own agenda and voice their own opinions. Their journalism is free from commercial and political bias—never influenced by billionaire owners or shareholders. This is a different breed of reporting. It means they can challenge the powerful without fear and give a voice to those less heard. At a time when factual information is both scarcer and more essential than ever, they believe that each of us deserves access to accurate reporting with integrity at its heart.

The *Guardian*, Grist, The Optimist, and Live Kindly provide good examples of factual information. They aren't afraid to discuss the complexity of environmental and social issues that we face while providing audience-centred and out-of-the-box alternatives to media giants controlled by corporate interests. They use language that recognizes the severity of the crisis we're all experiencing, and share in the loss and sadness we are feeling.

This kind of honest reporting is desperately needed to expose the taboos and myths surrounding critical issues like population and animal agriculture, and set straight the fairy tales of "green energy," "green growth," and "green technology." We must come to the realization that green growth no longer exists, and technology is not going to bail us out of this crisis.

Co-founder of Greenpeace International, Rex Weyler, puts the solution in a nutshell: "Embracing the crisis with technology feels good; it means growing our economy, advancing, having *more*, not giving up

502 Emma Marris, National Geographic, *To keep the planet flourishing, 30% of Earth needs protection by 2030*, Jan 31, 2019, https://www.nationalgeographic.com/environment/2019/01/conservation-groups-call-for-protecting-30-percent-earth-2030/

anything. This blind spot remains our deep, unspoken problem. We want to solve the ecological crisis and the humanitarian crisis with economic growth and "technology," without changing anything. **We want it all.** The problem is, we're avoiding the root problem and the genuine solutions. **All genuine solutions to our ecological dilemma must include a contraction of human scale.**"[503]

Weyler emphasizes that this solution is presently ignored by most people, governments, corporations, and **even most environmental organizations.** This is where the young activists of today are using social networking and street demonstrations to get the message across. Since most long-standing eco groups have betrayed their members by refusing to acknowledge the overwhelming impacts of overpopulation and animal agriculture, the youth have started up new groups to fill this void. This is a terrible travesty, since the existing eco groups had the framework and media connections already in place, and the new activists could have learned so much from these established groups. Instead, most are hiding their heads in the sand, and still refuse to tackle the root causes, leaving new groups like Fair Start Movement, One Planet – One Child, Center for Biological Diversity, and Birth Strike to tackle the tough issues of the day. **It is this new blood that will approach looming crossroads with a plan mapped out and the motivation to shift gears.**

With EIGHT BILLION people in our face and forcing most of our fellow species over the cliff, for goodness sake, why can't we talk about this taboo issue of overpopulation and how we feel about it?? Isn't it time that our population crisis went viral? Weyler notes, "We must abandon the idea that we can take baby steps. We require a **radical bold vision** ... fundamental shifts, and unprecedented adjustments."

Child Brides, Child Soldiers, Child Slaves—Transformed

As children are better spaced through family planning, it has a tremendous impact on the health, the education and the general well-being of children. And as child health and education—particularly of girls—improves markedly, more families actively seek family planning and have smaller families.

—James P. Grant, executive director of UNICEF[504]

We may find that we are leaving our children and grandchildren a planet greatly diminished in resources, beauty, and social benefits. Will the growing concern for children's well-being lead to a global movement to stabilize population and improve their quality of life? It is this great surplus in children that has contributed to child brides, child soldiers, and child slaves. It is only by preventing the 50 percent of unintended pregnancies on the planet that we can eliminate these most horrific industries.

But there is another interesting advent at play here at this 2020 crossroads, as we face the exploitation of children in the 21st century. Humanity needs to educate and support future populators in

503 *Rex Weyler,* Greenpeace, *The end of infinite growth,* Nov 15, 2019, https://www.greenpeace.org/international/story/26506/the-end-of-infinite-growth/

504 James Grant, UNICEF, https://www.unicef.org/french/publications/files/Jim-Grant-LR.pdf

preparation for this historic transition. Child brides, child soldiers, and child slaves must be transformed into the vibrant and dynamic leaders of tomorrow.

Existing Threats to Children

The rich require an abundant supply of the poor.
—Voltaire

In 2020, *The Lancet* tells us, "Despite dramatic improvements in survival, nutrition, and education over recent decades, today's children face an uncertain future. Climate change, ecological degradation, migrating populations, conflict, pervasive inequalities, and predatory commercial practices threaten the health and future of children in every country. In 2015, the world's countries agreed on the Sustainable Development Goals (SDGs), yet nearly 5 years later, few countries have recorded much progress towards achieving them."[505]

The rights and entitlements of children, *The Lancet* explains, are enshrined within the UN Convention on the Rights of the Child (CRC) ratified by all countries, except the U.S. The realization of these rights is the only pathway for countries to achieve the SDGs for children's health and well-being, and requires decisive and strong public action. Children and young people are full of energy, ideas, and hope for the future. They are also angry at the state of the world, and are protesting about environmental threats. We must find better ways to amplify their voices and skills if we are to meet the 2030 SDG target.

In 2020, there are more than 2.2 billion children on Earth. Nearly two billion of these live in a developing country. The most fundamental rights of minors are violated in most parts of the world, and even in developed countries, many are victims of violence, abuse, or discrimination.[506] Since studies show that three out of four assaults go unreported to police every year, that means that more than half the children on the planet have experienced violence or abuse in the 21st century.[507]

The UN Sustainable Development Goals highlight some of the critical issues affecting today's children, including human trafficking, sexual assault, children in detention, abuse, exploitation, torture, child brides, child soldiers, infanticide, neglect, war, and all other forms of violence and exploitation. And unfortunately, this has become a worldwide epidemic disproportionately affecting our most vulnerable.

> *Although **billions** of dollars are spent taking care of teenage mothers and their children, **only millions** would be needed to provide good prevention programs.*
> **—Pregnant Teen Help**

505 The Lancet, *A Future for the World's Children*, Feb 22 2020, https://www.thelancet.com/journals/lancet/article/PIIS0140-6736(19)32540-1/fulltext

506 Humanium, *Children in the World*, 2020, https://www.humanium.org/en/children-world/

507 Rape, Abuse & Incest National Network (RAINN), *The Criminal Justice System: Statistics*, 2020, https://www.rainn.org/statistics/criminal-justice-system

WHO points out that experiencing violence in childhood impacts lifelong health and well-being. It also diminishes the quality of life of the whole family, community, and society at large. The world's youngest citizens suffer disproportionately from poverty, malnutrition, and disease. Also, their young, growing bodies are most vulnerable to widespread environmental hazards, such as air, soil, and water pollution.[508]

> *Globally, it is estimated that up to 1 billion children aged 2–17 years, have experienced physical, sexual, or emotional violence or neglect in the past year.*
> **—World Health Organization**

In its 1994 report, "The State of the World's Children," the United Nation's Children's Fund (UNICEF) emphasized that **population stabilization is a prerequisite** to ensuring the health and well-being of children everywhere, noting, "There is an obvious and profound connection between the mental and physical development of children and the social and economic development of their societies."

In its 2019 report, UNICEF states, "Despite progress in the past two decades, one third of children under age 5 are malnourished—stunted, wasted or overweight—while two thirds are at risk of malnutrition and hidden hunger because of the poor quality of their diets. At the center of this challenge is a broken food system that fails to provide children with the diets they need to grow healthy."[509]

Could this lack of progress be because we have failed to address the underlying population crisis, adding another billion to our overburdened planet every 12 years—which equals to **adding a million people per hour** every day of every year? In fact, despite the billions of dollars spent to improve the poverty and abuse issues, the situation has actually gotten worse. One type of violent threat that WHO is concerned about can often be brought on by misguided use of the internet. Enough Is Enough points out, "When children go online, they have direct and immediate access to friends, family, and complete strangers, which can put unsuspecting children at great risk. Children who meet and communicate with strangers online are easy prey for Internet predators."[510]

Often, we have an image of sexual predators lurking around school playgrounds or hiding behind bushes scoping out their potential victims, but the reality is that today's sexual predators search for victims while hiding behind a computer screen, or are actually friends and family members.

Child Sexual Abuse in the 21st Century

Research conservatively indicates that over one-third of children will be sexually victimized before adulthood. We know that ongoing sexual abuse of the vulnerable is a form of sexual slavery, and each year, millions of pregnancies have resulted from these abusive sexual encounters around the world. In 2020, the Guttmacher Institute states, "Adolescents in LMICs [low- and middle-income countries] have an estimated 21 million pregnancies each year, 50% of which are

---

508 World Health Organization, *Violence against chidren,* June 8 2020, https://www.who.int/news-room/fact-sheets/detail/violence-against-children.

509 UNICEF, United Nations, *The State of the World's Children,* 2019, https://www.unicef.org/reports/state-of-worlds-children-2019

510 Enough is Enough, *Internet Safety 101,* 2020, https://internetsafety101.org/internetpredator

unintended."[511] This doesn't include the millions of pregnancies occurring in wealthier countries, and sexual abuse resulting in pregnancy only compounds the indignity and difficulty.

Child pornography offences have exploded by more than 200 percent in the last decade, a trend that advocates say is fueled by the easy spread of illicit material over the Internet. Technology, in particular the Internet and cellphones, has enabled sex trafficking and sexual exploitation to become the fastest-growing criminal enterprise in the world. The increasing misuse of technology is changing the nature of trafficking, and the public must work with authorities to address it. UNICEF warns that what fuels Internet predators is easy and anonymous access to children, risky online behaviour of youth, sketchy law enforcement, and easy access to child pornography.

Equality Now notes that 92 percent of all child sexual abuse URLs identified globally in 2016 were hosted in these five countries: Netherlands, United States, Canada, France, and Russia. (The U.S. ranked second.) Internet safety is now the fourth top-ranked issue in the list of health concerns for American children, according to the C.S. Mott Children's Hospital.[512]

According to the West Virginia Department of Health, the United States has the highest rates of teen pregnancy and births in the western industrialized world. Teen pregnancy costs the United States at least $7 billion annually.[513]

With so many unplanned pregnancies, abortion is also an important consideration when teens think about their future plans. Many teens do not want to enter parenthood alone, and they often do not feel that they can afford it. This kind of responsibility at such a young age is often overwhelming for any parents. However, abortion has been a taboo subject for years and the cause of ethical, political, and medical debates around the world. Fortunately, the new abortion pill is making it much easier for teens to find solutions to this problem in recent years. However, preventive action through safe contraception, vasectomy, and tubal ligation are by far the wiser choices.

We are advised by the UN that **solutions** must be global and multi-dimensional and supported by actors including governments, tech companies, civil society, and UN agencies. They must be informed by the experiences and perspectives of survivors. This is a global effort, as the Internet, human trafficking, and the rape epidemic have affected almost every country on the planet.

Ending the Child Rape Epidemic

This is a global reality today, made worse by COVID-19 restrictions and lockdowns, as girls are trapped at home with their abusers and often in fear of their lives. CNN reporter Bukola Adebayo notes the worsening situation in Nigeria, "The UN says the raft of measures deployed by governments to fight the pandemic have led to economic

If a sexually active teen does not use contraceptives, there is a 90 percent chance that she will become pregnant inside a year.
—Guttmacher Institute

511 Guttmacher Institute, *Adding It Up: Investing in Sexual and Reproductive Health 2019*, July 2020, https://www.guttmacher.org/report/adding-it-up-investing-in-sexual-reproductive-health-2019#

512 Equality Now, *The Rise of Technology in Trafficking and Sexual Exploitation*, 2020, https://www.equalitynow.org/stop_tech_exploitation

513 West Virginia Department of. Health & Human Resources, *Sex Has Consequences, The National Campaign to Prevent Teen Pregnancy*, https://www.wvdhhr.org/appi/psa/pages/facts.html

hardship, stress, and fear—conditions that lead to violence against women and girls. The police force says it has now deployed more officers to its stations across the country to respond to the 'increasing challenges of sexual assaults and domestic/gender-based violence linked with the outbreak of the Covid-19 pandemic.' And last week, governors across the country resolved to declare a state of emergency on rape, according to the Nigerian Governor's Forum (NGF)."[514]

Many of these terrorized rape victims end up pregnant or have contracted HIV, adding untold suffering and enormous financial debt for the country to cope with. Federal and state authorities are coming out with a united voice to condemn gender violence, which validates the outcry of women in the country and the scale of the problem in Nigeria, and highlights a widespread global rape epidemic.

According to the Guttmacher Institute, by the age of 19, 70 percent of U.S. teenagers have engaged in sexual intercourse. For the majority of these who are underage, this is defined as statutory rape, family rape, and sometimes date rape. So, parents have a unique window of opportunity, and a responsibility, to educate their children about sex and what qualifies as rape at an early age. While it can be uncomfortable, parents need to talk to them about contraceptives and sexually transmitted diseases—and what statutory rape is— and encourage safe and consensual sex.[515]

When one looks at the teen sex statistics, it is clear that most teenagers have sex, despite the intense moralizing by conservatives and the religious factions around the world. It also appears that education, the prevention of rape, and access to family planning options can help delay unwanted teen pregnancy and prevent the need for abortion.

UN Women points out that our rape culture is pervasive and shockingly common, and is always rooted in patriarchal beliefs, power, and control. They state, "**Rape Culture** is the social environment that allows sexual violence to be normalized and justified, fueled by the persistent gender inequalities and attitudes about gender and sexuality. Naming it is the first step to dismantling rape culture."[516]

> *It's in our hands to inspire the future feminists of the world. Challenge the gender stereotypes and violent ideals that children encounter in the media, on the streets, and at school.*
> **—UN Woman**

We must not justify rape by saying, "Boys will be boys," or "She was drunk," or other such platitudes. UN Woman adds, "Rape culture is allowed to continue when we buy into ideas of masculinity that see violence and dominance as 'strong' and 'male,' and when women and girls are less valued." So it is essential that we redefine masculinity. We know that there is far more to men than the violent stigma presently attached. Men need to have the confidence to portray their more gentle and nurturing

514 Bukola Adebayo, CNN, *She's on the frontline of a rape epidemic,* June 23 2020, https://www.google.com/search?client=firefox-bd&q=She%27s+on+the+frontline+of+a+rape+epidemic.

515 Guttmacher Institute Fact Sheet, *Contraceptive Use in the United States,* April 2020, https://www.guttmacher.org/fact-sheet/contraceptive-use-united-states

516 UN Women, *16 ways you can stand against rape culture,* Nov 18, 2019, https://www.unwomen.org/en/news/stories/2019/11/compilation-ways-you-can-stand-against-rape-culture

side without the fear of being discriminated against. There is no need to perpetuate the macho image with violent sports, hunting, war, rape, and so on.

We must all establish policies of zero tolerance for sexual harassment and violence in the spaces in which we live, work, and play. Leaders must be particularly clear that they are committed to upholding a zero-tolerance policy and that it must be practised every day. **Governments must also commit to ending impunity.** To end rape culture, perpetrators must be held accountable. By prosecuting sexual violence cases, we recognize these acts as crimes and send a strong message of **zero-tolerance**.

Trafficking of Children

According to the Anti-Slavery Society, the international trafficking of children is a growing phenomenon. The children are kidnapped or purchased by slavers, and sold into slavery in sex dens or as unpaid domestic servants. They may become **child brides or child soldiers**, or sold to brokers to become **child slaves**. Their lives are at the mercy of their masters, and suicide is often the only escape. Most of the surviving young girls end up pregnant, contributing further to their suffering and to the overpopulation crisis.

In the 21st century, human trafficking has replaced the drugs trade as the world's largest illegal trade, and is now a multibillion-dollar industry. The reason for this is likely that drugs can only be sold once. The Organization for Security and Cooperation Europe spokesman Valiant Richey states about the rate of prosecution, "It's a drop in the ocean. The current rate of prosecution means most traffickers never spend a day in a courthouse, let alone a prison cell. This needs to change. **Countering impunity and establishing a strong rule of law** must be a fundamental cornerstone in the fight against trafficking in human beings."[517]

UNICEF's Kul Gautum at International Symposium on Trafficking of Children in Tokyo believes that a combination of poverty, globalization, organized crime, and discrimination against women encourage the trade.

Child Labour and Incarceration

Every single day, tens of millions of children work in conditions that shock the conscience.... We must wipe from the earth the most vicious forms of abusive child labor.
—President Clinton

Is slave labor acceptable just because you create the slaves yourself?
—Madeline Weld, Population Institute Canada

Clinton spoke these words at an International Labor Organization conference in Geneva seeking to end the worst forms of child labour—slavery and bondage, prostitution and pornography, drugs and hazardous work. But since that time, little has changed. It has become imperative that we accelerate efforts to achieve the Sustainable Development Goal Target to eliminate child

517 ReliefWeb International, *Prosecute human traffickers and deliver justice to victims*, 20 Jul 2020, https://reliefweb.int/report/world/prosecute-human-traffickers-and-deliver-justice-victims-osce-alliance-conference

labour and slavery by 2030. Alliance is a global partnership for eradicating forced labour modern slavery, and human trafficking around the world.[518]

World Vision states, "As many as 152 million children aged 5 to 17 around the world are engaged in child labor, working in jobs that deprive them of their childhood, interfere with schooling, or harm their mental, physical, or social development. Nearly half of them—73 million children—work under hazardous conditions, such as carrying heavy loads on construction sites or digging in open-pit mines. By definition, child labor is a violation of both child protection and child rights."[519]

Poverty, says World Vision, is the primary reason children are sent to work. But sadly, child labour keeps children from getting the education they need to break the cycle of poverty. We also know that this poverty is compounded by the alarming rate of population growth that is undermining efforts to reduce child labour. According to the International Labor Organization, a U.N. agency, more than **two-thirds of child labourers are involved in family agriculture**. Others work long hours in factories or in domestic service. Millions of children are in forced labour, including child soldiers, child brides, and children exploited in the commercial sex trade.

The Anti-Slavery Society believes that childhood should be a time to grow, learn, play, and flourish in safety. But millions of children around the world don't have that chance. Instead, they wake up each day trapped in modern slavery. The society explains, "Some have been forced into back-breaking work in mines, brick kilns, sweatshops and private homes. Others are forced into sexual exploitation, begging or selling drugs. And it happens closer than we would like to think—children in the UK are often just as vulnerable to trafficking or exploitation."[520]

On the Ivory Coast of Africa, hundreds of thousands of small farms have been carved out of the forest. This is where about two-thirds of the world's cocoa supply comes from. *Washington Post* reporters Peter Whoriskey and Rachel Siegel tell us, "According to a 2015 U.S. Labor Department report, more than **2 million children** were engaged in dangerous labor in cocoa-growing regions. When asked this spring [2019], representatives of some of the biggest and best-known brands—Hershey, Mars and Nestlé—could not guarantee that any of their chocolates were produced without child labor."[521]

Antonie Fountain, managing director of the Voice Network, an umbrella group seeking to end child labour in the cocoa industry, said, "We haven't eradicated child labor because no one has been forced to. What has been the consequence ... for not meeting the goals? How many fines did they face? How many prison sentences? None. **There has been zero consequence.**"

The roots of the problem lie in poverty linked to overpopulation, the low price paid to cocoa farmers, children as young as ten being abandoned or overworked by their own desperate parents, child trafficking, corrupt governments, and lack of enforcement of both local and international law.

"Child labor in the cocoa industry will continue to be a struggle as long as we continue to pay farmers a fraction of the cost of sustainable production .... Fairtrade isn't a perfect solution," said Bryan Lew, chief

518 Global March Against Child Labor, *Alliance*, 2020, https://globalmarch.org/alliance-8-7/

519 World Vision, *Child labor: Facts, FAQs, and how to help end it*, 2020, https://www.worldvision.org/child-protection-news-stories/child-labor-facts

520 Anti-Slavery Society, *Child Slavery*, 2020, https://www.antislavery.org/slavery-today/child-slavery/

521 Peter Whoriskey and Rahel Siegel, Washington Post, *Cocoa's Child Laborers*, June 5, 2019, https://www.washingtonpost.com/graphics/2019/business/hershey-nestle-mars-chocolate-child-labor-west-africa/

operating officer for Fairtrade America. But, he said, the higher prices for certified cocoa and the group's efforts to organize farmer cooperatives are steps toward alleviating its root cause: **poverty**.

Paul Schoenmakers, a Tony's company executive, noted, "**Nobody needs chocolate**. It's a gift to yourself or someone else. We think it's absolute madness that for a gift that no one really needs, so many people suffer." Like the blood diamonds nobody really needs, this travesty should make one reconsider our society's priorities, traditions, and impacts. At this 2020 crossroads, it is the moment to vote with our dollar, examine these societal factors, and make change.

In 2020, the cocoa industry, which makes an estimated $103 billion annually, claims to have made major strides in eradicating child labour, including building schools, buying more ethical "certified" cocoa, and supporting agricultural cooperatives. Hershey's Cocoa For Good program promises to invest a half-billion dollars by 2030 to eliminate child labour, economically empower women, and tackle poverty and climate change.

Meanwhile, according to a 2016 Human Rights Watch World Report, worldwide **children languish behind bars**, wrongfully convicted of crimes, often victims of brutality and sexual violence. The United Nations Children's Fund, UNICEF, has estimated that more than one million children are behind bars around the world. Many are held in decrepit, abusive, and demeaning conditions, deprived of education, access to meaningful activities, and regular contact with the outside world.[522]

This global overuse of detention of children violates the International Convention on the Rights of the Child.
—Human Rights Watch

Children detained in the name of national security include captured, surrendered, or **demobilized child soldiers**, even though international standards call for states to treat them primarily as victims and offer them rehabilitation. However, a 2017 UN study promises to put **international focus** on the detention of children and hopefully result in more systematic monitoring of abusive practices.

Child Brides Reformed - #EndChildMarriage

In order to finally end this practice by 2030—the target set out in the UN Sustainable Development Goals—progress must be significantly accelerated. Global progress would need to be 12 times faster than the rate observed over the past decade.
— UNICEF

Globally, one in five women were married before their 18th birthday, and the United Nations stresses that we must end this unacceptable exploitation by 2030. When a girl is forced to marry as a child, this is a form of child and sexual slavery, and she faces immediate and lifelong consequences. Her odds of finishing school decrease while her odds of experiencing domestic violence increase. UNICEF notes that she is more likely to become pregnant during adolescence,

522 Michael Garcia Bochenek, Human Rights Watch, *Children Behind Bars*, 2016, https://www.hrw.org/world-report/2016/children-behind-bars

there are huge societal consequences, and higher risks of perpetuating intergenerational cycles of poverty.[523]

While over the last 25 years the prevalence of child marriage has slowly decreased globally, mainly in South Asia, it is still increasing in places like Africa—mainly due to population growth, stubborn cultural values and a lack of gender equality. Child marriage occurs in high-income countries too. In the United States, the majority of all 50 states have an exception in law that allows children to marry before the age of 18. Globally, the total number of girls married in childhood is estimated at 12 million per year.

The solution is simple. UNICEF advises that all countries ban child marriage, invest in education, and empower young people, families, and communities to bring about positive change. All sectors of society, and especially the media, need to use **strong public messaging** around the illegality of child marriage and the harm it causes if we are to see this required progress. Already, UNICEF is working to end child marriage in 64 countries globally. What we do today to address this issue will echo through generations to come. In the next defining decade, these child brides could be transformed into thriving, happy children, with a promising future ahead.

Successes and Solutions

In 2020, Girls Not Brides U.S. (GNB) celebrated a historic international milestone with the U.S. House passage of the Keeping Girls in School Act.[524] The strategy is the first by any government in the world to look at girls' lives holistically, with attention to rights, education, health, safety, and harmful practices such as child marriage.

"When girls are educated and empowered, we uplift communities and families, reduce poverty, and create a safer and more prosperous world," said Congresswoman Lois Frankel. "Today's bill passage is a big step towards helping girls around the world overcome the obstacles keeping them out of school, like child marriage and other forms of gender-based violence."

Over 130 million girls worldwide are not in school today. This bipartisan legislation focuses on closing the gender gap for adolescent girls and keeping them in school at the secondary level, a time when girls are most at risk of dropping out of school due to child marriage, early pregnancy, poverty, contracting HIV, lack of safety at school or travelling to school, and other harmful social and cultural norms.

In the last few years, child marriage has gone from a taboo topic to a prominent world issue. Girls Not Brides reporter Yasmin Mace tells us, "In 2016, child marriage was embedded within the UN's Sustainable Development Goals (SDGs). They're a set of ambitious and urgent goals and targets aimed at changing our world for the better."

Mace further notes that 1,300 organizations and 193 countries have committed to end child marriage by 2030: "Norway and Malawi banned child marriage and Indonesia raised the minimum

---

523 UNICEF, Fast Facts: *10 facts illustrating why we must #EndChildMarriage*, 11 February 2019 https://www.unicef.org/eca/press-releases/fast-facts-10-facts-illustrating-why-we-must-endchildmarriage

524 Girls Not Brides, Girls Not Brides USA celebrates the House passage of the Keeping Girls in School Act, Jan 30 2020 https://www.girlsnotbrides.org/category/uncategorized/

age that girls can marry from 16 to 19. In Latin America, a number of governments raised the minimum age of marriage to 18 without exceptions: the Dominican Republic, Honduras, El Salvador and Guatemala. Grassroots efforts have raised millions of dollars to stop child marriage through the Girls First Fund, and in 2018 Rebeca Gyumi was awarded the 2018 Human Rights Prize by the United Nations in recognition of her contribution to girls' rights."[525]

From Nigeria to Guatemala, Congo to Chad, Girls Not Brides joined the Restless Development initiative to motivate 94,000 global citizens to protest for an end to child marriage. More than ever the world needs **young people's leadership** to solve its greatest challenges, and make child marriage and overpopulation a thing of the past.

Child Soldiers Reformed

Theirworld tells us that in 2020 there are an estimated 250,000 child soldiers in the world in at least 20 countries. About 40 percent of child soldiers are girls, who are often used as sex slaves and taken as "wives" by male fighters.[526]

A child soldier is any girl or boy below the age of 18 who is recruited or used by an armed force or armed group outside government control, in any capacity. Theirworld adds, "A child soldier is not just someone who is involved in fighting. They can also be those in other roles such as cooks, porters, messengers, human shields, spies, suicide bombers, or those used for sexual exploitation. It includes children recruited and trained for military purposes, but not used in war."

Fifty countries still allow children to be recruited into armed forces, according to Child Soldiers International. Many non-state armed groups also recruit children. The UN Secretary-General's annual "name and shame" list for 2017 highlighted the armed forces of Afghanistan, Myanmar, Somalia, South Sudan, Sudan, Syria, and Yemen for recruiting and using under-18s for armed conflict. But non-state armed groups also recruit children in these and other countries. The report said there were at least 4,000 verified violations by government forces and over 11,500 by non-state armed groups in the 20 country situations it examined.

According to Theirworld, "Children become soldiers in different ways. Some are forcibly recruited. They may be abducted, threatened or coerced into joining, while others are enticed with money, drugs or in other ways. Being poor, displaced, separated from their families or living in a combat zone can make children particularly vulnerable to being recruited. Armed groups target children for several reasons. They are easier to manipulate, they don't need much food and they don't have a highly developed sense of danger."

Movements to End the Use of Child Soldiers

In the past two decades, thousands of boys and girls have been freed as a result of action plans mandated by the United Nations Security Council. Launched in March 2014, the campaign

525 Yasmin Mace, Girsl Not Brides, *10 ways the world got closer to ending child marriange in the last 10 years*, Jan 10 2020, https://www.girlsnotbrides.org/10-ways-the-world-got-closer-to-ending-child-marriage-in-the-last-10-years/
526 Theirworld, *Child Soldiers*, 2020, https://theirworld.org/index.php?p=explainers/child-soldiers#section-5

Children, Not Soldiers is working to galvanize support to end and prevent recruitment of children by national security forces in conflict. It is particularly focused on Afghanistan, Chad, the Democratic Republic of the Congo, Myanmar, Somalia, South Sudan, Sudan, and Yemen, all of which have signed UN Action Plans.

The UN Secretary-General's 2017 report on children and armed conflict said there had been several positive moves. That included armed groups in the Democratic Republic of the Congo and in the Philippines being delisted and the signing of a peace agreement in Colombia that included a special agreement on the release and reintegration of children. Measures are being taken around the world to prevent children becoming soldiers in the future. It is important to establish and **enforce 18 as the minimum age** for recruitment, and many governments have begun to do this.

Organizations such as War Child and Child Soldiers International are dedicated to rescuing and rehabilitating child soldiers who may have been desensitized to violence or traumatized by what they have been forced to do or witness.

Of course, much remains to be done. The Middle East Institute states, "The international community must take action beyond just signing protocols at the UN. Many of the signatories simply ignore their obligations. It's critical to understand the relevant international laws, force countries to sign these protocols through continuous isolation and scrutiny until they comply, and then hold them accountable for violating the protocols they signed."[527]

In 2020, Human Rights Watch reporter Jo Becker notes the 2000 United Nations treaty banning the use of child soldiers has seen some progress. She states, "Because of the treaty, governments—including the United States and United Kingdom, which used 17-year-olds in combat—have changed their deployment practices. More than 140,000 child soldiers have been released or demobilized. At least a dozen governments and armed groups have fulfilled formal agreements with the UN to end their use of child soldiers, including Chad, Cote d'Ivoire, Sudan, and Uganda. Commanders who once recruited children with impunity have **been convicted of war crimes** and received long sentences."[528]

Yet, at this critical crossroads, with only ten years to achieve the UN goals, there is a need to do more. Human Rights Watch calls on all governments to "investigate and prosecute commanders who recruit underage children, cut off support for the forces and groups that exploit children, negotiate more action plans to end the use of children in war, and to ensure that former child soldiers get the rehabilitation and support they need."

As an international community, we are not **demanding** *100 percent compliance, and that needs to change.*
—Middle East Institute

527 Mick Mulroy, Eric Oehlerich, Zack Baddorf, Middle East Institute, *Begin with the children: Child soldier numbers doubled in the Middle East in 2019*, April 14, 2020, https://www.mei.edu/publications/begin-children-child-soldier-numbers-doubled-middle-east-2019

528 Jo Becker, Human Rights Watch, *Stop Backsliding on Use of Child Soldiers*, May 25, 2020, https://www.hrw.org/news/2020/05/25/stop-backsliding-use-child-soldiers

The Dreamers of Tomorrow Take Action

Achieving and maintaining a sustainable relationship between human populations and the natural resource base of the earth is the single most critical long-term issue facing the peoples of the world and this issue will increasingly be the focus of international affairs for the foreseeable future.
—Russel E. Train, World Wildlife Fund

As more people come to understand the importance of this last decade to act, we will be seeing more population awareness programs in the near future—as people who truly care about child well-being refuse to remain silent. By targeting school-aged children, we will be setting the tone for the next generation of activists. Our children will grow up with the knowledge, the awareness, and best of all, the answers. Population awareness will become as much a part of their lives as growth, poverty, waste, and denial have become a part of ours. But first we, the adults they depend on for a safe and prosperous future, must face the reality!

Time reporter Elijah Wolfson claims that reality has already started to influence change. In 2020, after a decade of devastating fires, record-breaking storms, political turmoil, environmental destruction, and the shocking COVID-19 pandemic, some things are actually getting better.

Wolfson states, "Right now, it's hard to find reasons to be optimistic about human civilization's relationship with nature. Nevertheless, there have been some positive advances in the worlds of environment, climate, and energy in the past year. For example, the latest figures show that the global community is continuing to protect more and more parts of the Earth that are **essential to biodiversity**—and thus, long-term human survival—from exploitation. Investors are starting to rapidly divest from fossil fuels. An analysis undertaken by the environmental advocacy nonprofit 350.org found that assets of institutions now committed to divestment grew a ridiculous 22,000% from 2014 to 2019. Yes, you read that right. In addition, fewer people than ever before in recorded history are dying because of air pollution."[529]

We are also told that more children have access to health care and safe drinking water, the rates of global literacy and children completing high school are rising, and global gender equality is improving. Wolfson notes, "UN data show that fewer and fewer adolescent girls are being exploited by being married off and forced to have children in their teens, for example. So at least the trend is in the right direction."

Children around the world are also the focus for protection through the UN 1989 Convention on the Rights of the Child. This international law is the foundation of Equality Now's work. We can be assured that every day this committed group is making progress in protecting children from exploitation and trafficking, and in abolishing traditional practices that cause children harm.

In the last decade, numerous programs have united to forward family planning for young people, such as Thriving Together, Population Services International, Fair Start Movement, Jane Goodall's Roots & Shoots, and numerous others.

529 Elijah Wolfson, Time, *20 Ways the World Got Better in 2019*, Dec 23, 2019, https://time.com/5754155/global-advances-2019/

Population Services International advisor Amy Uccello reports that in 2018 their Family Planning 2020 pledge to reach ten million young people with contraceptives was exceeded two years early. PSI is catalyzing bold ways of driving health breakthroughs, including a **shift from a youth-centred to a youth-powered approach**, and they are just getting started.

Uccello states, "PSI recognizes that young people remain drastically underserved and under-represented in the health sector. We remain unwavering in our promise that all young people—regardless of age, marital status or parity—have access to the widest range of contraceptive options, where, how and from who they want. We're committed to keeping the momentum going."[530]

Now at this epic crossroads, Jane Goodall's **Roots & Shoots program** is working to empower young people to care for the world they inherit. It is leading a global movement in conservation by equipping an entire generation of young people to become activated and empowered conservation-minded citizens in their daily lives. Now nearly 100 countries strong and growing, Roots & Shoots is an unprecedented multiplying force in conservation and service-based learning, giving young people the knowledge and confidence to act on their beliefs and make a difference by being part of an initiative that includes reproductive health at its core.[531]

Jane Goodall notes, "Recognizing that **rapid population growth** coupled with chronic impoverishment was destroying the surrounding biodiversity, JGI Canada launched our *Delivering Healthy Futures* project in 2016." Their health care workers offer training on reproductive health and family planning, while simultaneously, villagers received training on **wildlife protection.** It is an elegant and effective arrangement.

The 2017 American documentary film, *Stopping Traffic*, was also produced to educate children worldwide, and motivate them to join the movement to end sex trafficking. Produced by a team of monks, this film traces the links between child sexual exploitation, pornography, social media, and sex trafficking.[532] One outcome that promoters of this documentary hope to achieve is a change in our legal system. Outdated sexist laws that discriminate against children have no place in the 21st century.

If children are to achieve some level of happiness, we should be looking to the happiest countries on the planet for guidance. From COVID-19 to climate change—in times of crisis, being happy promotes a more resilient and positive response.

530 Amy Uccello, Population Services International, *PSI Exceeds FP2020 Goal Early!*, May 28, 2019, https://www.psi.org/2019/05/psi-exceeds-fp2020-goal-early/

531 Jane Goodall Institute, *Roots and Shoots*, https://janegoodall.ca/our-work/roots-and-shoots/

532 Siddhayatan Tirth, Wikipedia, *Stopping Traffic*, 2017, https://en.wikipedia.org/wiki/Stopping_Traffic

Finding Happiness at This 2020 Crossroads

> *Listen, man, if you want to do something, don't buy yourself a hydrogen-fuel-cell car or an electric car or even a bus pass and think you're saving anything. If you really want to make a difference, the only thing you can do—the only thing—is to dedicate the rest of your life to population control as if the earth depended on it. Population control is ultimately the one thing that's going to save us, our kids and our kid's kids.*
> —Mark Vaughn, *Auto Week* magazine

"For the third year in a row, Finland has been named the happiest country in the world. Other Nordic countries dominate, with Denmark, Iceland, Norway and Sweden all making the top 10," states *Huffington Post* reporter Laura Paddison.[533]

Paddison notes that the report, now in its eighth year, ranks the happiness of 156 countries by how happy their citizens perceive themselves to be and is based on data from Gallup World Poll studies. Paddison adds, "In order to persevere through a pandemic, it's all about trust and supporting one another." This holds true for all of the critical world issues, such as extinction of species, overpopulation, political turmoil, and climate change.

Report co-author Jeffrey Sachs notes, "Time and again we see the reasons for well-being include good social support networks, social trust, honest governments, safe environments, and healthy lives." The Nordic countries tend to put a high value on these factors, as shown by their high environmental standards, strong societies, and low population levels. Societies that are paralyzed by overpopulation, polarization, and mistrust, on the other hand, fare badly in the face of huge disruptions—as is evident in the U.S. and other nations in turmoil.

Similarly, the **"Raising a Family Index"** will help you find incentive from the best countries for raising children in 2020—mainly Nordic countries. The index was based on the following six categories: safety, happiness, cost, health, education, and time. The top nine countries were Iceland, Norway, Sweden, Finland, Luxembourg, Denmark, Germany, Austria, and Belgium.[534]

Children are our future, and every child has a right to a fair start in life. This can only happen if every child is an intended child, and every parent is a willing parent. At this defining moment in our history, at this crossroads of epic proportions, it has never been more urgent that we place the welfare of Earth's children at the centre of our attention. By having fewer children, parents will have the time and energy to nurture and prepare each one for a bright and healthy future—a future free of exploitation and full of the beauty of nature.

---

533 Laura Paddison, Huffington Post US, *What The World's Happiest Country Can Teach Us About Surviving The Coronavirus Crisis,* March 20 2020, https://www.huffingtonpost.ca/entry/finland-coronavirus-pandemic-well-being-community_n_5e74ba30c5b6eab7794643b2?ri18n=true&ncid=newsltushpmgnewworld

534 Asher & Lyric Fergusson, *The Best (35) Countries to Raise a Family in 2020,* July 24 2020, https://www.asherfergusson.com/raising-a-family-index/

CHAPTER 13
TAKING ACTION

The only thing necessary for the triumph of evil is for good men to do nothing.
—Edmund Burke

The first **Earth Day** was a massive grassroots event, described by *American Heritage* magazine as "one of the most remarkable happenings in the history of democracy."[535] On this historic day in 1970, one of the biggest concerns of activists, student protesters, and Earth Day founder U.S. Senator Gaylord Nelson was human population growth and what it was doing to the planet. They joined to take action and remedy the overpopulation problem.

In 2020—just 50 years after those iconic days of the Beatles and the Vietnam War—we have doubled our population to almost EIGHT BILLION. The Center for Biological Diversity (CBD) points out, "Our appetite for energy, land and meat has skyrocketed; and the discussion of our runaway population growth and overconsumption has all but disappeared from the environmental movement."[536]

Reimagining Earth Day 50 Years Later

Activists worldwide want to reclaim Earth Day and put population back on the table. Now with a window of only ten years laid out by the UN to radically transform our world, we desperately need to bring population and the extinction crisis back into the Earth Day conversation. In this decade of change, a growing chorus of scientists and concerned citizens is unwilling to let this opportunity for change pass. But the million dollar question is, "What will it take to really make this a coordinated universal action?"

The Center for Biological Diversity is one organization leading the way with their Earth Day toolkit, Endangered Species Condoms toolkit, and new 2020 Crowded Planet resource library. These are all great **SOLUTIONS** to get us started. They state, "It's easy to be cynical about April 22 when you see Earth Day festivals sponsored by car companies and big box stores holding sales on yard accessories and mass-produced 'green' products. But the original Earth Day wasn't skewed toward just buying better products. We want to return it to its roots."[537]

535 What Is Earth Day?, Chelan Earth Day Fair, https://www.chelanearthdayfair.org/what-is-earth-day/
536 Center for Biological Diversity, *Reclaiming Earth Day: Putting Population Back on the Table,* April 18, 2014, https://www.biologicaldiversity.org/programs/population_and_sustainability/pop_x/pop_x_issue_41.html#one
537 Ibid

In 2020, Californians for Population Stabilization note the words of Earth Day founder, Senator Gaylord Nelson, "Central to the theme of the first Earth Day in 1970 was the understanding that U.S. population growth was a joint partner in the degradation of our nation's environmental resources."[538]

In the 1960s and '70s, population growth was widely and publicly connected to environmental ills by politicians, religious leaders, environment groups, and scholars. The legendary David Brower, director of the Sierra Club, urged Paul Ehrlich to write the controversial *Population Bomb* in 1968, which helped launch a worldwide debate that continues today. It is no coincidence that just two years later, the first Earth Day took place—which declared a challenge to the status quo, a recognition that our remarkable progress had come at too high a cost to the environment and the disappearing wildlife.

In 2020, many feel that Earth Day has lost its luster, become too commercialized and "greenwashed," promoting the deceptions of sustainable growth, green agriculture, clean energy, and green growth. Population has been swept under the rug, and corporate giants dominate the landscape. In 1972, cartoonist Wally Kelly observed, "We have met the enemy and he is us." It is humanity's obsession with growth that will be our undoing.

It is clear that we must change our attitude from one of arrogance, exploitation, and neglect to one of respect, humility, and cooperation if we have any hope of stopping population growth at EIGHT BILLION and starting to reduce our numbers to sustainability. We are all witnessing the heartbreaking tragedies playing out on Earth today as fires ravage the planet, species are disappearing at an alarming rate, and weather disasters drive millions from their homes—while our numbers continue to soar. Without action now, we will undoubtedly drive our own demise.

Negative Population Growth contributor Leon Kolankiewicz states, "In 1970 many Americans were alarmed that in just the previous two decades alone, between 1950 and 1970, some 52 million people were added to the U.S. population. They could see the dire consequences of this unchecked growth: sprawl devouring forests and farmlands, atrocious air pollution, rivers and lakes choked with filth and dead fish, disappearing wildlife. It was sickening to the senses and it confronted our collective conscience."[539]

Californians for Population Stabilization (CPS) points out, "Somewhere along the way, we stopped connecting the dots." CPS is working to formulate and advance a range of policies that promote sustainable, stable populations for our planet and nation—to ensure we save some America for future generations.

538 Californians for Population Stabilization (CAPS), *Earth Day at 50 Years*, 2020, https://capsweb.org/earthday/
539 Leon Kolankiewicz, Negative Population Growth, *EARTH DAY AND POPULATION:A MISSED OPPORTUNITY*, https://npg.org/library/forum-series/earth-day-and-population-a-missed-opportunity.html

The Only Way Forward

> *When family planning focuses on healthcare provision and meeting women's expressed needs, empowerment, equality, and well-being are the result; the benefits to the planet are side effects.*
>
> —Population Matters, Project Drawdown 2020

For centuries, we the people have looked to our political leaders and scientists to guide us and solve our most pressing problems. Unfortunately, neither have truly reacted to our critical world issues in an effective and timely manner. For over two decades—1992 to 2017—the Union of Concerned Scientists and most others in the science community were silent about the impacts of overpopulation, as were most politicians. That is what brought us to this appalling state of affairs—citizens need to be far more outspoken and involved in decision-making—especially women and youth. Most of us have been blinded to the root causes and the human factors involved here. Finally, a young Swedish girl came along to show the world that women's wisdom and youth activism could be much more successful at rallying the troops than a PhD and a flag.

Most of today's elite decision-makers are old enough that they will not experience the worst of our shameful legacy. What is needed now is to focus all of the world's funding allotted for further technological fixes, professional sports, military escapades, and space exploits, and invest it all into ending unintended pregnancies, now over 40 percent worldwide. We don't need any new scientific findings or technological fixes. We already know what the problems are and what the solutions are, as we hear today's population experts screaming it out at the top of their lungs. Why not focus on the greatest threat to the planet—that of population—first, and other world crises will all begin to diminish?

And that starts with **family planning** and education. I am **not** talking about academic education, although that is a basic human right that must be recognized. However, education about how to plan for children or be childfree is even more important and effective for creating smaller more manageable families. This is very evident in countries like America where school education is mandatory, **yet teen pregnancies are at epidemic levels**. School education on its own has little benefit to the state of the world when population and family planning are not incorporated into the process. In fact, religious and right-wing influences have prevented family planning initiatives in American classrooms, and this forced religious dogma has been a detriment to the family it claims to protect. There is little discussion of family planning taking place by parents or schools, the media rarely speaks of it, and most governments avoid the issue like the plague. It is little wonder kids today are suffering from eco-anxiety and depression, as we continue to ignore the root cause of our problems.

The only way out of this predicament is to admit our shortfalls, show some humility, and collectively decide to take immediate action with the **UN Sustainable Development Goals pointing to our True North. The most vital of these goals being the need for universal access to family planning.**

Science magazine reporter Sid Perkins warns, "The best way to reduce your carbon footprint is one the government isn't telling you about."[540] He uses the chart below to emphasize his point.

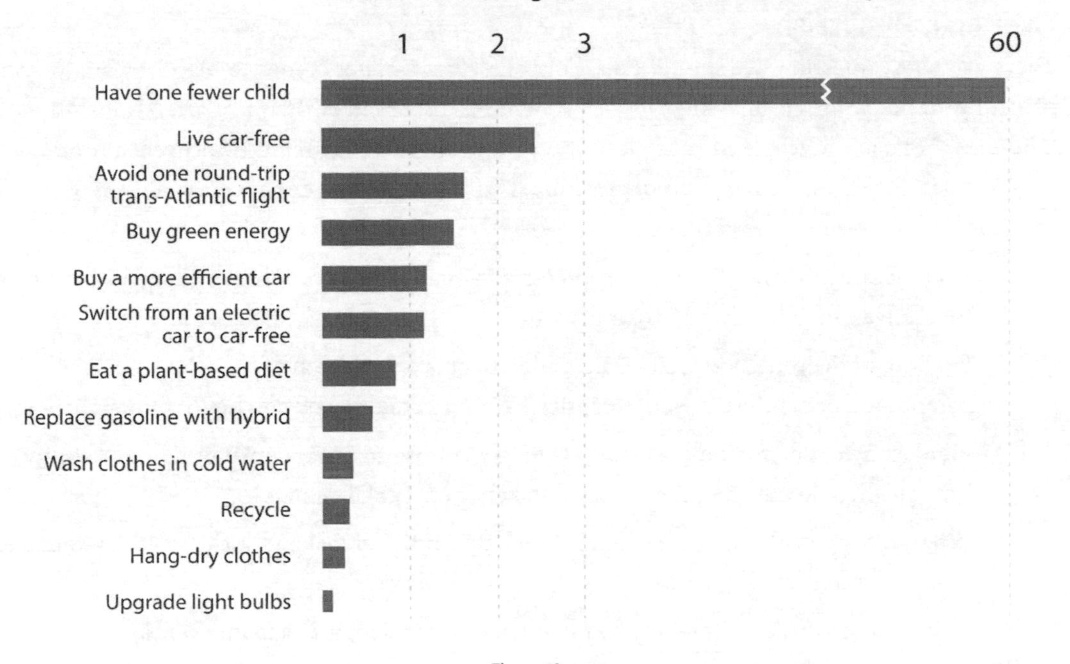

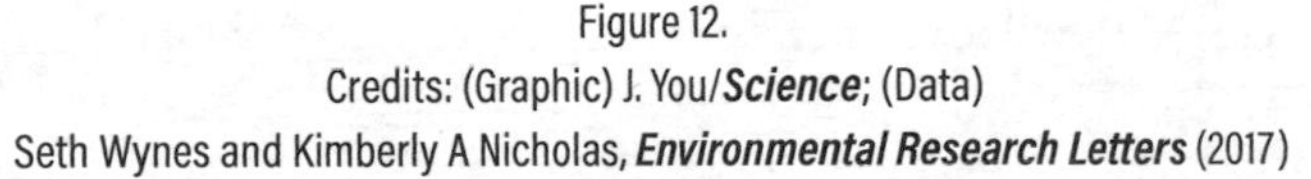
Figure 12.
Credits: (Graphic) J. You/***Science***; (Data)
Seth Wynes and Kimberly A Nicholas, ***Environmental Research Letters*** (2017)

We can see that having one fewer kid in our lifetime is by far the most important action we can take to reduce climate disruption, so we must demand that governments acknowledge this obvious fact, end the harmful population denial, and take action. We also know that current growth in consumption is greatest in the developing world, often in those same countries where population growth is highest.[541] So it is ironic that the world points only at countries of the north when blaming over consumers. We are all in this together.

The *Guardian* reporter John Vidal states, "So, of all the lifestyle choices you could make, having one less child is by far the best option. Research from Lund University in Sweden found that by choosing to have one less child, a parent would reduce their CO_2 emissions by 58.6 tonnes a year during their lifetime; over 25 times more than from any other action. Getting rid of the car, avoiding

540 Sid Perkins, Science Magazine, *The best way to reduce your carbon footprint is one the government isn't telling you about,* Jul 11 2017, https://www.sciencemag.org/news/2017/07/best-way-reduce-your-carbon-footprint-one-government-isn-t-telling-you-about

541 Population Matters, *It's Time to end harmful population denial,* Aug 28 2020, https://populationmatters.org/news/2020/08/28/its-time-end-harmful-population-denial

long-haul flights and going vegetarian are all well and good, but these actions, say the researchers, save very small amounts of CO_2 in comparison with having one less child."[542]

The Family Discussion

There are many questions we should be asking ourselves when it comes to planning a family or planning to be childfree, and these issues should be discussed early on in a relationship with a potential mate. The whole idea of planning your family size is meant to prevent unintended pregnancies, abortions, or coercion of any kind. If this planning is done responsibly, it will take into consideration a number of factors:

1. What is the minimum level of happiness and well-being a couple is willing to accept for themselves, their family, and their planet?
2. Do both of the parents actually want children, or is one or both of them being coerced into parenthood or deprived of contraception? Would a childfree option better fit their lifestyle?
3. How many children would a family willingly choose to have, considering their ability to support them in the present economic and environmental climate?
4. What impact would each additional child have on the global and local health system and food and water supply?
5. What impact would your family size have on other species' habitat and food?
6. What kind of burden will each family's size place on society—and on our social, physical and mental well-being?
7. What kind of birth control options will be required to meet your desired family size and the desired population targets set out by the United Nations?
8. How much does it cost to raise a child, and does every child deserve a fair start in life?

By considering the global context of these factors when doing a family planning exercise, we can prevent the vast majority of abortions and coercive programs that have caused so much concern in the past. According to a report released by the Population Reference Bureau in Washington, "Family planning is a lifesaver for millions of women and children in developing countries. It permits the healthy spacing of births, prevents the spread of sexually transmitted diseases like AIDS, reduces the number of low birth-weight babies, allows for longer breast feeding, prevents unsafe abortions, and averts death from childbirth."[543]

On the other hand, most methods of family planning have a low rate of failure if used correctly, and they are completely safe for the majority of users. So, family planning is a low-cost way to prevent many needless maternal and child deaths, prevent unintended pregnancies, as well as reduce population growth. This is a win-win solution.

542 John Vidal, The Guardian, *Well done, Prince Harry, for talking about population – but ditch the private jets,* July 31, 2019, https://www.theguardian.com/commentisfree/2019/jul/31/prince-harry-talking-population-royal-family-two-children
543 Population Reference Bureau, *Family Planning Saves Lives,* March 12 2009, https://www.prb.org/fpsl/

Family Planning

> *We can't go on like this. We can't push human population growth under the carpet. I would encourage every single conservation organisation, every single government organisation to consider the absurdity of unlimited economic development on a planet of finite natural resources.*
>
> —Primatologist **Jane Goodall** warns that human population growth is damaging the planet's future.

Every child should be a wanted child, and every parent should be a willing parent. If humanity could **accomplish this one goal**, it would do far more good than everyone converting to "renewable energy." In reality, no energy is truly "green energy" as it all requires something from Earth, and with EIGHT BILLION people, there is very little to spare.

The coronavirus is taking its toll worldwide not only as a health tragedy but also as a population tragedy—a baby boom! *The Phillippine Star* reporter Sheila Crisostomo tells us, "Following the warning of the United Nations Population Fund that the lockdowns may result in seven million unwanted pregnancies worldwide, the Department of Health (DOH) yesterday reminded couples to practice family planning during the enhanced community quarantine."[544]

Health Undersecretary Maria Rosario Vergeire said the DOH is working with local government units to promote reproductive health. She said many health centres in the Phillipinnes will remain open so that couples can receive condoms and contraceptive pills good for three months. Underprivileged countries like the Phillippines are often recipients of contraceptives funded by dedicated foundations like the Margaret Pyke Trust, Marie Stopes International, or the Bill and Melinda Gates Foundation. In 2012, the Gates Foundation, staunch supporters of family planning, announced that the foundation was doubling its investment in family planning to a total of more than $1 billion by 2020.

Despite the anti-contraception movement led by the religious right—conceived in the U.S., but also unfolding in Africa and elsewhere—champions like the Gates Foundation have stood strong. Although the Vatican managed to derail family planning initiatives in the United Nations, they have no power in these realms. "What we're doing is an enormous undertaking," Melinda Gates said.[545]

Her foundation's pledge was part of an ambitious $4.3 billion joint initiative to address a crisis in contraceptive access in the developing world that leads to roughly 80 million unintended pregnancies. The goal was to bring high-quality contraceptives to more than 120 million underserved women in the world's poorest countries. In 2012, the Gates Foundation became involved with the UN Family Planning 2020 campaign, and Melinda Gates quickly became the campaign's public face. *Harper's Bazaar* dubbed her "The Savior in Seattle."

---

544 Sheila Crisostomo, The Philippine Star, *Practice family planning amid ECQ, couples urged*, April 30, 2020, https://www.philstar.com/nation/2020/04/30/2010733/practice-family-planning-amid-ecq-couples-urged

545 Kathryn Joyce, Pacific Standard, *The New War on Birth Control*, Sept 16 2018, https://psmag.com/magazine/new-war-on-birth-control

We need more champions like this, as an overcrowded planet is making life harder for all of us, and many people are asking if this is fair ethically. Should couples really have the right to have as many children as they wish, when it causes such chaos, pressure, and conflict for everyone else? Just by preventing the 45 percent of unwanted pregnancies, a massive shift in population growth would be possible.

Fair Start Movement

In 2020, the population activist group Fair Start Movement tells us, "It's simple. The **most effective way to address climate**, and reduce our impact on the environment and other species, is by choosing **smaller families**. And when parents choose to have smaller families, they and their communities can invest more time, attention and resources in each child to ensure a fair start in life. The proven impacts are 10 to 20 times higher than traditional downstream attempts to protect kids, animals, and the environment."[546]

Very few environmental organizations publicly support an ethic and policy of smaller families for all. So Fair Start Movement is **issuing this challenge**: "We are urging environmental and animal protection organizations to adopt a public position on family planning, simply one that furthers their mission."[547]

Derived from dozens of published articles, and consistent with the leading thinking on human rights–based family planning, Fair Start Movement has developed the child-centred Fair Start family planning model. The United Nations warns that one million species face extinction thanks to human activity, and that rising inequality and accelerating climate change are affecting more children than ever before. It's time to think differently about having kids and every child's right to a fair start in life.

> *We must recognize that how we plan families, more than any other one thing, determines the quality of life on Earth.*
> **—Fair Start Movement**

In this model, each of us would delay having kids until we are ready to be good parents, and we would have small and sustainable families that put less demand on our shared resources and our world. Child-first family planning provides a clear human rights-based solution and path forward, addressing the biggest problems affecting kids such as poverty, inequality, and neglect. It's sustainable and equitable, and reflected in the worldwide trend towards smaller families that can invest more in each child and ensure a safer future.

This model shifts the focus of family planning from a narrow focus on just the parents to a broader one that includes the interests of the future child, the parents, and the community. This compels us to think out of the box and plan ahead, requires us to cooperate, and thereby helps ensure the best outcomes.

Fair Start Movement points out, "The poor family planning laws **designed to promote economic growth to benefit elites** also ensured countless children were born into poverty, abusive

546 Fair Start Movement, *Making An Impact: Sustainable Families,* 2020, https://havingkids.org/impact/
547 Ashley Berke, Fair Start Movement, *Earth Day Challenge,* Apr 19, 2018, https://havingkids.org/earth-day-challenge-environmental-organizations-to-address-family-planning/

homes, and misery over the past several decades. This, in turn, ensured the extinction of countless species, the creation of factory farming, the onset of the climate crisis, and other examples of non-human suffering and death that has only skyrocketed in our lifetimes. Human population growth, or the act of having kids, was the genesis of it all."

Fair Start Movement pursues change at legal, institutional, and cultural levels, making the most vulnerable—children and animals—their focus. All animals, including humans, have a right to nature, children have a right to a fair start in life, and no one has a right to have children in abusive or neglectful conditions. We must invest in future generations rather than exploit them, and protect the vulnerable from the powerful. They emphasize that if you care about giving kids the future they deserve, you'll care about family planning. It's everyone's business and everyone's responsibility.

This new model maintains, "When parents wait and choose smaller families who invest more in each child it makes for increased stability, less crowded communities, a smaller impact on the planet, wildlife, and farmed animals, greater investment in each child and more shared resources for everyone to get a fair start in life. It means thriving together, not just surviving. Who wouldn't want that?"

Fair Start Movement is working to liberate future generations by making them **self-determining**. They emphasize, "We have to create people with a 'sense of justice' who will protect animals, rather than relying exclusively on traditional animal protection laws that—especially when it comes to farmed animals—**are often not worth the paper on which they are written.**"

These priorities must be reworked and we also need to consider diet-focused veganism as a way to eliminate the confinement and suffering that humans inflict on non-humans in factory farms, and to improve human health and well-being at the same time. Otherwise, animal abuse will continue to increase as the population increases. Another key recommendation is to amend tax policies that encourage large families—to encourage delaying parenthood and planning smaller families. It is counter-productive to reward parents for having larger families.

We must also urge public figures to speak out, given the massive impact these role models could have in promoting sustainable family planning. Fair Start Movement notes, "Meghan Markle and Prince Harry's recent statement about their intent to keep their family small was exemplary."

Family Planning 2020s

The 2012 London Summit on Family Planning led to the formation of Family Planning 2020 (FP2020), today a coalition of four Core Partners, 40 civil society organizations, 55 government commitment makers, 23 target countries, and 7 foundations, all pledging to dramatically improve access to life-saving contraceptives. The core partners are the Bill and Melinda Gates Foundation, UK's Department for International Development, UN Population Fund, and U.S. AID.[548]

Hosted by the United Nations Foundation, FP2020 is based on the principle that all women should have universal access to sexual and reproductive health services by 2030, as laid out in UN Sustainable Development Goal #5.

548 Family Planning 2020, Overview, 2020, https://www.familyplanning2020.org/about-us

Many high birth-rate/low family-planning countries committed to making multiple forms of modern contraceptives more available, while wealthier nations and foundations pledged to increase funding for those efforts. After decades of lackluster support for global family planning initiatives, contraception was back as a cause. An article in the medical journal *The Lancet* declared it "**the rebirth of family planning.**"

Family Planning 2020 notes, "In 2015 alone, donors provided US $1.3 billion in bilateral funding for family planning. Five years later on the anniversary of the London Summit, the international community gathered for the 2017 Family Planning Summit for Safer, Healthier, and Empowered Futures to renew its commitment to the FP2020 partnership, growing to include more than 120 commitment-making partners."

FP2020 was a massive effort, and the Gates Foundation—and Melinda Gates in particular—quickly became the campaign's public face. *Harper's Bazaar* dubbed her "The Savior in Seattle."

Is child-bearing a right or a privilege? Iran has one of the world's more aggressive programs to encourage smaller families: to get a marriage licence there, **you must take a course in family planning.** In fact, it worked so well that their government is now promoting larger families again, but that's not working— citizens have now experienced the many benefits of small families, and there is no turning back.

Contraception

Women must come to recognize there is some function of womanhood other than being a child-bearing machine.
—Margaret Sanger, founder of Planned Parenthood Federation

The Binary Birth Rate

0101010101010101010101010101010101010101

The binary number system is made up of only two numbers, 0 and 1, and is the basis for all **binary code** used to write computer instructions.[549] This same elegant design would also be applicable for an enhanced reproductive system for the 21st century. The **number** system represents electricity being allowed to flow through a circuit. The **reproductive** system represents flow of genetic material from the male body to the female reproductive system. Either system can be turned on or off to control the desired outcome. In the case of reproduction, the desired outcome would be to trigger a reduction in fertility rates and ultimately human population through the use of a wide range of voluntary family planning services.

Binary choices encouraged:

1. Have one child as promoted by the humane One Planet – One Child education campaign. Because we have neglected the population crisis for far too long, more radical solutions are

549 Computer Hope, *Binary*, 2020, https://www.computerhope.com/jargon/b/binary.htm

now needed to achieve a sustainable population as recommended by the UN Sustainable Development Goals. We must stop growth at EIGHT BILLION.

2. Choose to be childfree and perhaps join one of the many Childfree, Fair Start Movement, or GenderStrike groups around the world. This is also an opportunity to volunteer with one of the numerous oranizations striving to make the planet a better place to live for all sentient beings.
3. Make adoption an option. According to UNICEF (the United Nations Children's Emergency Fund), there are roughly 153 million **orphans worldwide**. Every day, an estimated 5,700 more children become **orphans**. Children are often relinquished due to war, natural disaster, poverty, disease, stigma, and medical needs. Every child deserves to feel happy, healthy, safe, and most importantly, loved.[550]

One reason there are so many orphans, teen births, and unintended pregnancies is because, globally, parents play a minimal role in educating their children about sexual reproductive health. We need to get parents to **talk about sex education** with their children, including birth control options and impacts on the environment. A 2019 Geopoll report shows that parents have failed to educate their children, especially girls, on sexual reproductive health, leaving the responsibility to the Internet or TV. Leaving this education to sources like the Internet can seriously skew attitudes about healthy sexuality and lead to more performative as well as demeaning expressions of sexuality based on exposure to pornography for both girls and boys. While the majority of youth would be interested in sex education, less than 20 percent of parents have provided it.[551]

Why Have Unwanted Children?

Besides the cases of child marriage, sexual abuse, and lack of contraception mentioned above, there are many more reasons for unintended or neglected children:

1. Many couples won't give up until they have a son. This practice often produces far more children than the parents planned on having or can actually afford. This kind of gender discrimination is common in many countries around the world besides China.
2. Many women have children to fill a void in their life. One such case was that of **octomom** Nadya Suleman, who had 14 children trying to fill this void, and then told Dr. Phil she couldn't cope and was having a nervous breakdown. Octomom is now often referred to as the human equivalent of a Hummer.
3. Men are not held accountable. Global opinion and law are far too lenient on men who feel free to impregnate as many women as they like, often with no consequences. It is the women and/or our social systems (the taxpayers) that are usually held accountable and left with the responsibility of caring for the child.

550 Adoption.org, https://adoption.org/many-orphans-worldwide

551 Business Diary, *Parents to blame on sex responsibility, Report,* June 25, 2019, https://teddyleting.com/business/parents-to-blame-on-sex-responsibility-report/

4. Couples may have a child to hold a marriage together. You probably know someone who was having marital problems and thought that maybe having a child (or another child) would bring the couple closer together. However, most often this tactic backfires, they end up divorced anyhow, and now the child has only one full-time parent. Using a baby as a pawn in a marital relationship is cruel and senseless, and being a parent is much more difficult than most people imagine. This is especially true in today's environmentally crippled, COVID-19 stricken world.
5. Having a child acts as a security blanket in old age. Parents may want a large family to have someone to take care of them when they get old. In poor countries, this may be because they have no safety net, such as pension or social security, to depend on in old age. However, as more young people move to cities, governments implement better social services, and grown children increasingly move away from their families, this perceived need is diminishing.
6. Families fall for the "happily ever after" illusion. Mainly due to a lack of family planning education globally, most families have failed to consider the impacts of having unintended children or large families and can't cope. This is one reason that about half of marriages fail. Family law specialist Travis Krepelka warns that in 2020, the global pandemic will increase our already high divorce rates of 40 to 50 percent—depending on location. He states, "We will see a spike in new divorce filings in the latter half of 2020. Some of these divorces may not otherwise have happened, or at least not yet, if the pandemic had not occurred. But, the global financial and health crisis has created an enormous amount of economical and interpersonal difficulties for relationships."[552]

I think we can all agree that parenting is a very demanding job with a lifetime of responsibilities, and not everyone is suited to it. Yet it is by far the most neglected. Unlike most other jobs, there is no accountability for parenthood except in very extreme cases of abuse or neglect, and then only if authorities are informed. Also, unlike most other jobs, there is no training or experience required. Even the simple act of driving a car demands more training and accountability than raising a child. That is one reason I believe **Family Planning 101** should be a requirement in all high schools and colleges.

If you're in the market for a new way to ***keep your body baby-free****, this cheat sheet is for you.*
—Health.com

We all seem to have this preconceived, and often unrealistic, notion about having a perfect family like *The Partridge Family* or *The Brady Bunch*. We think that we won't make the same mistakes that our parents made, but it rarely works out as we planned. For some parents, it is a rude awakening and a great disappointment. I can't tell you how many parents have confided in me that they wished someone had been honest with them about how hard it is, and if they could only do it over again, they would make a different choice.

552 Travis Krepelka, Business Journal, *Divorce rush looms as Covid-19 strains arriages beyond repair*, May 7 2020, https://www.bizjournals.com/sanjose/news/2020/05/07/divorce-rush-looms-as-covid-19-strains-marriages.html

Similarly, I'm sure we have all heard children shout, "I never asked to be born" or "I wish I had never been born" in a fit of rage. Of course, we just brush it off as words uttered in anger. But what if we were to look at it a little deeper, and try to actually understand how they see life? It may not be a pretty picture, but this could change. If we were to create a world of two billion citizens living an ecocentric lifestyle, we could all be celebrating our magical planet.

So Many Options for Birth Control

Let's make the 2020 decade a time of discovery. Schools, doctors, governments, and the media should be major players in encouraging parents to get involved with sex education. For example, a UN Population Fund report noted that Kenya had the highest number of teen pregnancies, so their organization extended the conversation online. This has been found to be a major source of sexual reproductive health education, through a campaign dubbed WIWIK (**What I Wish I Knew**). This kind of program should be available globally and watched by all teens with their parents.

When it comes to contraception education, Power to Decide provides data on how to prevent unplanned pregnancy. They believe that all young people deserve the opportunity to decide if, when, and under what circumstances to get pregnant and have a child. And that means knowing all there is to know about birth control.[553]

Another helpful site to find the most effective birth control ranked is Health.com. This site notes, "Since effectiveness is so important, it's a good idea to use **effectiveness rates** as a guide for what birth control to go with. We've laid out the rates (based on Centers for Disease Control numbers) for every method out there—from the most reliable to the iffy."[554]

Health.com reporter Louise Sloan tells us whether it is pills, rings, patches, shots, vasectomy, tubiligation, or morning-after pills, there are so many different ways to get birth control into your body now:

- The oral **birth control pill** comes in many variations now, with different kinds and ratios of hormones (estrogen and progestin). This pill may also be available for men around the world in the near future, as it has already been approved in many countries. Gender equality requires more focus on male contraception **to bring birth control into the 21st century**.
- Then there are the **hormonal methods** that don't involve swallowing anything: implants, hormone-infused vaginal rings, and birth control patches. Contraceptive implants, like Jadelle, are progestin-based, highly effective, and rapidly reversible methods of contraception that require little of the user and have few side effects. Over the past 35 years, they have been approved in more than 60 countries and used by millions of women worldwide.[555]

553 Power to Decide, *Find Your Method*, 2020, https://powertodecide.org/sexual-health/your-sexual-health/find-your-method
554 Sarah Klein, Health.com, *The Most Effective Birth Control, Ranked*, October 05, 2017 https://www.health.com/condition/birth-control
555 Valerie French, Glowm.com, *Implantable Contraception*, July 2015, https://www.glowm.com/section_view/item/398/recordset/18975/value/398

- **The patch** is ideal for women who don't want to take the pill every day due to the negative effects of hormonal birth control, and it has a higher success rate than the pill (you have to remember to take the pill). The cervical cap is also available, but less effective.
- For **men** who are eager to step up and take responsibility in family planning, the **vasectomy** is the preferred choice. Otherwise known as male sterilization, the vasectomy involves cutting or blocking the tubes sperm travel through from the testicles. This is a simple procedure taking only minutes, and is one of the most reliable methods of permanent birth control. This procedure can also be done without an incision, as seen on the video, *The Vasectomist*. More about this below.
- Female sterilization (or **tubal ligation**) means blocking or cutting the fallopian tubes, which prevents an egg from travelling to the uterus and being fertilized by sperm. This is a much more complex surgery than vasectomy, more risky, and more expensive. This option is ideal for those sensitive to the negative effects of hormonal birth control.
- For **males**, a trendy choice has been the **condom**, which is convenient and can also help prevent the spread of AIDS. However, it can be put on incorrectly, slip off during intercourse, or even tear—so it is not as reliable for birth control as some other options.
- The introduction of the drug **mifepristone (aka RU 486)**, which gives women the option of having safe early induced **abortions** in private locations instead of public clinics, is a welcome addition to birth control. It also prevents anti-abortion stalkers from terrorizing those who have chosen abortion as an option.
- For emergency birth control, the **morning-after pill (MAP)** is available. The active ingredients in this pill are similar to those in birth control pills; the MAP simply contains higher doses. The morning-after pill is designed to be taken within 72 hours of intercourse, with a second dose taken 12 hours later. According to the manufacturers, the MAP (also known as plan B or emergency contraceptive pill) is more than 80 percent effective in preventing pregnancy.
- A revolutionary option to tubal ligation is an oral form called **quinacrine sterilization (QS), which is presently being blocked for use in some countries**. In 2015, *Science Direct* reporter Jack Lippes stated, "Like all methods of family planning and especially permanent methods, QS should always be offered in human rights frameworks of fully informed consent. Much is now known about QS. It is nonsurgical, there is no need for anesthesia, and studies have shown that it is safe. With the improved Hieu insertion technique, and two treatments, QS effectiveness (failure rate of 1.2% after 2 years) compares favorably, with surgical tubal ligation (0.7% at 2 years). The cost is low, and QS can be performed by non-physicians. It is **time to reexamine** the epidemiological and clinical data on the use of QS and approve its use in both developed and developing countries."[556]
- With the high prevalence of unintended pregnancies and rape, it is reassuring to know that **legal abortion** is an ethical, safe solution to a destructive situation (which any pregnancy

556 Jack Lippes, Science Direct, *Quinacrine sterilization (QS): time for reconsideration*, Aug 2015, https://www.sciencedirect.com/science/article/pii/S0010782415002322

that is unwanted or places the female in a state of uncertainty/anxiety is). The *decision* represents a powerful and positive perspective for young people who think life has more to offer than raising children. Part of the population control argument should be to help relieve the existing stigma from the act of choosing an abortion. Having the choice to have an abortion emphasizes the mental, physical, and lifelong liberating consequences that are associated.

Sloan points out that this wide selection is great news for women and men because it gives them the opportunity to find a "method of delivery" that works best for their bodies and lifestyles. Of course, this may suddenly find anti-abortion **protesters all dressed up with nowhere to go**—and no one to terrorize. The religious right's 40-year campaign of violence against U.S. and Canadian medical clinics has resulted in numerous murders, bombings, arsons and kidnappings—and hundreds of death threats and assaults. Many of these protesters claim that family planning means promoting abortions and coercion. In fact, the opposite is true, since if all women had access to contraception, there would be no need for abortions, and coercive tactics would be a thing of the past. In reality, **coerced parenthood is a far greater threat** today, with a more deadly outcome.

Human Life International, a hardline U.S. anti-abortion group that maintains a network of global affiliates, is supported by the religious right, and has gone to great lengths to block many forms of birth control. These attempts are often successful, and often with government approval. It is little wonder that the political environment and COVID-19 have caused a boom in **at-home abortions**, especially since they are safer and more affordable. In fact, CTV reporter Rachel Gilmore notes that in Canada, the 2020 epidemic is causing a shortage of the abortion pill named Mifegymiso, which is manufactured in Europe. However, the Government of Canada has ensured citizens that they are taking measures to make sure it's equipped to meet this demand.

Gilmore explains, "A **medical abortion** is when a person takes a series of pills to terminate a pregnancy. With the medical option, the patient doesn't need to head to a clinic or a doctor's office the way they would for a **surgical abortion**. If these pills aren't available, women seeking to terminate a pregnancy would have to opt for the surgical option instead—an option that providers have ensured remains available. This new option of drug-induced medical abortions is putting more abortion providers back in service. This medical abortion gives women the dignity of going through the process in the privacy and comfort of their own homes, rather than having to hunt down a clinic, face a hysterical mob, and endure a painful and invasive surgery."[557]

Of course, all forms of abortion are stressful for a woman. Voluntary family planning is a low-cost alternative that would prevent much of the suffering women endure on this planet every day due to unwanted pregnancies. Reproductive choice also gives a female control over her life. By exercising this choice, a girl or woman is better able to take advantage of opportunities for education and employment and contribute more to her community.

Contraception is especially important in cases where the mother has HIV or an addiction to drugs or alcohol. We often see cases on TV where an addicted mother is not sufficiently aware to

---

557 Rachel Gilmore, CTV News, *Canada experiencing shortage of abortion pill amid COVID-19 outbreak*, March 30, 2020, https://www.ctvnews.ca/health/coronavirus/canada-experiencing-shortage-of-abortion-pill-amid-covid-19-outbreak-1.4874666

take precautions and often gets pregnant with an unwanted baby. When that baby is born, it often has fetal alcohol syndrome or a drug addiction, and this is an ever-increasing scenario. These babies often end up in the system for government agencies to care for at the taxpayers' expense, or they may be left in a single-parent situation, quite possibly with a lifelong disability. These situations are leading to an alarming number of unplanned pregnancies and babies that grow up to a life of misery.

When it comes to the alarming increase in opioid births, Fair Start Movement reporter Ashley Berke believes that Congress is overlooking the obvious solution. She states, "Only about 6% of women with an opioid use disorder report using Long-Acting Reversible Contraceptives (LARC), such as an IUD or implant. Because of this massive hole in family planning, children undergo physical trauma at birth and the lifelong emotional and health consequences that result."[558]

A staggering 90% of pregnancies among opioid-maintained women are unintended.
—Ashley Berke, Fair Start Movement

As Berke notes, what is needed is a public health response from the U.S. government that would focus on prevention instead of treatment after the fact. Colorado's successful proactive contraceptive program serves as a model, as Fair Start Movement works to replicate these efforts in other states. Indeed, this must become a global initiative in the next decade if the UN goals are to be met by 2030.

Berke adds, "We can help women *and* children by removing the hurdles to accessing long-acting and effective birth control. By doing so, we can improve lives and better help those struggling with addiction to overcome their dependency without the added burden of parenting before they are ready. **The obvious solution to preventing infants from becoming additional victims in this crisis is through a conscious public effort for family planning.** And what about cost concerns? While an IUD can cost about $1,090, the cost of caring for an infant with neonatal abstinence syndrome (NAS) can be upwards of $60,000. Therefore, it makes fiscal sense for the public to support family planning."

We can prevent babies that are born addicted from enduring agonizing withdrawal symptoms and often lifelong learning disabilities, by simply preventing these unintended pregnancies.

Men and Contraception—Changing the World

How can you be prepared to cut back on your car emissions and your plane emissions but not on your baby emissions? Can you really celebrate the pitter-patter of tiny carbon-footprints?
Mifegymiso –Journalist Johann Hari

Mr. Condom - Meanwhile, in 2017, Mr. Condom is still causing quite a stir in Thailand. Mechai Viravaidya is a former politician and activist who began his family planning program in the '70s

558 Ashley Berke, Fair Start Movement, *Congress Overlooking the Obvious Solution to Opioid Births*, Oct 11, 2018, https://havingkids.org/skyrocketing-opioid-births-family-planning-crisis/

to make his home country a better place for life and love. *Bangkok Post* reporter Andrew Biggs tells us that since then Mechai has been affectionately known as "Mr. Condom."[559]

In his Ted Talk, Mechai Viravaidya explains that in the '70s, families were having seven children, and there was great poverty in Thailand, so Mechai could see that family planning was desperately needed. Reporter Jonathan Stack noted that "Mr. Condom" first started out promoting condoms and family planning to get the neighbourhood women on board and then the nurses and midwives. He used creativity, fun, and humour to win over his audiences and volunteers, showing people that there was **no reason to be shy about discussing sexual health**. His program had an extraordinary impact, handing out condoms at both small roadside stands and the biggest stores—and **teaching everyone about family planning.**[560]

He made his priority the children, then the mothers, and then family planning, and it was all done in stages. Although there were few doctors in Thailand, his team was able to make the pill readily available for women and condoms available for the men—as well as vasectomy festivals later on. They teamed with Coca-Cola booths to supply pills and condoms even in remote villages throughout the country. Mechai convinced the Catholic Church and Buddhist monks to bless the pills and condoms with holy water, and recruited the floating markets to hand them out. A picture of this blessing event was widely circulated to encourage participation.

Mechai jokes, "Wherever you find people, you'll find contraceptives in Thailand. You need everybody to be involved in trying to provide whatever it is that makes humanity a better place. So we went to the teachers. Over a quarter of a million were taught about family planning with a new alphabet —A, B for birth, C for condom, I for IUD, V for vasectomy. And then we had a snakes and ladders game, where you throw dice and promote family planning on every move. In taxis, you get condoms. And also, in traffic, the policemen give you condoms—our **cops and rubbers** programs. So, can you imagine New York policemen giving out condoms? Of course I can. And they'd enjoy it immensely; I see them standing around right now, everywhere. Imagine if they had condoms, giving out to all sorts of people. And these were the condoms that we introduced. One says, **Weapon of mass protection.**"

However, in the 1980s and 1990s, HIV/AIDS struck, threatening to divert attention and funding from their program. When the government refused further assistance, Mechai went to the military to borrow 300 radio and television stations—and they agreed. This was an elegant turn of events, since the condoms needed to control AIDS also worked to reduce population growth. Then the next government increased funding 50-fold, and there was backing for the entire package—and Mr. Condom became a huge success. Since his contraception and HIV/AIDS campaigns began in 1974, the average number of children in Thai families has decreased from 7 to 1.5 today. According to the UN, new cases of HIV have declined by 90 percent. And so that's the remarkable case in

559 Andrew Biggs, Bangkok Post, *Mr. Condom is still causing a stir*, Jul 16 2017, https://www.bangkokpost.com/life/social-and-lifestyle/1288059/mr-condom-is-still-causing-a-stir

560 Mechai Viravaidya, TED Talk, *How Mr. Condom made Thailand a better place for life and love*, https://ted2srt.org/talks/mechai_viravaidya_how_mr_condom_made_thailand_a_better_place

Thailand of everyone, especially men, joining in. They didn't have a strong government. They didn't have lots of doctors. But everybody united to change attitudes and behaviour.

Population Matters, a UK-based charity that addresses population issues, believes we should all recognize World Vasectomy Day (WVD), held in November. It is a worldwide movement and celebration of men who are stepping up for their partners, their families, and their future. Men are often ignored in the family planning landscape, but in the last decade, they have been increasingly sharing the responsibility by using condoms and having a vasectomy. An additional option will soon be available in the form of a male birth control pill, which will reinforce the notion that both men and women can be responsible for contraception.[561]

The YouTube story ***The Vasectomist*** highlights an as noble and successful a movement, as Dr. Doug Stein travels the world inspiring one person, one family, and one community at a time. The *Daily Star* notes that WVD was founded in 2012 by the American filmmaker Jonathan Stack while he was working on a documentary about the decision of having a vasectomy. The underlying goal was to involve men in family planning decisions and educate them about vasectomies as a simple way of taking responsibility for birth control.[562]

The event first took place in 2013 when together Stack and Stein set out to inspire 100 doctors in 25 countries to do 1,000 vasectomies. They achieved the goal within 24 hours. The following year, over 400 physicians in 30 countries, including the U.S., Australia, India, Kenya, China, and Colombia, performed almost 3,000 vasectomies, making it the largest male-focused family planning event in history.

According to Stack, "Sadly, while men waffle on the sidelines, more than 300 million women have had tubal ligations. That's six times the number of men having vasectomies, even though tubal ligations are more invasive, costly and risky. While both procedures are almost equally effective, tubal failures can result in ectopic pregnancies, a leading cause of maternal mortality."

Vasectomy is a surgical procedure for male sterilization or permanent contraception, especially suitable for those who feel their families are complete or those who cannot afford more children. The two-week WVD event continues to spread and grow, a product of the year-round collective effort that builds awareness through communication and community.[563]

Childfree and Thriving

I'm completely happy not having children. I mean, everybody does not have to live in the same way. And as somebody said, "Everybody with a womb doesn't have to have a child any more than everybody with vocal cords has to be an opera singer."

—Gloria Steinem, acclaimed journalist and women's rights activist

In the 21st century, millions of women around the world will reach 40 and still be childfree. For

561 Population Matters, *World Vasectomy Day*, Nov 17 2019, https://populationmatters.org/events/world-vasectomy-day-0

562 The Daily Star, *World Vasectomy Day*, Nov 11 2018, https://www.thedailystar.net/health/news/world-vasectomy-day-2018-1658671

563 Jonathan Stack, YouTube, *The Vasectomy Series*, May 16, 2013, https://www.youtube.com/watch?v=c7H4VrY-7Qo

some, it will be due to infertility; for a growing number, it will be a conscious choice; and others are simply still on the fence about parenthood. The main difference between these groups is that the childfree by choice don't see themselves as lacking anything—quite to the contrary, studies show they are happier than those with or desperately wanting children.

I'm sure most of us have heard the children's nursery rhyme, the KISSING song, which starts, "First comes love, then comes marriage, then comes the baby in the baby carriage." It's true that this is what we are taught as children, and it's a common expectation for most newlyweds, but more and more young couples are exercising their individual reproductive freedom, and remaining childfree. This is a **success** we can be proud of.

What is Truly the Selfless Act?

> *The world might, perhaps, be considerably poorer if the great writers had exchanged their books for children of flesh and blood.*
> —Virginia Woolf

Many who would like to perpetuate population growth, including the Pope, say that choosing to be childfree is morally a selfish act. But is it really? At a time when overpopulation is threatening our security, resources, and environment, perhaps it is having a large family that should be viewed as an act of betrayal, selfishness, and excess!

A government study indicated that Canadian families with children are greatly subsidized for simply breeding, but not childfree families. The study found that Canadian families benefit from public services worth an average of about $41,000 a year, or 63 percent of the median family income.[564] Most of this goes towards child-rearing expenses. In 2020, on average, families benefiting from the Canada Child Benefit alone will receive around $6,800 to help with the high cost of raising kids.[565]

Everyone's taxes pay for these services even if they don't have children, and childfree couples pay a disproportionately larger percent of taxes than those with children because they are allowed fewer income tax deductions and are put in a higher tax bracket. This privileged status for parents is accomplished by the government providing daycare services, the Canada child benefit, income tax deductions allowed for each child, school tax, paid maternity leave, health insurance, welfare, subsidized housing, and so on, which discriminate against childfree citizens. Childfree taxpayers are actually being forced to pay for breeders to add even more consumers to an already overburdened planet, even though many of us strongly oppose population growth, as do many organizations such as the United Nations.

Then there are the non-monetary contributions that childfree citizens make to the children of our society. Valerie Bell makes this point in her book *Nobody's Children* when she notes that

564 Howie West, Policy Alternatives, *Scapegoating Canada's public sector*,2015, https://www.policyalternatives.ca/Harper_Record_2008-2015/04-HarperRecord-West.pdf

565 Government of Canada, *The Canada Child Benefit*, March 2020, https://www.canada.ca/en/employment-social-development/campaigns/canada-child-benefit.html

loneliness is a tragedy that millions of children live with on a day-to-day basis. It sends them off to school in the morning and greets them at the end of the day. For an increasing number of children, the need for love and attention from the adult world is unmet. Because these kids feel so emotionally detached, on that level they are indeed "nobody's children." Ms. Bell says, "To whom can children look in the quest for belonging? Who will pick up the emotional slack too often unrecognized by their own parents? It's a job for any sensitized adults who are 'there' with enough time to make a child feel that he belongs."[566]

Many childfree people take up this slack for parents who don't have adequate time to give their children. I know that I certainly put in a lot of hours helping my sisters out with their children. I enjoyed taking them hiking, teaching them to swim, and often babysitting for a weekend to give the parents a break. I did this gladly, as I am sure many others who are childfree do. So those who choose not to breed often make great contributions to those who do breed, which I would call a **very selfless act indeed!**

In her book *Why Have Kids?* Jessica Valenti states, "If parenting is making Americans unhappy, if it's impossible to 'have it all,' if people don't have the economic, social, or political structures needed to support child rearing, then why do it? New parents find themselves struggling to reconcile their elation with the often exhausting, confusing, and expensive business of child care. When researchers for a 2010 Pew study asked parents why they decided to have their first child, nearly 90 percent answered, for 'the joy of having children.' Yet nearly every study in the last ten years shows a marked decline in the life satisfaction of those with kids." [567]

As a mother, Valenti explores this disconnect between would-be parents' hopes and the day-to-day reality of raising children—revealing all the ways mothers and fathers are quietly struggling. Perhaps this new reality is why so many millennials have decided to **forego bringing more babies into this world**. As a young childfree woman in 1971, the responses I received were pity, hostility, and patronizing quips meant to change my mind. However, in 2020, my decision sounds increasingly sensible and is becoming more widely accepted, especially in light of climate change and increasing pandemics.

The childfree movement actually started around 1960 when the birth control pill was approved. Then in 1973, abortion was legalized with the landmark *Roe v Wade* legal victory, giving couples even more control over parenthood. This gave rise to numerous childfree groups.

In 2020, groups like Childfree International, No Kidding, BirthStrike, and National Organization for Non-Parents are thriving around the world, since they realize that parenting is not for everyone. They don't feel that it is fair to the child if they cannot give it their full attention and the kind of love that every child deserves.

Childfree states, "'Childfree by Choice' is a group of adults who all share at least one common desire: we do not wish to have children of our own. We are teachers, doctors, business owners, authors, computer experts—you name it. We choose to call ourselves 'childfree' rather than

566 Valerie Bell, *Nobody's Children*, https://www.biblio.com/book/nobodys-children-bell-valerie/d/35129282?aid=frg&gclid=EAIaIQobChMIr5XfvcnC8gIV0Rt9Ch2Nzg6yEAQYASABEgJCu_D_BwE

567 Jessica Valenti, Goodreads, *Why Have Kids?*, https://www.goodreads.com/book/show/13594583-why-have-kids

'childless,' because we feel the term 'childless' implies that we're missing something we want—and we aren't. We consider ourselves childfree—free of the loss of personal freedom, money, time and energy that having children requires."

Childfreedom, according to the Happily Childfree website, is choosing not to create or raise children. It's about wanting to devote one's life to other objectives. For some, that is their career; for others, their hobbies (or volunteer work). Most childfree people recognize that parenting is a stressful, time-consuming, and often thankless job, and would rather do something else with their lives. As they point out, making the choice to be childfree is not something that is chosen lightly. In many cases, childfree people have thought longer and harder about why NOT to have kids than their counterparts have thought about why TO have kids.

No Kidding is a childfree group that was founded in Vancouver, British Columbia, in 1984 and now has more than 700,000 members in over 92 chapters across the globe. They are currently growing by more than 1,200 subscribers per day. No Kidding states, "r/childfree is one of the fastest growing communities on Reddit and it's for people who do NOT want children. They don't want to be told why they should have them, how much they'll regret it if they don't, and how 'selfish' they are for not contributing to society."[568] Turning that notion on its head, many people are choosing to create rather than procreate.

One *No Kidding* member points out, "And watching all of my friends have children, now I can see how exhausted they are financially, emotionally, how depleted their relationships are, because it's so difficult for them to be able to be a part of a partnership as well as try to raise these tiny humans." Others say that they **"came out" as childfree** because they don't want to subject their kids to environmental destruction—or take part in environmental destruction by having kids. They say the realization that you don't have to become a parent has been liberating, and they don't have to follow the stressful "life-script" trap of past generations.

Fair Start Movement points out that even though most couple's decision not to have children isn't based on finances, they've always enjoyed more disposable income—around $233,000 to raise a child through age 17, according to the United States Department of Agriculture.

There are many famous people who agree with this ideology. Celebrities like Renee Zellweger, Jay and Mavil Leno, Oprah Winfrey, Lily Tomlin, Dolly Parton, Betty White, Anjelica Houston, and Stevie Nicks have all chosen the childfree path. In 2020, *Slice* magazine quotes some of these progressive thinkers:

- Elizabeth Gilbert, author of *Eat Pray Love*: "I have come to believe there are three sorts of women, when it comes to questions of maternity. There are women who are born to be mothers, women who are born to be aunties, and women who should not be allowed within ten feet of a child. It's really important to know which category you belong to... And there is a curious rush of joy that I feel, knowing this to be true—for it is every bit as important in life to understand who you are NOT, as to understand who you ARE. Me, I'm just not a

568 Josh Swartz, No Kidding, *The Childfree Movement Hits Close To Home*, Dec 06, 2019, https://www.wbur.org/endlessthread/2019/12/06/no-kidding-childfree-movement

mom... Having reached a contented and productive middle age, I can say without a blink of hesitation that I wouldn't trade my choices for anything."

- Ellen DeGeneres: "Honestly, we'd probably be great parents. But it's a human being, and unless you think you have excellent skills and have a drive or yearning in you to do that, the amount of work that that is and responsibility—I wouldn't want to screw them up! We love our animals." *People*, March 2013
- Ashley Judd: In her 2011 memoir *All That Is Bitter and Sweet*, the Harvard grad wrote, "I figured it was selfish for us to pour our resources into making our 'own' babies when those very resources and energy could not only help children already here, but through advocacy and service, transform the world into a place where no child ever needs to be born into poverty and abuse again. My belief has not changed. It is a big part of who I am."[569]

It is likely that most of these women would not have pursued their successful careers had they chosen to have children. In 2019, insight therapist Noam Shpancer PhD, in *Psychology Today*, confirms past studies that despite lingering stigma, childfree women are often happier.[570]

Shpancer states, "Like all choices, though, this one is taken inside a social context. That context is quite heavily tilted in a 'pronatalist' direction. Our culture views parenting as an essential part of achieving fulfillment, happiness, and meaning in life, and as a marker of successful adulthood. Remaining childless by choice (AKA childfree) is still an outlying path, a move that raises questions and is met with prejudice and even moral outrage. This is particularly true for women, whose gender identity and social value have long been tied to fertility and motherhood. Thus, women who decide to not have children are commonly viewed unfavorably."

Whether to have children is one of the most consequential decisions we get to make.
—Noam Shpancer PhD,
Psychology Today

In 2015, 7.4 percent of 15- to 44-year-old women were voluntarily childless, according to Centers for Disease Control data. So what are the motives for child freedom? Shpancer believes that in most cases, they involve an interaction between macro-social forces—such as the availability of contraceptives, declining marriage rates, and women's increasing labour force participation—and micro-level considerations, including the desire for greater opportunity, personal freedom, self-fulfillment, and mobility.

Does research confirm that childfree women are happier? Shpancer states, "The broad answer appears to be a qualified **yes.** For one, research has consistently shown negative correlations between having children and marital and life satisfaction. Having children is associated with reduced happiness, particularly for women, and particularly in the U.S., and that link appears to sustain over the long term. Moreover, the money, time, and energy spent on raising children may be used instead to acquire sufficient wealth to allow one to purchase competent help in old age."

---

569 Slice magazine, *28 Famous Childfree Women Who Chose Not To Have Kids*, August 12, 2020 https://www.slice.ca/20-famous-women-who-chose-not-to-have-kids/

570 Noam Shpancer Ph.D., Psychology Today, *Why so many are satisfied being childless by choice*, Dec 29, 2019, https://www.psychologytoday.com/us/blog/insight-therapy/201912/why-so-many-are-satisfied-being-childless-choice

The literature suggests that most childfree women are not lonely, miserable, or odd. In fact, the opposite is more often the case. Shpancer adds, "The fact that the childfree path is less traveled does not make it less worthy or legitimate. Those who choose it may contribute to the social good in ways other than reproduction and cultivate new forms of meaning for themselves, untethered to motherhood."

So if you decide to be chidfree, you may want to be prepared for the cliché phrases that will be thrown at you:

- "It's different when it's your own child."
- "You'll forget all about the pain of labour."
- "Don't you like kids?"
- "Who will take care of you when you're old?"
- "Your biological clock is ticking."
- "The children are our future."
- "Don't you want your parents to have grandchildren?"
- "What if your child grows up to discover a cure for cancer?"
- "What if your parents had decided not to have kids?"
- "Parenthood is the most important job in the world."

Childfree couples must make it clear that they have given their decision a lot of thought, and they may prefer to own a pet, want a clean house, appreciate personal and financial freedom, and enjoy a career, a peaceful quiet house, and a good night's sleep. A person is not shirking their responsibilities for choosing this equally meaningful lifetyle.

BBC News reporter Ted Scheinman notes, "UK-based **BirthStrike** shares the goal of shifting the focus from population to carbon divestment and related policy action. Blythe Pepino, an activist and musician who founded BirthStrike at the end of 2018, says the group aims to push governments and corporations to drastically lower emissions while offering better humanitarian support to would-be mothers."[571]

The BirthStrikers have decided they can't bring children into a world facing an ecological breakdown, and where scientists predict climate change will bring bigger wildfires, more droughts, and food shortages for millions of people. Since children born in 2020 will be inheriting a world far worse than ours, **we are gambling with their lives**.

In 2019, CNN reporter Stephanie Bailey states, "US congresswoman Alexandria Ocasio-Cortez told her 3 million Instagram followers, 'there's a scientific consensus that the lives of children are going to be very difficult... is it still ok to have children?'"[572] In addition to fears surrounding the quality of life for future generations, some BirthStrikers don't want to have children because of the extra emissions that their kids, and their descendants, will produce. With a ticking clock, perhaps we should reconsider having children until the future looks a lot more promising.

---

571 Ted Scheinman, BBC News, The couples rethinking kids because of climate change, October 1, 2019, https://www.bbc.com/worklife/article/20190920-the-couples-reconsidering-kids-because-of-climate-change

572 Stephanie Bailey, CNN, *BirthStrike: The people refusing to have kids, because of 'the ecological crisis'*, June 6, 2019, https://www.cnn.com/2019/06/05/health/birthstrike-climate-change-scn-intl/index.html

In 2020, to be successful, to be valued, to be a global champion, it is favourable to forego having children and become an eco-activist. So instead of blindly following societal expectations, stop and think—why do you want that baby? You will be introducing the child to a world groaning at the seams, with melting icecaps, increasing pandemics, and warming temperatures. Can you bring them up healthfully? Is that ethical and just?

It's too bad that more couples don't think about their choice to be a parent as much as childfree couples do. Those deliberately childfree are making an intentional, honest, responsible, respectful, fair, and for a growing number, right decision. Beyond that, many considering the childfree lifestyle have observed parenting up close, and don't like what they see—how unhappy, stressed, financially depleted, or neglectful many of the parents are. Many children also look so sad, stressed, confused, or depressed—subjecting them to a world greatly diminished seems unjustified and wrong.

In the *Solutions Journal*, Robert Engelman notes, "If world population peaked at close to 8 billion rather than 9 billion, along the lines described in a low-fertility demographic projection published by the UN Population Division, the model predicted there would be a significant emissions savings: about 5.1 billion tons of carbon dioxide by 2050 and 18.7 billion tons by century's end. What if we could prove wrong the popular conviction that a future with 9 billion people and a growing population is inevitable? Suppose we could demonstrate that world population size might peak earlier and at a lower level if government policies aimed not at reproductive coercion but at individual reproductive freedom?"[573]

Amazing Success Stories

Society is always taken by surprise at any new example of common sense.
—Ralph Waldo Emerson

When I talk to fellow Canadians about overpopulation, they usually respond that Canada isn't overpopulated, and the problem lies in Asia and Africa with the fastest population growth. Well, that isn't always the case. Ironically, there seems to be a baby boom in my region of Canada in 2020, and yet many countries abroad have fertility rates below ours—even in parts of Africa and Asia.

The Overpopulation Project: the case of **Tunisia, Africa**, with the first population policies implemented on the African continent. In the early 1960s, Tunisia significantly improved women's status when they launched a voluntary national family planning program, and now it is one of the most progressive Arab countries in terms of women's rights. In 1962, Tunisia's national Economic and Social Development Plan recognized **the need to reduce population growth in order to achieve economic and social improvement**. The Overpopulation Project reporter Patricia Derer tells us, "Today Tunisia has some of the most progressive family planning (FP) policies in Africa, and it is the most progressive of all Arab countries in terms of gender equality and women's rights.

573 Robert Engelman, The Solutions Journal, *An End to Population Growth: Why Family Planning Is Key to a Sustainable Future,* May 2011, https://www.thesolutionsjournal.com/article/an-end-to-population-growth-why-family-planning-is-key-to-a-sustainable-future/

As a consequence, its fertility fell from 7 children/mother in the 1950's to around replacement level by the early 2000's, allowing Tunisia to harness the benefits of its demographic dividend."[574]

Despite having limited natural resources, Tunisia's recent growth and development performance have been notable relative to its geographic neighbours and other countries at similar levels of development. Under President Habib Bourguiba, Tunisia's almost entirely Arab and Muslim population experienced rapid development in almost every area.

Derer notes, "Shortly after independence, he began a remarkable liberalization of the law, inducing social changes outstanding at the time, especially for a Muslim country. The president, educated in Paris, recognized that empowering women was essential for the development of the country. He gave women the right to vote, to remove the veil, and to divorce. Uniquely in the Arab world, he forbade polygamy and child marriage. In addition, Bourguiba made female sterilization and abortion legal, considerably improving women's legal status in a very short period of time. These measures, including limiting government subsidies after the fourth child, are considered the first modern population policies in the Arab world and on the African continent."

Bourguiba's unique legislation was based on a modern interpretation of Islam, and was generally well received by both religious leaders and the public. Tunisia's family planning services were free and voluntary, and their liberal abortion law is still unique in the Arab world. We can only hope that eventually all countries in Africa will develop these same time-tested policies.

Despite some fears of aging at the beginning of the national family planning program, the program did not change its direction **and continues today**. In 2014, Tunisia had the second strongest family planning effort in Africa (after Rwanda) in terms of policies, access to contraception and services, and evaluation.

The Iranian Miracle: Iran has experienced a similarly dramatic fertility decline. When discussing the population crisis, *Mother Jones* contributor Julia Whitty states, "In Iran, the fertility pendulum has gone the other way in recent years. From a high of 7.7 in 1966, total fertility fell to 6 during the Shah's reign, spiked to 7 during the Islamic Revolution (when marriage became legal for 12-year-old boys and 9-year-old girls), then plummeted 50 percent between 1988 and 1996, continuing down to 1.7 today."[575]

> *That plunge, known as the "Iranian miracle," was one of the most rapid fertility declines ever recorded.*
> **—Julia Whitty,** *Mother Jones*

Whitty notes that Iran's demographic reversal was swift, uniform, and voluntary. She stated, "In this traditional society ruled by Islamic law women of all childbearing ages in urban and rural parts of the country simply began to have smaller families practically overnight. Demographer Mohammad Jalal Abbasi-Shavazi of the University of Tehran writes that the feat was engineered through a mobilization between **government and media**: Information was broadcast nationwide

---

574 Patricia Derer, The Overpopulation Project, *The first population policies implemented in Africa: the case of Tunisia,* July 29, 2019, https://overpopulation-project.com/the-first-population-policies-implemented-in-africa-the-case-of-tunisia/

575 Julia Whitty, Mother Jones, *The Last Taboo,* June 2010, https://www.motherjones.com/environment/2010/04/population-growth-india-vatican/

about the value of small families, followed up with education about birth control, implemented with free contraceptives." Also, population education became part of the curriculum at all educational levels; even **university students had to take a two-credit course on population and family planning**.

Population Reference Bureau contributor Farzaneh Roudi-Fahimi states, "In 1993, the legislature passed a family planning bill that **removed** most of the **economic incentives for** large families. For example, some allowances to large families were cancelled, and some social benefits for children were provided for only a couple's first three children."[576]

Iran also extended social security and retirement benefits to all parents so that they would not be motivated to have many children as a source of old age security and support. It was mandatory for both men and women planning to marry to participate in family planning classes before receiving their marriage licence, increasing male involvement in family planning. In recent years, the government has encouraged a higher fertility rate, but here again, Iranians aren't buying it.

The Overpopulation Project reporter Patricia Derer points out that Iran's program had three goals:

- Encourage birth spacing intervals of three to four years.
- Discourage pregnancy among women younger than 18 and older than 35.
- Limit family size to three children, but encourage family size of two or less.

"Iran stands out for lowering its fertility in a very short time without coercion. In rural areas it dropped from 8.1 to 2.1 in only one generation, which in comparison, took European countries 300 years."[577] Similar spectacular fertility reversals have occurred in Costa Rica, Cuba, South Korea, Taiwan, Thailand, Tunisia, and Morocco—**as quickly as in China but minus the one-child policy**.

Bangladesh is a country neighbouring India, and on the southern coast an enormous mangrove forest shared with Eastern India is home to the royal Bengal tiger. In 1976, the villages in the Matlab region of Bangladesh committed to increase access to family planning by training 23,000 women health workers to provide free doorstep reproductive health services. By 1996, the f**ertility rate had been cut by half through this voluntary program,** a remarkable achievement which set an example for other regions.

Family Planning 2020 adds, "In Bangladesh, family planning remains one of the top priorities in the 4th Health Sector Programme 2017-2021, as a path toward achieving the Sustainable Development Goals.... Bangladesh reiterates its commitment to end child marriage. For these, the Government of Bangladesh will mobilize USD$615 million for the family planning programme over 2017-2021, which is a 67% increase from the previous program."[578]

---

576 Farzaneh Roudi-Fahimi, Population Reference Bureau, *IRAN'S FAMILY PLANNING PROGRAM*, https://www.prb.org/wp-content/uploads/2016/09/IransFamPlanProg_Eng.pdf

577 Patricia Derer, The Overpopulation Project, *The Iranian miracle: The most effective family planning program in history?*, March 21, 2019, https://overpopulation-project.com/the-iranian-miracle-the-most-effective-family-planning-program-in-history/

578 Family Planning 2020, *Bangladesh - Commitment Maker Since 2012*, 2020, http://www.familyplanning2020.org/bangladesh

In 2019, India's overpopulation emergency is prompting a trend in couples voluntarily choosing to be childfree. Why? *Times of India* reporter Spardha Pandey explains, "The reasons for going child-free in India are not limited to financial constraints, high level of responsibility or lack of family support. Ensuring the safety of the child and rising pollution levels were also key reasons."[579]

One couple in India shares, "I feel scandalised when I see the news of kids being molested in their schools, kidnapped for ransom, and four-year-old girls being brutally raped. I can't think of bringing up a child in such an environment; I would be in a constant state of stress. Also, what about the child's health? The pollution levels are going crazy. And there is always a question of whether we would be able to afford the child's expenses or not."

The World's Wealthiest Give Back

There may be help on the way from the planet's elite. For decades, wealthy philanthropists have benefited from industries that contributed greatly to climate destruction, the use of slave labour, the glorification of the war machine, and the list goes on and on. Now they are finally set on giving back and reversing some of the damage that has been done.

It all started in 2009, as *WorldNetDaily* explains, "Some of the richest men and women in the world met secretly recently in New York to conspire on using their vast wealth to bring the world's population growth under control. The meeting included some of the biggest names in the 'billionaires club,' according to the *London Times*—Bill Gates, David Rockefeller, Ted Turner, Oprah Winfrey, Warren Buffet, George Soros and Michael Bloomberg."[580]

Everyone in the group agreed that the population crisis is something so nightmarish that it needs big-brain and well-funded solutions. They felt that the group needed to be independent of government agencies, which seem unable to head off the disaster we all see looming. They felt it was necessary to join forces to **overcome political and religious obstacles** to change.

Fast forward to 2019, and this giving pledge is renewed by this same billionaires' club, but organizers Bill Gates and Warren Buffett have some concrete goals in mind. The idea was to persuade their fellow billionaires to pledge at least 50 percent of their wealth to charity, and many have. *Vox News* reporter Kelsey Piper states, "Ten years later, we are in a position to evaluate some of the effects of the Giving Pledge. Hundreds of billionaires have signed on, with Bloomberg and other big names like Mark Zuckerberg, Elon Musk, and Mackenzie Bezos among them. **More than $500 billion have been pledged,** and money is already being donated to a range of causes."[581]

Gates and Buffett, who had pledged half their fortunes to work on global health and education, have set a remarkable example. There are important moral priorities that U.S. and other governments are not pursuing and will not pursue, from advocacy for civil liberties to **reproductive health care.** Piper points out that **philanthropy can help fill that void.**

579 Spardha Pandey, Times of India, *Why these Indian couples are opting to ot have kids*, Dec 3, 2019, https://timesofindia.indiatimes.com/life-style/parenting/getting-pregnant/why-these-indian-couples-are-opting-to-not-have-kids/articleshow/64135670.cms

580 WorldNetDaily, *Secret Billionaire Club Seeks Population Control*, May 24 2009, https://www.wnd.com/2009/05/99105/

581 Kelsey Piper, Vox News, *The Giving Pledge, the campaign to change billionaire philanthropy*, Jul 10, 2019, https://www.vox.com/future-perfect/2019/7/10/18693578/gates-buffett-giving-pledge-billionaire-philanthropy

Forbes reporter Will Yakowicz tells us that Bill Gates has also partnered with two African billionaires in a venture to ease Africa's population burden. He states, "Bill Gates, the second richest person in the world; Aliko Dangote, the richest man in Africa; and Mohammed "Mo" Ibrahim, a U.K. billionaire born in Sudan, spoke about ways Africa can reach its potential in the coming decades."[582]

Africa today has the biggest gap in what people want in family planning, and what's available to them. Melinda [Gates] is trying to close that gap, because then everything gets easier—education, food, stability, jobs.
—Bill Gates

When discussing how to provide jobs and feed Africa's hungry hordes, they were in agreement that family planning is essential to Africa's future. Ibrahim commented, "When our economy is growing by two percent, we're running on a treadmill. Why are we Africans unwilling to talk about family planning?" He explained that unemployment rates are already high, especially among the youth in countries like South Africa.

"Africa today has the biggest gap in what people want in family planning, and what's available to them. Melinda [Gates] is trying to close that gap, because then everything gets easier—education, food, stability, jobs," said Gates.

According to the United Nations, Africa's population is expected to reach 2.4 billion people by **2050**—double that of the population in 2016. Gates said it is estimated that 50 percent of all newborns worldwide will be born in Africa by the end of the century. However, if the billionaires make population their focus, these numbers could be greatly reduced. Meanwhile, many billionaires have **not** signed on to the Giving Pledge yet. Piper adds that there are now 607 billionaires in the U.S. and more than 2,000 worldwide. So just imagine the possibilities!

Oxfam tells us that in 2020, over 80 millionaires from seven countries around the world called on governments to implement a permanent tax increase for the super-rich. The "Millionaires for Humanity" wish to contribute towards COVID-19 and other social issues related to the pandemic.[583]

The letter highlights the role that the richest people in society can play in helping to **rebalance the world economy**. The group urges governments to **raise taxes on millionaires and billionaires "immediately, substantially and permanently."** Millionaires For Humanity is a project by Bridging Ventures, Club of Rome, Human Act, Oxfam International, Patriotic Millionaires, and Tax Justice UK.

The millionaires, mostly based in the US, the UK, Germany, Canada, and New Zealand articulate in the letter that increasing the taxes will adequately fund health systems, security, and schools.
—Business Today

Pearl, chairman of the Patriotic Millionaires, said, "The COVID-19 crisis has revealed the fragility of our

582 Will Yakowicz, Forbes, *Bill Gates and Two African Billionaires Say Family Planning is Essential to Africa's Future,* Sept 24 2019, https://www.forbes.com/sites/willyakowicz/2019/09/24/bill-gates-and-two-african-billionaires-say-family-planning-is-essential-to-africas-future/#6825c64d20d1

583 Oxfam, July 13 2020, *Over 80 millionaires around the world call for higher taxes on the richest to help COVID-19 global recovery,* https://www.oxfam.org/en/press-releases/over-80-millionaires-around-world-call-higher-taxes-richest-help-covid-19-global

system and shown that no one—rich or poor—is better off in a society with massive inequality and a failing social safety net. We must reset our tax structure to one that values the contribution of labor as much as the contribution of capital."

But this isn't the first such entreaty. Even before the COVID-19 pandemic upset public finances, the Patriotic Millionaires pressed for a more forward-looking taxation system. This is especially encouraging news, since citizens have been condemning the unjust imbalance of wealth for decades.

Grassroots Taking Action

But it isn't just the filthy rich that are in the mood for giving. Live Kindly reporter Kat Smith says, "Swedish environmental activist Greta Thunberg is donating her 1 million Euro (over 1.5 million Canadian Dollars) Gulbenkian Prize for Humanity to climate charities."[584] Smith adds that the Greta Thunberg Foundation, created by the Swedish vegan activist, will donate the prize money to SOS Amazonia in Brazil, and €100,000 to the Stop Ecocide Foundation. This Netherlands-based charity aims to make mass destruction of nature an international crime.

Global Population Speak Out is taking the first step in providing a platform for individuals or groups to speak out about their ideas and actions to raise public awareness. It is the brainchild of the Population Media Center, and started by Dr. John Feeney in 2009. Speaking out is the missing link in addressing the population crisis, so check this site out for ways that you can participate. This initiative has recruited substantial numbers of world-class scientists, academicians, opinion-leaders—and thousands of lay environmentalists and concerned citizens—to help bring international attention to the crises posed by population growth.[585]

The Center for Biological Diversity is a world leader in tackling the root causes of the world's woes, most importantly population and animal agriculture. In 2020, they created the Crowded Planet Library to provide a variety of resources on population issues, including ecology, conservation, demography, psychology, and policy. The center points out, "By bringing them all together in one place, we can better understand the wide-ranging effects of human population pressure as well as the solutions that can benefit endangered species, human rights and the health of communities and the planet."[586]

---

584 Kat Smith, Live Kindly, *Greta Thunberg is Donating €1 million Award to Climate Charities*, July 21, 2020, https://www.livekindly.co/greta-thunberg-donating-award-climate-charities/?goal=0_8051ea5750-32d3cbfd0b-136030931&mc_cid=32d3cbfd0b&mc_eid=834df3ebaa

585 Population Media Center, *Global Population Speak Out*, 2020, https://www.populationmedia.org/projects/global-population-speak-out/

586 Center for Biological Diversity, *Crowded Planet Library*, 2020, https://crowdedplanet.org/?emci=1c7f4a1c-08dc-ea11-8b03-00155d0394bb&emdi=b7359d22-95e3-ea11-8b03-00155d0394bb&ceid=1885004

Science in Action

We declare clearly and unequivocally that planet Earth is facing a climate emergency.
—Union of Concerned Scientists, "World Scientists Warning to Humanity – A Second Notice"

Chapters 5 and 7 provide ample scientific evidence of the climate and biodiversity emergencies humanity has created, and the solutions available to rectify these problems. This forewarning is in line with the "World Scientists Warning to Humanity – A Second Notice" and the UN Sustainable Development Goals. In 2020, world influencers have come to an agreement that the window of opportunity is ten years, the threat is catastrophic, and our response to date has been totally inadequate. WE MUST ACT NOW OR IT WILL BE TOO LATE.

The *Guardian* reporter Damian Carrington notes, "The statement was a collaboration of dozens of scientists and endorsed by further 11,000 from 153 nations. These scientists say the urgent changes needed include **ending population growth**, leaving fossil fuels in the ground, halting forest destruction and slashing meat eating."[587]

Scientists feel they have a moral obligation to clearly warn humanity of this imminent threat saying, "The climate crisis has arrived and is accelerating faster than most scientists expected. It is more severe than anticipated, threatening natural ecosystems and the fate of humanity. Despite 40 years of global climate negotiations, with few exceptions, we have largely failed to address this predicament." Since these scientists had dropped the ball on the population issue for a couple of decades, it is good to see that it is finally back on the radar in 2020.

These 11,000 scientists set out a series of **urgently needed actions**:

- Use energy far more efficiently and apply strong carbon taxes to cut fossil fuel use.
- Stabilize global population—currently growing by 200,000 people a day—using ethical approaches such as longer education for girls.
- End the destruction of nature and restore forests and mangroves to absorb CO_2.
- Eat mostly plants and less meat, and reduce food waste.
- Shift economic goals away from GDP growth.

"**The good news** is that such transformative change, with social and economic justice for all, promises far greater human well-being than does business as usual," the scientists said. The recent surge of concern was encouraging, they added, from the global school strikes to lawsuits against polluters and many nations and businesses starting to respond with positive change.

One 2020 case of legal action comes from a group of Portuguese youth who are suing 33 countries over the climate crisis. Grist reporter Jonathan Watts tells us, "Young activists from Portugal have filed the first climate change case at the European court of human rights in Strasbourg, demanding 33 countries make more ambitious emissions cuts to safeguard their future physical and mental wellbeing. The crowdfunded legal action breaks new ground by suing multiple states

587 Damian Carrington,The Guardian, *Climate crisis: 11,000 scientists warn of 'untold suffering'*, Nov 5 2019, https://www.theguardian.com/environment/2019/nov/05/climate-crisis-11000-scientists-warn-of-untold-suffering

both for the emissions within their borders and also for the climate impact that their consumers and companies have elsewhere in the world through trade, fossil-fuel extraction, and outsourcing."[588]

This kind of large-scale environmental devastation is legally called ecocide: the damage, destruction, or loss of ecosystems, whether by human agency or other causes. This is the legal definition proposed to the UN by international environmental lawyer, Polly Higgins, in April 2010.[589] Higgins anticipates that this crime will be included in international law by 2025.

Live Kindly is excited to highlight the YouTube video, *ENDGAME 2050,* which has included population as a driving force behind global turmoil. This scientific expose adds to the dire warning to humanity. *ENDGAME 2050* warns, "Humanity has backed itself into an ecological endgame as we approach the year 2050."[590]

A **Cornell University study** of population trends, climate change, increasing pollution, and emerging diseases concludes that "life on earth is killing us." The study notes that an estimated **40 percent of deaths around the world can now be attributed to various environmental factors,** especially chemical pollutants. Such environmental degradation, coupled with the *growth* in world *population,* are major causes behind the rapid *increase* in human *diseases* worldwide.[591]

A **2017 Mount Sinai Medical report** concurs, adding, "Environmental pollution—from filthy air to contaminated water—is killing more people every year than all war and violence. More than smoking, hunger or natural disasters. More than AIDS, tuberculosis and malaria combined."[592]

The report notes that one of out every six premature deaths in the world in 2015—about nine million—was attributed to disease from toxic exposure, according to a major study released Thursday in *The Lancet* medical journal. The financial cost from pollution-related death, sickness, and welfare is equally massive, the report says, costing some $4.6 trillion in annual losses—or about 6.2 percent of the global economy.

One Example of Corporate Response

When the last tree has been cut down, the last fish caught, the last river poisoned, only then will we realize that one cannot eat money.
—Cree Indian proverb

This Cree proverb is interesting in that it reminds us that the Indigenous lifestyle, not centred around cities, is much more connected to the earth and much less likely to promote income

588 Jonathan Watts, Grist, *Portuguese youth are suing 33 countries over the climate crisis,* Sep 6, 2020, https://grist.org/climate/portuguese-youth-are-suing-33-countries-over-the-climate-crisis/?

589 Polly Higgins, Stop Ecocide, 2010, https://www.stopecocide.earth/polly-higgins

590 LIVEKINDLY, *ENDGAME 2050,* https://www.endgame2050.com/?utm_source=LIVEKINDLY+Mailing+List&utm_campaign=6c742298b3-

591 Cornell University, Science Daily, *Pollution Causes 40 Percent Of Deaths Worldwide, Study Finds,* Aug 14 2007, https://www.sciencedaily.com/releases/2007/08/070813162438.htm

592 CBC News, *Pollution causing more deaths worldwide than war or smoking: Lancet,* Oct 20, 2017, https://www.cbc.ca/news/health/pollution-worldwide-deaths-1.4363613

disparity. Perhaps we have a lot to learn from this concept.

Every city in an "overdeveloped" nation has within it a slum of malnutrition, homelessness, and unemployment. Every city in a "developing" nation has within it a super-rich district of high tech, high fashion, and high finance. Seeing the cities of the world as globally connected—all run by the super- rich, yet all including a component of homelessness and poverty—changes the stereotype perspective and invites all cities of the world from Paris to Calcutta to unite in this decade of change. In 2020, protesters are demanding that we eliminate poverty and tax the wealthy enough to substantially close this gap by 2030.

One company that is taking today's youth protesters seriously is Unilever, which is based in the U.K. and the Netherlands and owns brands such as Surf, Dove, and Breyer's. In 2020, they pledged to drop fossil fuels from their cleaning products by 2030 to reduce carbon emissions, as urged by Generation Z activists. BBC News notes that the consumer goods giant said it would invest €1 billion (£890 million, $1.2 billion) in the effort to replace petrochemicals with ingredients made from plants, and marine sources like algae.[593]

The move is part of the corporation's "Clean Future" initiative, which aims to bring Unilever's carbon emissions to net zero by 2039. The company plans to halve its use of new plastic in packaging by 2025 and also wants all its product formulations to be biodegradable by 2030. "People want more affordable sustainable products that are just as good as conventional ones," said Peter ter Kulve, Unilever's president of its home care division. This is only one company's enlightened response to the global crisis—a call to action for all corporations to think of the common good and generations yet to come.

> *We must stop pumping carbon from under the ground when there is ample carbon on and above the ground if we can learn to utilise it at scale.*
> **—Unilever**

Media Working for Change

Fortunately, there are organizations trying to encourage honest and effective media involvement in population coverage. The *Population Institute* is an international educational organization that was established in 1969. Headquartered in Washington D.C., it now has members in 172 countries.[594]

The Population Institute's Global Media Awards are designed to encourage greater media coverage of population and development issues. They honour those who have contributed to creating awareness of population problems through their outstanding journalistic endeavours. The awards serve to encourage editors, news directors, and journalists to acquire a more in-depth knowledge of population issues to stimulate high standards for journalism. There are 12 categories for awards, including magazine, TV shows, radio, news, editorials, online, and films.

In 2017, the Population Institute announced the winners of its 38th annual Global Media Awards. In a slight departure from previous years, that year's winners were selected for their

593 BBC News, *Unilever to cut fosil fuels from cleaning brands*, Sep 2 2020, https://www.bbc.com/news/business-53994319
594 Population Institute, *Population Institute Announces 2017 Global Media Award Winners*, February 6, 2018, https://www.populationinstitute.org/our-news/population-institute-announces-2017-global-media-award-winners/

research, writing, or reporting on topics related to the escalating political assault on reproductive health and rights in the U.S.

In announcing the awards, Robert Walker, the president, noted that, "Last year [2017 was an extraordinary year, especially at the federal level, where President Trump and his allies in Congress, sought to target Planned Parenthood, restrict abortion rights, repeal the Affordable Care Act, eliminate funding for family planning, terminate the Teen Pregnancy Prevention (TPP) program, and stack the federal judiciary with judges who are hostile to reproductive rights." Given these unprecedented attacks on reproductive health and rights, Walker said, "We felt it was important to acknowledge the individuals and organizations that have done the most to keep the public informed on these issues and their public policy implications."

Despite all of these efforts to highlight the "big picture" and include other species in the equation, the myth still exists that when we choose between feeding the hungry and conserving nature, people must come first. After all, we wouldn't sacrifice starving African children in order to save the threatened black rhino, would we? As a result, in 2020, the black rhino is classified as critically *endangered—on the brink of extinction,* and human population is at nearly EIGHT BILLION. Ideally, we should and could be saving both if we included population in the conversation.

Governments and people around the world spend billions of dollars on art, music, and entertainment to improve our quality of life. Half of this money could, in fact, be going to save those species most at risk—why not consider nature as important a commodity as art? After all, it is often the catalyst that inspires art!

Landmark Policies

In 2017, the Government of Canada took a step in this direction. Open Canada reporters Gilian Barth and Sandeep Prasad stated, "Last month, the Government of Canada announced its first feminist international assistance policy. This landmark policy pledges that by 2021/2022, at least 95 percent of Canada's bilateral international development assistance investments will either target or integrate gender equality and the empowerment of women and girls. The new policy is accompanied by a $CAD650 million commitment over three years to support comprehensive sexual and reproductive health care in development contexts."[595]

Barth and Prasad note that this funding is intended to "fill the gaps" created by Canada's former policies to restricted funding for sexual and reproductive health and rights, including contraception and safe abortion. Furthermore, in March 2017, Canada pledged $20 million towards the She Decides movement, which seeks international cooperation to fill the global funding gap created by the United States' Mexico City (global gag) policy.

By coordinating its investments with other donors, Canada could ensure global access to sexual and reproductive health services in crisis-affected areas—and in a cost-effective manner. These crises are becoming more prevalent and intense as pandemics, political turmoil, and population

595 Gillian Barth, Sandeep Prasad, Open Canada, *In the world's worst crises, access to sexual and reproductive health and rights is paramount,* July 7, 2017, https://www.opencanada.org/features/worlds-worst-crises-access-sexual-and-reproductive-health-and-rights-paramount/

disasters plague the planet. So Canada's leadership and contribution could help fill a crucial and growing gap in some of the harshest landscapes on Earth.

Heavy Hitters

As I see it, humanity needs to reduce its impact on the Earth urgently and there are three ways to achieve this: we can stop consuming so many resources, we can change our technology and we can reduce the growth of our population.
—Sir David Attenborough

Throughout the book I have discussed how neither reckless consumption nor irresponsible population growth are sustainable, and both demand immediate action. In 2012, at the UN Rio Earth Summit, two of the world's iconic heavy hitters fought down the Vatican and religious right at the conference to move forward with empowering women and promoting family planning. In 2020, these same powerful women are still fighting for sexual equality in a new initiative: *The Elders.*

Population Growth reporter Suzanne York states, "**Gro Harlem Brundtland**, former president of Norway and now the UN Special Envoy on Climate Change has been around big international and environmental negotiations for decades. A great concern, she said, is that many of the sensitivities linked to population issues are still with us. In Brundtland's opinion, at Rio and beyond, the focus must be on what women want and need, and we must put a stop to all discrimination against women and girls. We can only succeed by enabling young women to avoid early pregnancy, breaking the inter-generational cycle of poverty, and by providing education."[596]

Brundtland is also known for having chaired the Brundtland Commission, and has been involved in The Elders' initiative on child marriage, including the founding of Girls Not Brides: The Global Partnership to End Child Marriage.

York added, "**Mary Robinson**, former president of Ireland, stressed the importance of bringing the message of supporting family planning and reproductive health to a much wider audience and link it with women's issues. At the Rio+20 text negotiations, Robinson said it is very worrying that these issues are not on the agenda. '**Who is trying to prevent this**?' she asked. The former Irish president challenged the audience that this is a fight that still has to be fought and we have to be prepared to fight it."

Robinson is also former UN High Commissioner for Human Rights and Chair of The Elders; she is a passionate, forceful advocate for gender equality, women's participation in peace-building and human dignity. She asserted, "We need leadership on this issue. We need to connect the dots and integrate issues, and the essential element is the economic empowerment of women. And the removal of discrimination, especially of the girl child, has to be front and center."

One remarkable form of leadership came from **Nelson Mandela**, who stepped up to champion women's equality. Robinson noted, "When Nelson Mandela brought us together as Elders, he did

596 Suzanne York, Population Growth.org, The Heavy Hitters of Population Growth & Women's Rights, Jun 19 2012, http://populationgrowth.org/the-heavy-hitters-of-population-growth-womens-rights/

so in the belief that together we are stronger, that change happens when people collectively take action to make our world a better place."[597]

Who are the members of the Elders? Present at the 2007 launch were Kofi Annan, Jimmy Carter, Graça Machel, **Nelson Mandela**, Mary Robinson, Desmond Tutu, Muhammad Yunus, and Li Zhaoxing. Members who were not present at the launch were Ela Bhatt, Gro Harlem Brundtland, Lakhdar Brahimi, and Fernando Henrique Cardoso. They agree that what the world needs now is a nucleus of wise elder statesmen and women to grapple with seemingly intractable global issues that governments and international institutions overlook or have failed to correct.

Retired from public office, the Elders say they need no longer fear political fallout from positions they take. They bring diverse knowledge and experience from every region on a broad range of issues, from political and economic to social and environmental. Archbishop Tutu noted as others have that **some persistent customs palmed off as religious are not part of religion at all**. This point was made forcefully at the Rio Earth Summit when Mary Robinson refused to allow religious entities to derail progress on sexual and reproductive health services that would transform the lives of women and girls globally.

One of the world's most prominent organizations has been **taking action** for decades to reduce carbon emissions by reducing population growth. In 2020, Population Connection celebrated World Population Day, which was established in 1989 by the UN in the wake of the world's population reaching five billion people. Population Connection states, "The purpose of this day is to bring attention to the health, development, and environmental impacts of rapid population growth."[598]

In 1968, the United Nations made it a basic human right for each woman to decide the number and spacing of her children. However, access to contraception, especially in rural parts of Africa and Asia, has not kept up with these legal changes. Recognizing this tremendous shortfall, 151 organizations have now joined forces to create the **Thriving Together** campaign, which is spearheaded by the London-based Margaret Pyke Trust. Together, they spend £8 billion each year on family planning and environmental work in 170 countries globally.

The focus of this year's World Population Day is "the sexual and reproductive health needs and vulnerabilities of women and girls during the pandemic."
—Population Connection 2020

Among the 151 organizations are Greenpeace, Marie Stopes International, the United Nations Population Fund, Conservation International, and the Bill & Melinda Gates Institute for Population & Reproductive Health. The campaign will focus on **getting the environmental organizations to talk about family planning** and vice versa in a coordinated approach. They have launched their campaign on World Population Day and plan to table a motion at the 2020 World Conservation Congress to get it on the agenda of national governments worldwide. Also recognizing the wisdom of getting environment groups involved, Paul Ehrlich brilliantly declared, "You don't have a conservation policy unless you have a population policy."

597 Mary Robinson, Th e Elders, https://www.theelders.org/profile/mary-robinson
598 Population Connection, *World Population Day: July 11, 2020*, 2020, https://www.populationconnection.org/world-population-day-2020/

School First #BabyLater

Through this program designed to break the mould, Population Services International (PSI) encourages couples to think about their financial goals and how to make choices aligned with those life goals, including decisions about when to start a family. Most importantly, PSI notes, it's working—four in five girls who interact with PSI's Smart Start take up a contraceptive method.[599]

These initiatives that recruit youth as co-creators allow them to take the reins and influence the project early on. More than 25,000 married girls chose to use a contraceptive method as a result of Smart Start since January 2018. By 2024, PSI is on track to reach 500,000 girls with contraception. They have made some bold commitments to achieve faster results and revolutionize the way adolescents access contraception

Prominent population group, **Growthbusters**, is also working to get the attention of governments and the media. They warn, "Scientific estimates of the population the Earth can sustainably support range from under 1 billion to 5 billion. The good news is that overpopulation is solvable. We can, and must, voluntarily and dramatically reduce births so that total numbers drift back down to a truly sustainable level. By averaging 1.5 children, total world population would—instead of rising to 11 billion—decline to about 5 billion by 2100. A one-child average would reduce total population to below 3 billion—cutting our total human consumption by more than half from its current levels."[600]

599 Population Services International, *School First #BabyLater*, Aug 28 2019, https://www.psi.org/2019/08/babylater/
600 GrowthBusters.com, *Overpopulation*, http://www.growthbusters.org/population-growth/

PART 5

FAST FORWARD TO PROSPERITY

The opposite of wealth is not poverty but sufficiency. This is critical. Sufficiency is not a matter of sacrifice and deprivation. It is a means of working out different ways of achieving satisfaction in our own lives.
—Jonathon Porritt, Friends of the Earth

Ancient civilizations predicted a time when we would come to a crossroads, a time when humankind would have to decide to make the necessary changes to save our planet—or not. Many today see Easter Island as a metaphor of the modern world, giving fair warning of what a runaway population can do. In 2021, this milestone reminds us that our planet is finite, and the insane rate of population and economic growth we have been promoting cannot continue. Part 5 will explore the exciting benefits of a decreasing population and give a sampling of what this GOOD NEWS will look like. It will instill in every reader that sense of power we each possess—to make this dream a reality.

CHAPTER 14

THE DAY THAT GROWTH STOOD STILL

Population is the multiplier of everything we do wrong.
—Dr. Martha M. Campbell, Berkeley School of Public Health

We will recognize the day when humankind's collective wisdom begins to pay off because that will be the day that population begins to decrease—**World Depopulation Day**. That **one special day when growth stands still**—the population clock on my website will hesitate for a moment, and then the numbers will actually start to reverse direction. When that momentous day actually arrives, it will give the world reason to celebrate as never before. As I realized long ago, that is my purpose on this planet—to help bring about this historic event in my lifetime. The good that could come from this one unprecedented phenomenon would finally put humanity on the right side of history. How about setting a date for **World Depopulation Day—July 11, 2025**. That day, July 11, is appropriate since it is the day the UN chose for World Population Day. And the countdown begins as humanity plans for the most important day of the 21st century. Check out my population clock at populationinsync.net.

When do **you** think that day will come? The United Nations estimates a population of ten billion by 2050, and if that comes to pass, humans will be one of the few species remaining on our depleted and crippled planet. So what if we blow that number out of the water, and stop at EIGHT BILLION instead? We know that this is possible, and we have the technology to achieve this. A chorus of voices is already promoting this bold endeavour, and a swell of funding is beginning to materialize to make this happen within the next decade. Can you imagine? The benefits would be extraordinary!

The **quality of life of all species on the planet would soar in ways we can't even imagine**—energy demands and pollution would begin to decrease, our real income would increase, the state of our environment would begin to improve as we began to reclaim the land, we could begin to remove threatened species from the Red List ... and the benefits go on and on. We would have all the advantages of modern technology but little, if any, environmental deterioration.

This is our last chance, since we have passed up all the other opportunities that were given us. What if we could rewind, and the government had taken Malthus, Ehrlich, or the Union of Concerned Scientists seriously? What if Martin Luther King's 1966 speech on family planning had been heeded and used to rewrite history? What if Gandhi had led a population revolution after his political revolution? Imagine how these actions would have furthered the greater good. Now India's population is predicted to surpass that of China's in this century. How different things could have been.

What about technology? Rex Weyler, co-founder of Greenpeace, notes, "Some economists imagine that computer chips or nanotechnology will save us from the laws of nature, but every technical efficiency in history has resulted in more consumption of energy and resources, not less. Remember when computers were going to save paper? That never happened. Computers increased paper consumption from about 50 million tons annually in 1950 to 250 million tons in 2009. Meanwhile, we lost 600 million hectares of forest."[601]

The more desperate the overpopulation problem becomes, the more people are looking to technology to bail us out. Yet, as technology flourishes and we continue to make spectacular breakthroughs, the number of people living in poverty remains over a billion. There are a million species on the brink of extinction, and 40 percent of births are still unintended. So what would make us think that technology could solve the problems of the future, when it falls so far short of dealing with the problems of the present? The Union of Concerned Scientists has warned that we cannot expect technology to save us from the problems caused by overpopulation. We can toss around ideas like the Green New Deal or converting to alternative energy, but they are meaningless unless they address the most fundamental problem facing us: overpopulation.

There is no magic bullet or imaginary god that is going to make the overpopulation problem disappear, so we should stop fooling ourselves and get down to the business of solving it ourselves. We already know what measures are needed, and we have the resources available to begin the process of stabilizing world population. This is not a technical problem, but a political one. We must cast off the entrenched mindset of the nay-sayers and create a bold new vision.

What Are We Waiting For?

World Population Balance tells us that if all countries followed the lead of countries with the lowest fertility rates—including Thailand, Japan, Taiwan, Poland, South Korea, Tunisia, Iran, Greece, and Italy—we could reach a global population of less than four billion by 2100! Their billboard campaign promotes the OnePlanetOneChild.org motto.[602]

Figure 13. OnePlanetOneChild.org

World Population Balance states, "But we cannot focus only on overconsumption and ignore the population multiplier in the sustainability equation. **The small-family solution to**

601 Rex Weyler, The Tyee, *Idea #10: Biophysical Economics*, Jan 2 2009, https://thetyee.ca/Views/2009/01/02/Economics/

602 OnePlanetOneChild, World Population Balance, About the Billboard Campaign, 2020, https://oneplanetonechild.org/#!/billboard

overpopulation is compassionate, voluntary, and ethical. It's the most loving thing we can do for the children of the world."

In our modern world, we can forget how connected we once were to our wild spaces and how these places protect the very things we hold dear. We may not realize that the birds singing in our backyards migrate every year to the disappearing rainforests. We may not understand that the rivers that provide our drinking water are being threatened by contamination and over-use on their way from our boreal forests to the ocean.

Time reporter Harry Campbell interviewed Jane Goodall, and she reflected, "I've stood with Inuit elders by a great ice cliff in Greenland as water cascaded down and icebergs calved. It never used to melt, the elders told me. I've witnessed the shrinking of a Mount Kilimanjaro glacier. I've watched wildfires rage in Africa and in California. And I've seen the carcasses of animals who have died in droughts. As I travel around the globe, people tell me how the weather patterns have been disrupted and the worst kind of hurricanes, typhoons and cyclones are getting more destructive and more frequent.... How is it possible that the most intellectual creature ever to walk the earth is destroying its only home? There has been a **disconnect between our clever brains and our hearts.** We do not ask how our decisions will help future generations, but how they will help us now, how they will help our shareholders, etc."[603]

It would be easy, in our busy lives, to say that we don't have time to worry about our population or our life-support systems. Let someone else worry about it—after all, that's what our governments are paid to do, right? Well, that kind of passive approach may have been acceptable at one time, but it isn't good enough in 2020 in our global emergency. We all need to take action to protect our special places, and now there is a **climate clock** that can steer us towards sustainability. A New York clock that once told time now tells the time remaining.

New York Times reporter Colin Moynihan states, "For more than 20 years, Metronome, which includes a 62-foot-wide, 15-digit electronic clock that faces Union Square in Manhattan, has been one of the city's most prominent and baffling public art projects. Now, instead of measuring 24-hour cycles, it is measuring what two artists, Gan Golan and Andrew Boyd, present as a critical window for action to prevent the effects of global warming from becoming irreversible."[604]

Moynihan adds that the **MCC Carbon Clock** shows how much CO_2 can be released into the atmosphere to limit global warming to a maximum of 1.5 degrees Celsius, as recommended by the UN. In 2020, messages including "The Earth has a deadline" began to appear on the display. Then numbers—7:103:15:40:07—showed up, representing the years, days, hours, minutes, and seconds until that deadline. This number is based on calculations by the Mercator Research Institute on Global Commons and Climate Change (MCC) in Berlin, and will be displayed during Climate Week. The creators say their aim is to arrange for the clock to be permanently displayed, there or

603 Harry Campbell, TIME, *These 4 Issues May Not Seem Related to Climate Change. But They Are and We Need to Solve Them Now*, September 12, 2019, https://time.com/5669043/jane-goodall-climate-change/

604 Colin Moynihan, New York Times, *A New York Clock That Told Time Now Tells the Time Remaining*, Sept 20 2020, https://www.nytimes.com/2020/09/20/arts/design/climate-clock-metronome-nyc.html

elsewhere. Golan said, "This is our way to shout that number from the rooftops." Mr. Golan said just before the countdown began. "The world is literally counting on us."

A Little Good News—Yet to Come?

Better family planning is the most effective and comprehensive way to protect kids, animals, and the environment, and to make our communities stronger.
—Fair Start Movement

When considering solutions to the population crisis, I can't help but think of Anne Murray's song "A Little Good News." Just imagine how nice it would be to turn on the radio and hear these breaking headlines:

- The Red List has just been revised to indicate that every species on the endangered species list has now been safely recovered due to increased protection and habitat recovery. As the human population continues to decline, humankind is gradually giving back the land we stole from other living species. A new appreciation for nature is now being realized, along with an increased civility toward the animal kingdom.
- The Family Planning 2020 program has announced that every woman worldwide now has access to safe and effective contraception services. This was accomplished by reallocating one week's budget from global military spending, since peace has now been reached in the Middle East. Public outrage and a dwindling supply of expendable soldiers were also contributing factors to ending these military exploits.
- The Vatican announced last week that they have heeded the wishes of the majority of Roman Catholics and have decided to support family planning programs to reduce the poverty, misery, and spread of AIDS they have caused globally. Due to public protest, the Vatican has also given up their veto power in the United Nations. This will truly bring the United Nations into the 21st century—and align the decision-making process with the environment, not religious affiliation.

- The World Health Organization has confirmed that great strides have been made in reducing the number of women and children being abused through the slave trade, child soldiers, child labour, child marriage, and various forms of rape. This has greatly reduced the number of unwanted pregnancies and improved the well-being of women and children worldwide.
- The United Nations International Court announced today that they have included the crime of ecocide as a war crime, and offenders will be punished as such. This law is defined as "Making the destruction of our planet a crime … A CRIME AGAINST PEACE." Until now there was no law that addressed crimes of this magnitude against nature.
- Most of the world's wealthiest millionaires and billionaires have now agreed to give back half of their net worth to beneficial causes, with a focus on population reduction and environmental restoration, a project started by the Bill and Melinda Gates Foundation. This

achievement will ensure that the 40 percent of unintended pregnancies worldwide will be averted.

- The board of education has announced that family planning and parenthood classes have been approved in schools and colleges around the world. This will better prepare students to make decisions regarding family function that will greatly benefit their futures.
- The Union of Concerned Scientists is pleased to announce that there are positive indications that global citizens have taken their "Warning to Humanity" seriously and are taking measures to reduce the human footprint on this planet. There are now signs of reduced population and consumption in many countries, and greater prosperity on every continent.

- Economists worldwide are now reporting that the genuine progress indicator (GPI), developed by Redefining Progress, has now replaced the gross domestic product (GDP) as a tool for shifting the economy toward sustainable practices. Besides economic health, the GPI also measures indicators that really matter to people—a sustainable population, good health, safety, a clean environment, and the social well-being of a nation.
- The United Nations has announced the creation of the World Population Organization to address the population crisis in the context that it is linked to environmental, economic, health, social, and political issues. This new initiative is expected to radically change the international community's approach to population governance and to be as effective as the World Health Organization and World Trade Organization. A major component will be youth-led, using their amazing leadership abilities and the passion displayed in climate protests since 2019.
- The Global Media Network has launched a multimedia strategy dedicated to population outreach and partnership building. Thousands of media strategists have been appointed to market and promote the human aspect of a conscious, intelligent, and creative population vision.

- Feminist organizations on every continent have joined forces and have launched a global campaign to eradicate the degradation of women, ensure access to contraception globally, and renew their demand for equal rights.
- Sustainable Population Australia has rallied the medical profession worldwide to address the issue of overpopulation, as it greatly affects the public's health and well-being and diminishes our quality of life. As the rate of population growth continues to decline, statistics are showing enormous reductions in communicable diseases and stress-related illnesses.
- The United Nations recently confirmed that every country now has a population policy in place created by the people—to guide the people to sustainable lifestyles. These policies reflect a country's carrying capacity, and include guidelines for reducing immigration numbers.

- Governments around the world are developing incentive programs that encourage voluntary population reduction and small families. These include tax incentives and various reward programs that have proven very successful in countries that already have a declining population.
- The World Wildlife Fund has noted that the numerous "Plant a Tree" projects being implemented in most of our world's great forests are creating new habitat for wildlife and reducing CO_2 emissions. As the declining human population retreats from wilderness areas, we are able to reclaim and replant our lost forests.

- Population groups on every continent are now issuing Population Report Cards that indicate the level of participation each country has reached in achieving a sustainable population. Due to a concerted effort through the United Nations, and involvement of women and youth at all levels, great progress is being made. The Thriving Together campaign has become a driving force behind this movement.
- Right to Die groups around the world have reported that government policies to allow death with *dignity* are being implemented for those with terminal illnesses and reduced quality of life. This marks a milestone for citizens petitioning for humane and compassionate treatment for those living in constant misery.
- Oxfam recently reported that they have created the International Women's Panel on Population, consisting of women from every corner of the planet to participate in decision-making at every level of government. This panel has been recognized by the United Nations and will replace the Vatican presence at UN conferences.
- The World Health Organization has announced that increased awareness of the value of population reduction coupled with increased access to adoption has encouraged couples to forego fertility treatments. It is no surprise that a decrease in racial conflict and more cooperative international communication have occurred as a result of adoption negotiations.
- The Pentagon has just announced that it will be leading a global campaign to withdraw all forces from foreign countries in an effort to re-evaluate the "permanent state of war" mentality, which clearly is not working. Due to public outrage, a new strategy of cooperation, relationship building, and conflict resolution will be designed to end the senseless profiteering related to the war racket. As the population continues to decline, there has also been a substantial decline in conflict over natural resources, bringing peace to many regions.
- UNESCO has just announced a substantial increase in protected and park land to provide citizens with opportunities to experience uncluttered wild spaces, encounters with wildlife, and places of natural beauty. Studies have shown that these kinds of primal encounters provide a sense of connectedness and well-being that is sadly missing in our modern lives. Eventually, more than half of the earth will be returned to other critters we share this beautiful planet with.

- The media, governments, the medical community, the education establishment, and non-government organizations are making population the buzzword through TV ads, billboards, magazine ads, brochures, and most importantly, dinner discussions. The Internet, which has often been faulted for its unsavoury uses, has now become the catalyst for population consciousness and environmental stewardship.
- And the good news just keeps on coming, as a heady mood of competition has taken hold in countries from Iceland to Uganda. There are now reports of acts of civility and global stewardship coming in from every continent. The number of confirmed cases of childfree couples has reached a record high, and children of the world are being treated with a newfound respect. It seems that the "pay it forward" effect (as depicted in the movie *Pay It Forward*), has manifested throughout the land, and a new sense of optimism has been documented, the likes of which has never been seen on this planet before.

Wow! Can you imagine the impact of a truly responsible and honest commitment to the Declaration of Human Rights? But wait a minute—this isn't all just a pipedream. Many countries like Canada are already at an enviable level of fertility that would allow a declining population—**if** their immigration levels were reduced. If we truly wanted to, we could stop the growth factor in these countries tomorrow, and allow the healing to begin.

Japan, Thailand, Bhutan, Ukraine, Serbia, and parts of Europe are leading the way for the world, with populations that are already declining. They are setting an example of how education and government incentives can work without any coercive measures being used. It is interesting to note that in these cases where citizens are given honest information and access to family planning, they will freely choose the binary system—zero or one child. There is always a transition period for this kind of revolutionary positive change, but future generations will thank these countries for their wisdom and foresight.

The 21st century must be a feminists' century. We desperately need to evolve to a gentler, more nurturing behaviour toward our planet and fellow beings if we are to have a world that is tolerable for all inhabitants. Women must get the feminine adrenaline flowing to be heard in today's testosterone-driven landscape. Women must have a stronger voice in the dialogue taking place about population issues so that female leaders can emerge to guide the way. Coerced parenthood must be a thing of the past, and women must never again be so shackled by discrimination and a patriarchal society.

Men also need to be involved in promoting gender equality and ensuring women's reproductive health, especially since they make many of the decisions affecting women's rights and often shape public opinion. Fathers and partners often play a key role in reproductive decisions affecting women, so it is imperative that they step up and protect women from oppression and sexual abuse. We must **achieve a sense of cooperation between genders to solve this problem**. Men must share family planning responsibility by using condoms, having vasectomies, and using the new male birth control pill, gel, and injection that are becoming available around the world.

Global Strategy for 2020 and Beyond

In order to create an effective global strategy for overpopulation, we must recognize the advantages to declining numbers and realize that falling fertility often stokes economic growth. Population Institute Canada contributor J. Anthony Cassis notes, "National strategies to encourage falling birth rates have been a factor in improving human well-being in South Korea, Thailand, and China among others. Falling birth rates provide what is called a '**demographic dividend**' when having fewer dependent children allows more adults to participate in the workforce, increasing productivity and prosperity."[605]

The prospect of adding several more billion people to our planet over the next 50 years as predicted is nightmarish. This would have a devastating effect on ecosystems already reeling under current human demands. If we fail to act and bring human population growth under control, we will undoubtedly face increasing conflict over resources, devastating social and economic collapse as well as an environmental disaster that is impossible to repair. To think that over 40 percent of births are unintended is a crime against humanity that these and future children will pay for with their lives. However, there is nothing fated about a planet of ten billion humans in this decade or the next. We must collectively decide to stop population growth at EIGHT BILLION.

> *All governments should show real leadership by designing institutions that can help humanity shrink its way to sustainable prosperity.*
> **—Population Institute Canada**

Cassis adds, "The international community needs to come to an unequivocal agreement that a significant reduction of human population is a desirable goal." In 2020, the clock is ticking, since the UN tells us we have only ten years to make this happen. Global wilderness explorer and author, Kevin Casey, gives us some advice on how to get started:

> What are the solutions to an overcrowded planet? Firstly, to stop getting sidetracked by the climate-change industry and recognize that the problem is our sheer numbers and blatant disregard for the planet's health—not the climate. We must replace political and economic agendas and warped ideologies with better education (especially in science). We need more global promotion of family planning, more female empowerment, and government incentives to have fewer children—not more. And sadly, we should have been proactive about all this stuff at least sixty years ago, instead of just waking up to our self-inflicted predicament now. While it's reassuring that today's young people are increasingly aware of the seriousness of their environmental plight, they are protesting up the wrong tree. They should direct their passionate attention

605 J. Anthony Cassils, Population Institute Canada, *A Strategy for a National and International Population Policy for Canada*, August 2009, https://populationinstitutecanada.ca/wp-content/uploads/2018/03/Strategy-for-Nal-Interntl-Popn-Policy-Cassils.pdf

> to the real enemy—a greedy, arrogant, two-legged species that's in furious denial and has become far too adept at making excuses for the inexcusable.[606]

However, this is not an impossible problem—there is a very simple solution. We could have the future that we dream about if we could only return our numbers to the carrying capacity of the planet. That has become my dream—to live to see the day when we make that U-turn. This must become a global effort, the focus of every level of government, all foreign aid groups, as well as the reluctant environmental sector and media. As this book has worked to argue, we have reached the monumental moment when we can no longer continue on this course without complete collapse. We must all get involved with spreading the word to everyone living on this **jewel of a planet.**

606 Kevin Casey, Free Enquiry, Volume 40, No. 4, Why Climate Change Is an Irrelevance, Economic Growth Is a Myth, and Sustainability Is Forty Years Too Late, June / July 2020, https://secularhumanism.org/2020/06/why-climate-change-is-an-irrelevance-economic-growth-is-a-myth-and-sustainability-is-forty-years-too-late/

INDEX